# The Inventor of Stereo

ALAN DOWER BLUMLEIN 1903–1942

# The Inventor of Stereo:

## The Life and Works of Alan Dower Blumlein

Robert Charles Alexander

Focal Press
Taylor & Francis Group

NEW YORK AND LONDON

First published 1999
Paperback edition 2000

This edition published 2013
by Focal Press
70 Blanchard Road, Suite 402, Burlington, MA 01803

Simultaneously published in the UK
by Focal Press
2 Park Square, Milton Park, Abingdon, Oxon OX14 4RN

*Focal Press is an imprint of the Taylor & Francis Group, an informa business*

Notices

Practitioners and researchers must always rely on their own experience and knowledge in evaluating and using any information, methods, compounds, or experiments described herein. In using such information or methods they should be mindful of their own safety and the safety of others, including parties for whom they have a professional responsibility.

To the fullest extent of the law, neither the Publisher nor the authors, contributors, or editors, assume any liability for any injury and/or damage to persons or property as a matter of products liability, negligence or otherwise, or from any use or operation of any methods, products, instructions, or ideas contained in the material herein.

**British Library Cataloguing in Publication Data**
A catalogue record for this book is available from the British Library

**Library of Congress Cataloguing in Publication Data**
A catalogue record for this book is available from the Library of Congress

ISBN 13: 978-0-240-51628-8 (pbk)

Typeset by Avocet Typeset, Brill, Aylesbury, Bucks

# Contents

# Foreword

It is a privilege for me to write this foreword to Rob Alexander's book on Alan Blumlein because it enables me to pay tribute to a man with whom my acquaintance was short but historic. In the later chapters of this book readers will find an account of the problems facing Bomber Command in the early stages of World War II and of the circumstances that led to the urgent development of the long-range bombing aid which bore the code-name H2S. In the last days of 1941 I had been ordered to form a group to develop this system and had been informed that the finest group of electronic engineers in the country at EMI had been allocated the task of transforming the experimental system to an operational state.

I had already heard of Blumlein and of his skills in dealing with the embryonic Air Interception (AI) 1½-metre radar for night fighters, but I did not meet him professionally until early in 1942. His immediate impact on the elementary electronics of our first experimental H2S system lives in the memory. Alas, this impact was to be short lived. On a Sunday afternoon, 7 June 1942, Blumlein and his team from EMI were killed in the crash of a Halifax bomber when they were receiving a demonstration of the performance of our only H2S system using the, then, highly secret cavity magnetron.

A host of problems faced the further development of H2S to an operational state and it was the end of January 1943 before it was first used by Bomber Command over enemy territory. The receiver and the display unit contained many of Blumlein's ideas and it is tantalising to reflect on how much superior that early system would have been had Blumlein survived.

Alan Blumlein lived and worked in the thermionic-electronic age of the twentieth century and this book reveals his immense impact on those systems. It needs little imagination to reflect on how he would have grasped the post-war age of the solid state, microchip systems. He was one of the greatest electronic engineers of his time and in these few words it is an honour for me to pay my humble tribute to one whose tragic loss is forever seared in my memory.

Sir Bernard Lovell
July 1999

# Preface

Alan Dower Blumlein – a British genius?

Who was Alan Dower Blumlein? Incredibly, very few people have heard of Alan Dower Blumlein. Other than those whose fields of interest cross into the realms of his work, he is almost unknown. Yet this man is one of the twentieth century's foremost scientists, inventors and engineers. His achievements were enormous and many of them still affect our lives, nearly sixty years after his death.

If you were to ask most schoolchildren, 'who do you *think* invented television, or radio, or perhaps the jet engine?' even if they do not know the exact name of the person, I feel reasonably sure that they will have heard of John Logic Baird, Gugliomo Marconi or Frank Whittle. Yet none of their text books will mention Alan Blumlein, or any of his achievements.

And what achievements they were: numerous microphones and the techniques for their use; the lateral cutting system which made 'modern' record production possible – this method continues, despite the compact disc, well into the 1990s; telegraphic networks that linked countries and continents; submarine cables; much of the infrastructure of the 405-line high definition television system which, until 1986, was still being broadcast in Britain; improved radar systems, some of which helped win the Second World War and again, were still in use until very recently; and then there is stereophonic sound, a masterpiece of invention so advanced that neither the company he worked for, nor many of the colleagues who helped him perfect it, fully understood its complexities until well over fifteen years after Blumlein had died, and nearly thirty years since it had been conceived.

Were Blumlein to have 'invented' just one thing in his life, perhaps our history books might have treated him differently. It would seem however, that his very volume of work – 128 patents in all – might actually have gone against him. Surely, no one man could have achieved so much in so short a period of time? Yet the Americans have no problem in lauding over the achievements of Thomas Alva Edison who has over 1000 patents to his name. Why is it that Edison, with nearly 60 biographies written about him, is singled out, and Blumlein is not?

There may be an answer. There is an element of mystery and secrecy that surrounds certain events of the life, and of course the death of Alan Blumlein; we have, for over half a century now, been excluded from understanding exactly what happened throughout the key parts of his work. This is especially true of the period during the Second World War, and leading up to his premature death, at the age of just 38, in a mysterious plane crash in June 1942.

Part of the reason why so little has been written about him is buried, along with countless secret documents, in our Ministry of Defence vaults. These locked files, are only now, at the turn of the millennium, finally being opened and looked at by investigative writers, keen to bring into the light a compelling story of achievement and tragedy.

It is without doubt that had Alan Blumlein lived, he would have gone on to achieve even more quite extraordinary things. This is not to say that his achievements as they stand are not outstanding already, but one cannot help imagining the incredible engineering feats that might have come about earlier, or perhaps at all, had he turned his hand to them? What amazing inventions could he have designed had the tragic events of Sunday, 7 June 1942 not taken place?

We are left with only the 128 patents as a lasting legacy to the man. They span a mere fifteen years of his life. With the benefit of hindsight and modern understanding, we can attempt to look a little into this mind that was able to conceive such incredible visions; visions which in some cases, even ten, fifteen years after his death, were not fully understood or commercially exploited.

I first heard about Alan Blumlein from my father who, like Blumlein, held a fascination for engineering, electronics and the practical application of the human mind. It was my father who, upon leaving the RAF as a radar operator after the war, had gone to work for EMI in Hayes. This was where the memory of Blumlein's work was still fresh in the minds of the colleagues he had left behind. When I later became an audio engineer myself, I was bemused by the fact that so little had been written about a man that I had been hearing about since I was a child. Whose principles, in my particular field of interest were, and still are, used on a daily basis.

It was during my time as a lecturer in audio engineering that the first thoughts of this book were planted in my mind. As I taught about microphone theory and application, I was constantly asked by my students: 'Where can we read more about this chap Blumlein?', to which my natural reply would be, 'Well, in a technical library of course'; but, contrary to such auspicious databases as *Who's Who*, there are no such books. None.

Over the years since his death however, there have been no shortage of attempts to write about Blumlein; indeed, several so-called biographies have been written, and have been promised for decades. Yet none of these books have ever seen the light of day.

There have been quite a few so-called biographers of Alan Blumlein, some working with the blessing and co-operation of his family, and others without. In nearly every case these people seem to have lost the real reason that they began the task: to record the events that surrounded the work of such an astounding man, his achievements, his life, his colleagues and friends. In every case, like Blumlein himself, inexplicably, the 'book' that is promised, vanishes into the haze once again.

This situation has gone unchecked for decades.

Doreen Blumlein, Alan Blumlein's wife, died recently in the firm belief that a biographical volume of her late husband's work would appear. Yet, despite the years of help that she gave, allowing access to her private family archives and documents, the book never appeared; year after year she waited, always promised 'next year, next year'. Doreen Blumlein never did live to see that work in print.

Blumlein's eldest son, Simon, has, for an equal number of years, given his time and support to ensure that his father's work could be documented for the good of historians, and for the nation – so important were elements of Alan Blumlein's achievements. In return, he too has received nothing for his efforts. In more recent times, certain individuals have taken to publicly damning those who have called themselves 'biographer of Alan Blumlein'. And yet, despite the constant pressure from established academic and scientific bodies and numerous requests from researchers, libraries, and historians not to mention the Blumlein family, still no book has appeared.

This volume will hopefully, bring to light some (though I am the first to admit not all), of the accounts of the life and works of Alan Dower Blumlein.

It is not my place to stand in judgement on those who have, for so long, promised to deliver his biography, and who have done nothing. That is for others and history to do. However, I would say that this book is so long, long overdue. Were it not so incredibly important to document this life and these achievements, if only for history's sake, then the circumstances that have surrounded the publication or, more accurately the lack of publication of a biography of Alan Blumlein, it would almost be comical.

My only regret is that many of those who entrusted their faith in the work of individuals, who made such shallow promises, have now, sadly, passed away. Of the many that knew and loved Alan Blumlein today, but a handful remain. Had that which had been promised for so long been done, then many of his colleagues, his friends and of course his family, would have lived to see a biography in print.

Above all this book will set the record straight for future generations. As fully as research allows, given the distance in time, the full catalogue of Alan Blumlein's achievements is contained herein. In spite of the shroud of secrecy which, even now at the end of the century, still covers part of the work he undertook (and will continue to do so for some time to come), I have attempted to discover and understand the kind of mind that is capable of conceiving a new patent every 46 days of his working life.

It may not yet be possible to explore every avenue of his work. However, while there are still those alive who knew this incredible man, and who were close to him in life, I feel it is imperative that an account of his work be published.

Only then can he take his rightful place among the very best of this century's scientists and engineers.

# Acknowledgements

I should like to thank the large number of people who have given their enormous assistance to me during the many years of research that I have devoted to this book. Without them, once again, we would still not have the biography that should have been published decades ago:

Barry Fox for his encouragement and help throughout. Ruth Edge, Chief Archivist at EMI Music Archives and her staff for all their time and for allowing me complete access to the huge EMI archives. Reg Willard at EMI Music Archives whose knowledge of the company spanning 57 years, was an invaluable source of information for which EMI should be most grateful and proud, and without whom much detail would have been lost. Kate Callaway at EMI Music Archives for the preparation of many of the photographs that are contained in this book.

Simon Vaughan, Archivist of the Alexandra Palace Television Society and his new wife Marta, for allowing me to invade their home and rummage through innumerable documents and photographs for this book right in the middle of their preparations for their wedding, and again for allowing me access to a quite extraordinary source of archival information.

The entire staff of The British Library, Science and Reference Library, Patent Library, at Southampton Buildings, London (where many hours have been spent in dark, dusty old patent vaults).

Peter Trevett and the staff of the library at DRA, Malvern, an establishment which despite the security surrounding such an institution holds invaluable links to TRE and TFU documentation, which, with his guidance I was given access to. William Sleigh who compiled at RSRE the definitive report which finally shed light on the reason why Halifax V9977 crashed in 1942, his subsequent help and pre-publishing reading has been invaluable. Ernest Putley for his encouragement and recollections of TRE and TFU and for reading my material to check for any factual errors.

The staff of the library at the Imperial War Museum, Kennington, London, who again allowed complete access and gave me much help during several days spent there. The staff of the Royal Air Force Museum at Hendon for all their help. The staff of The British Library, Sound Library for access to its archive material. The staff of The British Library, Newspaper Library at Colindale for their help during many hours spent there. The staff of Westminster Library for their help in tracking down many of the government and legal documents that have been referred to in this book.

The staff of Waterstone's bookshop in Watford who for many months gave me enormous help tracking down and then ordering the dozens of books that were needed for reference material for this book. Patrick Vaughan and Jerome Vaughan for allowing me access to their land and to Douglas Kirby and Peter Kirby for accompanying me. Kim Templeman-Holmes who, while at CRL, gave me some of my first photographic material on Blumlein. Alastair Sibbald at CRL who, despite being very busy the day I visited, took time out from his schedule to copy the Blumlein binaural film archive material on to VHS for me.

Colin Greenaway at SD Post, London, who transferred the video material via the AVID system onto a Zip disk in order that the stills in this book could be reproduced. Wayne Thomson for his photographic skills on a far from sunny day at Goodrich Castle. Neil Lucas at AM Publishing for his original graphic design work and file preparation, Mandy Silk for the typing and tape transcribing, also at AM Publishing. Mike Skeet for his inexhaustible knowledge of microphones and microphone technique, which helped me clarify this area of my research. To Karen Newton and Hilary Tugwood for the sake of their fingers, as it was they who typed out most of the 128 patents, a job that I didn't relish, and which tested our friendship (at times) to the limit. To http://www.jayjaybee.com for the website.

Anthony Timson for keeping my computers going when they didn't feel like it and for finding me the software that he knew would get certain jobs done when I didn't have a clue. John Andrews, Christian de Haas, Andrew Bailey and especially Antony Askew for taking the time to read this book or part of it before it went to my publishers. Without their expertise and specialized help many details would have been missed and I am most grateful to them for their help. My great friend Jayson Chase for putting up with long, tedious hours of photocopying and research, helping me with parts of this book that would have taken me so much longer had I had to do it on my own, and for his never-ending encouragement throughout.

To my publisher, Margaret Riley, for knowing so much about Alan Blumlein when we first spoke of this book and thereby making my task of selling the project to her so much easier.

Special thanks to Felix Trott, Maurice Harker and Eric White, without whom I would have had little or no first-hand information that I could check and cross-reference against. Not only were they very special and memorable occasions for me meeting these three gentlemen, all now in their late eighties and early nineties, but the encouragement and enthusiasm which they instil is the kind of rare and magical inspiration that any investigative writer can usually only dream of finding. I am most grateful to have been given the opportunity to access their personal and private documents for use within this work and can only marvel at their extraordinary memories so long after the events took place. I hope that they find this book represents some small token of our thanks for the quite extraordinary achievements that they and their colleagues completed all those years ago.

Special thanks must also go to Simon Blumlein and to Sir Bernard Lovell. Simon Blumlein for his help and time. He finally believed that a biography of his father would appear in print after all the years of waiting. Sir Bernard Lovell for his kind words and also for the vivid recollections of what must have been a most inspiring and also tragic time in his life.

Finally, to my family for believing in me.

# Picture credits

| | |
|---|---|
| Front Cover | EMI |
| Back Cover | Wayne Thomson and English Heritage |
| Frontispiece | EMI |
| EMI | Figures 1.1, 1.3, 3.1, 3.2, 3.3, 3.4, 3.25, 3.26, 3.27, 3.28, 3.30, 3.31, 3.32, 3.34, 3.35, 4.7, 4.8, 5.1, 5.2, 5.3, 5.5, 5.6, 6.3, 6.4, 6.5, 6.10, 6.11, 6.18, 7.3, 7.4, 7.8, 7.11, 7.12, 7.13, 7.14, 7.15, 7.16, 7.17, 7.18, 7.19, 7.20, 7.21, 8.8, 8.9, 8.10, 8.11, 8.12, 9.3, 9.4, 9.5, 9.6, 9.7, 9.8, 9.9, 11.1, 11.4, 11.5, 11.6 |
| Eric White | Figures 1.2, 3.48, 5.4 |
| Maurice Harker | Figures 3.29, 3.37, 3.38, 3.42, 3.43, 3.44, 3.45, 3.46 |
| APTS | Figures 4.3, 6.7, 6.8, 6.14, 6.15, 6.16, 6.17, 7.1, 7.2, 7.10 |
| Felix Trott | Figures 6.2, 7.6, 7.7, 10.10 |
| Sir Bernard Lovell | Figures 9.1, 10.3 |
| William Sleigh | Figures 10.5, 10.8, 10.9, 10.11, 10.12, 12.1 |
| English Heritage | Figures 12.2, 12.3, 12.4 |

# Introduction

S hortly after 4.20 p.m. on Sunday, 7 June 1942, a glorious summer's day, clear skies, warm sunshine and perfect visibility for flying, a Halifax bomber crashed into the shallow hillside of a field just north of the River Wye near the village of Welsh Bicknor in Herefordshire. All eleven occupants aboard were killed in the enormous fire, which engulfed the aircraft on impact.

The consequences of that crash were, at the time, considered grave enough for none other than Winston Churchill to shroud the events in secrecy. Family members were withheld from publishing obituaries, no mention of the crash appeared in the press, even the bodies of those who had died were held for several days, before finally being released for burial.

These were the darkest days of the Second World War, and Great Britain, in desperate trouble, stood, practically alone, against the might of the marauding German armies. Even though the British propaganda machine had swung into action, trying as best as possible to keep the morale of the hard pressed British people going, the truth, the real truth of the situation however, was known by only by a few; and the information they had was not encouraging. Britain's defeated armed forces had been kicked out of Europe in June 1940, at Dunkirk. Nearly starved by the continued loss of vital shipping to the U-boat fleets; fought to near extinction in the skies over the Channel during the Battle of Britain, London had then been bombed to a shell during the Blitz of the winter of 1940/41.

The only way Britain could fight back, to restore some of the morale, to give the public the victory that they so badly needed, was in the skies over Germany. By the summer of 1942, the bombing of German cities and industrial sites took priority over almost every other operation. Yet, for those who were privy to the most secret of secrets in this, the darkest period of British history, the plain truth was that Bomber Command were falling woefully short of expectations. Even with its new squadrons of heavy, four-engined bombers, Bomber Command were only placing one bomb in five within ten miles of the target their bomb-aimers thought they were.

If a way, any way, could be found of directing the aircraft to a designated target with some degree of accuracy, then it must be found, at all costs.

That method was called H2S radar, and it was an early experimental form of the H2S equipment that the Halifax that crashed at Welsh Bicknor on 7 June 1942 was carrying. More critically perhaps, of those who died aboard the aircraft that day, the majority of the civilian and military development team who were perfecting the radar also perished, before H2S could be delivered to the Bomber Command squadrons.

The loss of such vital personnel, all at once, may well have had far reaching consequences for Britain. It may have even lost the war, had it not been for the dedication of the few, left behind, driven on by constant personal intervention by Churchill to complete the project and carry on with the work of the team who had died.

Of the scientific personnel who died that day, Alan Dower Blumlein stands out as possibly the greatest loss. 'A national tragedy', one of his colleagues would call it, for Blumlein was, without any doubt, at a time when scientific genius was at its foremost, one of the most brilliant engineers of the twentieth century.

So who was Alan Blumlein? And what exactly did he achieve? Certainly there are very few references to him in textbooks, and almost none of the history books seem to have remembered him. He does not appear in '*Who was who?*' for example, nor does he have an entry in most of the written recollections of the war years. For a man whose work was so influential, Alan Blumlein has been dealt a pretty rough hand by the game of history.

For the purposes of simplicity here, I have split his works into four general fields of interest: telephony, audio, television and radar, though he undoubtedly contributed greatly to other categories, such as circuitry and measurements; these are dealt with in detail later. It was these four areas however, all inextricably linked to each other, that has made it relatively easy to trace the route of development taken by Blumlein as he progressed from one subject to the next, often overlapping his work.

It was, as I have said, his secret work with radar in the Second World War that caused him to lose his life. This was something that he was aware could happen, and he had even discussed the matter with his wife Doreen. Alan Blumlein knew that the Germans had some idea he was part of a major British scientific team working on secret war projects. Though whether they were able to infiltrate the tight security surrounding such establishments as TRE, the Telecommunications Research Establishment, at Malvern, where much of the work was being carried out, is unknown (and frankly, unlikely).

It was however, as a direct result of the crash that an even tighter blanket of secrecy came down and covered all aspects of the work being carried out by those who had lost their lives.

The work continued of course, the H2S radar system was perfected and used to a very effective degree throughout the remainder of the war. Indeed, in various different versions, H2S was used continually until the end of the Falklands War in 1982.

Somewhere along the line however, the work that had been carried out by the men who had died, never became de-classified. It remained, until quite recently, very secret, and as such the people who had paid the ultimate price, made the ultimate sacrifice for their country and the work they believed in, were forgotten by all except those who knew and loved them.

This is the story of all the areas of interest that engaged Alan Blumlein during his working life. It spans a quite extraordinary era in human technological development covering a period from just before the first powered-manned flight was made, to a time just prior to the nuclear age. While I have endeavoured to place events in the context of Alan Blumlein's life – specifically, his surroundings, work, family and friends – it is also important to point out that many of his patents are co-written with colleagues who were extremely talented in their own right. Two of those who died with Blumlein in the Halifax had been friends and co-workers; they had, for many years, laboured to come through the trials and tribulations that developed the high definition television system with Blumlein before the war.

Yet other aspects of his life touch upon projects and people he may not have known personally, but who are mentioned here in the context of the achievements they made, achievements which ultimately contributed in some way towards his work.

This book then is not so much the story of one man, though it centres on the achievements of Alan Blumlein; it is, rather, the legacy of many. Just as inventions are not always the work of an individual but the achievements of the many, so Alan Blumlein's work can be attributed to the influence of the many that went before him. An extraordinarily modest man, I am certain that Blumlein himself would be the first to ensure that the work carried out by his friends and colleagues was recognized for its value in the context of his own achievements.

And yet, despite all that, there are fascinating elements of his work that simply did not exist before he imagined them. His mind worked in way that made problem-solving a daily challenge, a goal to attain, and he took up these challenges with a relish which most ordinary people would find hard to understand, or even believe.

It is this very individuality, this ability to create alongside the work of others that I believe, truly justifies the use of the word genius.

Robert C. Alexander

Watford, June 1999

# Chapter 1
# Earliest days

Alan Dower Blumlein was born on 29 June 1903, at 31 Netherhall Gardens (just off the Finchley Road), in Hampstead, London. A modest house, it was leased by his father Semmy Joseph Blumlein, who, originally a mining engineer, had recently enjoyed some success as a merchant and was beginning to prosper. Semmy Blumlein was a naturalized British subject having been born in the Alsace region of Germany in 1864. His wife, Jessie Edward Dower, was the daughter of a Scottish missionary, the Reverend William Dower, who had been working in South Africa where Jessie had been born. Semmy Blumlein had left Alsace in 1871, to avoid the conflict of the Franco-Prussian War and possible conscription into the German army. He travelled to England to meet with family friends and ended up for a while in Manchester, where he would probably have found his new world very strange. Manchester being a large, dirty, industrial town, was not completely unlike the Alsace part of Germany which is also an industrial and mining region.

There is no evidence to suggest that Semmy spoke much if any English when he arrived in Britain, though he certainly learned to speak the language in his newly adopted country soon after. He seems to have survived quite well, even thriving and would, in time become a successful businessman directing several public companies. Some years after his arrival in England, and after he had made enough money to travel, Semmy Blumlein decided that his prospects were better in the British colony of South Africa where massive European development was taking place. He packed up everything he had and travelled to South Africa to investigate, and hopefully exploit, the mining prospects in the region. These were undergoing extensive development at the time and offered a man of Semmy Blumlein's resolve an opportunity he could not pass up. It was while there in South Africa that he met and married Jessie.

For a while Semmy and Jessie Blumlein prospered in South Africa, but as political unrest escalated in the years before the Boer War, they decided, perhaps for safety's sake, perhaps for business and economic reasons, to return to England. Taking with them the not inconsiderable amount of money that they had acquired, plans were made for the passage home. When the Blumleins returned to England, they set up home in north-west London and it was here that Alan was born on 29 June 1903, and where he would spend the formative years of his childhood. At the time of his son's birth, Semmy Blumlein gave his profession on Alan's birth certificate as a financial merchant and, having decided to make England his permanent home, finally became a naturalized British subject in 1904.

At the age of five, Blumlein was placed in a local preparatory school by his parents called Gothic House, which was in Belsize Road. School would be one of the few places during his childhood where Blumlein was truly happy and his association with this part of London would last well into his adult life. So would Gothic House where he would later meet his wife Doreen, a preparatory teacher there. Blumlein's education during his time at Gothic House seems to have been of a somewhat liberal nature.

Alan Blumlein was a slight child with mousy brown hair and vivid, interested green eyes. He was a clever child, able to understand most things, but at the same time could be quite introverted. His parents, who were liberal-minded for their time, allowed the boy to work as he pleased, and it would seem that he did just this. His attention span was limited only to subjects that interested him, and he practically ignored those which he felt he had no use for.

One of the best examples of this during these formative years is his reading ability, or rather the lack of it. In most of the narratives that have been written about Blumlein, much has been made of the fact that until the age of about twelve, Alan Blumlein could not read, and very probably could not write. There is no direct evidence for this as none of his work at Gothic House has survived (assuming of course, Semmy and Jessie Blumlein even had it to keep in the first place). However, much later in life, Doreen Blumlein recalled that Alan had told her something of the kind of world in which he lived between 1908 and 1914.

> 'He couldn't read properly at all until he was about twelve, and he could never spell. Though he did get to read library books in the end. I think through me, suggesting that he did. However, it was no good if he read any sort of poetry or anything like that, he couldn't read it. It meant nothing at all to him.
>   Of course he *had* been reading up to then, as I said to him one day, because his nurse used to read it all. But because he hated reading, he went to a crammers when he was about twelve, and he must have got to it. So, I said to Alan one day, "You couldn't read properly till you were twelve?" and he said "No, but I knew a hell of a lot of quadratic equations." He knew what he wanted to know, and the rest he didn't bother with.'

This was not an entirely unusual situation for the time. Semmy and Jessie Blumlein were affluent enough to afford a nurse to look after their son. As was the popular belief in late Edwardian times, much of the formative education of the boy, during the hours that he was to spend at home, would therefore be left to her. Children were to be seen and not heard, and only seen when they were wanted. While there is no indication that Semmy and Jessie ran a particularly strict household, it would only be natural for them to take a distant interest in their son's education. Young Alan Blumlein would have been read to by his nurse. Stories, adventures, poems and rhymes. Perhaps as he got a little older Shakespeare or even the classic novels, but he himself found little interest in learning to read which, after all was being done for him whether he learned or not.

Semmy Blumlein's influence on his son, despite the normally strict edict of Edwardian life, seems to have been quite important as it was almost certainly he who first introduced Alan to the knowledge of electrical engineering. Though Semmy was in principle a mining engineer and a businessman, his interests were wide and varied, as is the case with many self-made men. Somewhere around the age of five or six, Semmy Blumlein began to notice that his son had a keen aptitude for knowledge of subjects that were of a distinct interest to him. Only when Alan was interested would he pay enough attention to learn. It was due as much to Semmy's awareness of this fact, as the boy's interest, that Alan Blumlein was first introduced to the wonders of electricity.

The first direct knowledge that we have of Alan's growing ability to study the application of electrical engineering is a small piece of paper. Essentially, it is an invoice written out to his mother in 1910, when he was just seven years old. It seems that a doorbell had broken. Alan had either been asked or, far more likely, had taken it upon himself to fix it. Having done so, he presented the bill to his mother stating the sum that he wanted for fixing the doorbell and signed it 'Alan Blumlein, Electrical Engineer'.

Over the next few years Alan began an almost total immersion in the study of electronics and mathematics which, bearing in mind his age, was a quite extraordinary thing for him to undertake as study subjects. He was undoubtedly encouraged by Jessie and Semmy Blumlein who, even at this tender age, must have seen great promise in the bright eyes of their young son. Like most boys, Alan Blumlein developed a healthy interest in steam trains, something which would remain with him throughout his life. He used to walk to a vantage point near Swiss Cottage and watch the trains going by for hours at a time. Having the kind of inquisitive mind that he did, just watching the trains was not enough for Alan Blumlein. So he set about studying the mechanics of how they worked, which in turn gave him a grounding in mechanical engineering that would serve him well in his adult life and during his later careers.

On 28 July 1914, Semmy Blumlein, by now a director of several public companies, died at the age of 50 from thrombosis of the pulmonary artery. Jessie Blumlein was at his side when he passed away. It was a great shock to the young Alan Blumlein, who determined that he would help his mother by condensing his education and making something of his life. The problem was that Alan Blumlein, despite all his expertise in mathematics and the inquisitive mind, which had served him so well up until now, could still hardly read at the age of eleven. As a child, he liked to have stories read to him about trains, anything and everything about trains. It may well have been this interest, coupled with the death of his father, which finally drove him to the realization that in order to progress he would have to get down to some serious reading. Alan Blumlein therefore finally learned to read.

It is a testimony to the kind of mind that he had, for within two years of not being able to read anything other than rudimentary passages from text books, Alan Blumlein could read well enough to win not one, but two scholarships. He was offered places at both Aldenham and Highgate Grammar Schools. It seems that his mother had also realized that her son had some catching up to do if he was ever to get into higher education. She sent Alan to a 'cramming' school called Ovingdean which was near Brighton where several years worth of catching up on reading and writing were packed into a period of about twenty months. Alan learned to read and write, though for the rest of his life his spelling ability wavered between plain bad and atrocious.

In later years Alan would recall for Doreen, his wife, how unhappy this time was for him. Though he knew only too well how important it was to study and cram his education, the time spent at Ovingdean was during the First World War. Alan Blumlein, having a Germanic sounding name was subject to much taunting and bullying by his fellow pupils. This was quite common in Britain during this time, with many people of German origin picked on or even attacked because of their names. It was during this time that the British Royal Family, who had mostly originated from Germanic ancestry, changed their name from Saxe-Coburg to Windsor because it sounded British.

The hatred of the words levelled at a completely innocent Alan Blumlein during this time caused him to became quite reclusive and introverted. He tended to bury himself in his work, rarely taking part in activities such as group sports. Throughout the remainder of his life, Alan Blumlein would show little interest in team sports such as cricket, soccer or rugby. He did swim however, a solitary activity. Blumlein found swimming exercise enough to satisfy the needs of a growing teenager, and secluded enough to escape the jibes and taunts of his peers. It is ironic that his name made him the target of such attacks. It was because of these that Blumlein made so much time available for study, and goes some way to explaining how he managed to catch up so fast, in so short a time.

Having won the two scholarships, Alan now had to decide to which of the two schools he wanted to go. He chose Highgate Grammar School because it placed a greater emphasis on science than Aldenham did. Highgate School was in fact one of Britain's oldest schools having been founded in 1565, during the reign of Elizabeth I, by Sir Roger Cholmeley. It is his name that has now been given to a small region of Highgate which straddles Highgate Hill and Archway Road. When Alan Blumlein joined as a boarder in January 1918, at the age of fourteen and a half, there were

Figure 1.1
Alan Dower Blumlein.
(*Courtesy of EMI*)

about 500 students at Highgate School. The headmaster was Dr J.A.H. Johnston MA, DSc (who had become headmaster in 1908, and who would remain as such until his retirement in 1936).

Dr Johnston had instigated several programmes at the school, all of which were directed towards a greater development of scientific education for the students. Fifth and Sixth form students at Highgate could choose from an array of scientific studies: physics, which included electricity and magnetism, chemistry, astronomy and biology. In every way, Highgate School was the right choice for Alan Blumlein and it did not take him long to settle down to what became a very happy part of his life (when he was studying).

Blumlein now immersed himself in almost total study. A contemporary of his, C.P. Fox, recalled later, 'He was a voracious reader who saw the manual instruction hours of the school time-table as a welcome outlet for his practical abilities as well as a happy alternative to aimless boredom which many of the other boys suffered. He seemed to need little if any instruction I remember, and while other pupils were starting to build model boats or bookshelves, but seldom finishing them, he designed and built for himself an (electronic) balance cabinet of quite professional excellence.'

Blumlein matriculated and entered the science sixth form of Highgate School where he stayed until July 1921, at the age of eighteen. That summer, having graduated from Highgate, Blumlein had to decide where he should go next. His first choice was City and Guilds but it was an expensive college and so an entrance examination would have to be sat. It was an external exam rather than an internal one. Things evidently went rather well as Blumlein not only succeeded in passing the test, but was admitted directly to the second year of the course having been granted a Governor's scholarship.

City and Guilds College in London, which was one of the constituent colleges of Imperial College, was one of the foremost centres of scientific learning in Britain at the time and perfect for the mind of somebody like Alan Blumlein. Some years later, again describing this time to his wife Doreen, he said that it was 'The happiest day of his life until he was married, when he first went to college, engineering. All day!'

That first day in October 1921, Blumlein met Ivan L. Turnbull who had joined City and Guilds from Battersea Polytechnic. Turnbull was also entering the second year directly and suggested to Blumlein that they should form a laboratory partnership and work together as all the other students who had gone through the first year's work had already formed theirs. Blumlein agreed and the two of them worked in the laboratory for about three or four months until the beginning of 1922, when Turnbull had to terminate the partnership because Blumlein was working far too fast for him. Turnbull said that, 'While I was still reading the instruction notes for the experiment that we might be working on, Blumlein would have gone ahead so that by the time I had finished reading and worked out what we were supposed to do, he would have completed the work.' Turnbull would later graduate from City and Guilds obtaining a MSc. He would renew his acquaintance with Alan Blumlein as a member of the EMI research team, of which Blumlein was one of the leading members.

During his time at City and Guilds Alan Blumlein proved a popular and likeable student, though 'He did not suffer fools', according to Turnbull. During this time, he swam regularly, obtaining colours for Imperial College for swimming, which remained his one passionate sporting activity throughout his life. Some years later, he swam for the Otter Club. He even won two cups for them, but as Doreen would recall: 'He gave up the Otter Club because, you see, Alan wasn't one that would go into the bar and be hearty afterwards, you see? He enjoyed the swimming.'

He completed the second and third years of his course, though these terms were not without their difficulties. At City and Guilds, at that time, the courses tended to emphasize heavy engineering, while steering away from mathematics and pure mathematics. Although Blumlein had been working with quadratic equations since he was a child, his mature mathematic ability often left him labouring. Typically an engineering course such as that which Blumlein had engaged in only went as far as calculus, and not very advanced calculus at that. This caused problems for Blumlein for a while and he needed special tutoring for a time, though he crammed enough information again to get him through his two years of the course. Mathematics however, would remain, for the rest of his life, one of Alan Blumlein's weaker subjects, and he would happily leave the calculations to somebody else wherever he could.

In June 1923, he obtained his Associate of the City and Guilds Institute (ACGI), with his first class honours BSc (in electrical technology, distribution and utilization, electrical machine design and electrical generation). His degree was awarded one month later in July 1923, when he had just turned twenty. Blumlein was by now beginning to show his brilliance, and it had not gone unnoticed. Following his graduation, he was invited by Professor Edward Mallett of City and Guilds to join him as an assistant demonstrator. The position, while junior, was nevertheless a great honour for Blumlein (again one must bear in mind his age at the time). So, along with another third year graduate, A.R. van C. Warrington, Blumlein elected to join Mallett who was running a postgraduate course in telephony and wireless telephony.

The decision to work with Professor Mallett might at first seem an unusual choice for Blumlein. He did have the option to carry on with his studies in postgraduate courses such as electrical machines and transformers, or electrical machines and traction, both of which presumably, his knowledge would be better suited to. On the other hand, the early 1920s were exciting times in the field of telephony and wireless telephony. It may well have been the lure of a relatively new and exciting science that gave Blumlein the incentive to choose the path that he did.

That Blumlein chose to work with Professor Mallett may well have changed the very course of his life. He was in good company to say the least as fellow research students in the electrical engineering department run by Mallett included A.H. Reeves, who would go on to join the Western Electric Company, and subsequently invent pulse code modulation (PCM), and Geoffrey F. Dutton, who had already begun his work on an investigation into the properties of the electrical reproduction of sound. Many years later Dutton would work in conjunction with Henry Clark to write a paper for the Institute of Electrical Engineers explaining the principles of stereophonic sound entirely based upon Alan Blumlein's work; it may well have even have been knowledge of the work that Dutton was undertaking that gave Blumlein the incentive to change careers in 1929, when he left Standard Telephones & Cables to go and work for Columbia on their recording and reproduction system.

Figure 1.2
Geoffrey F. Dutton was already at City and Guilds when Blumlein decided to work there as an assistant to Professor Mallett. They would be reunited some years later when both worked for EMI. (*Courtesy of Eric White*)

Alan Blumlein worked as an assistant demonstrator with Professor Mallett from September 1923 until September 1924, and during this time he met with and helped several students who would later work with him at the Columbia Graphophone Company, and who subsequently became part of his research and development team at EMI. One such student was Eric Nind who recalled Blumlein as:

'A delightful man, very human indeed who had a good sense of humour and was very good at explaining everything. He didn't get exasperated if you did not understand and would go through a point again and again. He had plenty of inexhaustible patience and had a great facility for converting quite complicated mathematics into very simple circuit elements.'

During the year that they were working together, Blumlein and Mallett devised a new method of high-frequency resistance measurement and prepared the work as a paper to be read before the Institute of Electrical Engineers (Mallett was already a member, and Blumlein joined in 1923). The paper took some time to prepare, especially considering the fact that Blumlein's command of English, and spelling in particular, still left a lot to be desired, with much of the mathematics probably done by Mallett rather than by Blumlein. However, it was finally ready in August 1924, and was originally received by the IEE on 29 September that year, at which point the IEE's referee commented that the paper 'appears to have been somewhat hurriedly put together; the authors should be asked to go carefully through it, to rectify errors and improve the English.'

Corrections were made, and a final form of the paper given to the IEE on 5 November 1924, with the paper read before the Wireless Section of IEE on

Figure 1.3
Eric Nind first met Alan Blumlein when a student at City and Guilds and Blumlein was an assistant to Professor Mallett. Nind and Blumlein would later work together, co-writing several patents at EMI. (*Courtesy of EMI*)

7 January 1925. The paper was considered of some importance and this was demonstrated by the awarding of a 'Premium' by the Institution. The work subsequently became part of a series of articles which appeared in *Wireless World* between October and December 1925. It was the first occasion (one of only three times), that Blumlein would be published, other than patent specifications, during his lifetime. The first instalment of the article appeared in the 21 October 1925 issue of *Wireless World*, and subsequent instalments ran each week (*Wireless World* later became a monthly periodical), until 9 December 1925 (with a break during the issue of 25 November 1925, for reasons which are unknown), a total of seven parts.

However, by the time the paper had been read before the IEE in January 1925, Blumlein had, some months earlier, already left his position as assistant demonstrator with Professor Mallett at City and Guilds to take up a position at International Western Electric (IWE), a division of the Western Electric Company of America (soon to become International Standard Electric Corporation and following that Standard Telephones & Cables). Mallett had evidently seen the promise in his young assistant and realized that in order to further his career, Blumlein needed to be 'out in the field' so to speak. In the autumn of 1924, Mallett approached a friend of his at IWE, R.A. Mac, who was the head of a new engineering department which had just opened, and had been looking for any bright young engineers that might suit.

Mallett suggested to Mac that he should consider Blumlein for a position at IWE as he had showed great promise at City and Guilds, and he felt sure that he would suit the needs of the new engineering department. At first, Mac refused to consider

Blumlein as he had a Jewish name. For various reasons, the Western Electric Company in America would not engage Jews. As it happened this turned out not to be the case, and Western Electric were indeed employing several Jews in America and elsewhere at that time. Mac replied to Mallett thanking him for his offer, but turned Blumlein down. However, Mallett contacted Mac again and explained to him that if he did not take on Blumlein he would lose one of the best men that they had ever had at City and Guilds on that type of work.

International Western Electric were, by 1924, deeply engaged in the research and development of international telephone lines and exchanges, and the need for good engineers in this field was becoming paramount to their continued success. Telephone engineering had by then become a science in its own right, with the systems approach dominant and, though the science itself was not exactly new (much of it having been invented in the nineteenth and early twentieth centuries), the application of many associated discoveries such as AC theory, were being used as routine tools to design the telephone systems required.

These systems, in turn, were growing into networks which were spreading all across Europe, with national networks being linked internationally through undersea cables and long telephone lines. It was the very length of these telephone lines which were causing much of the problem then associated with telephony as overall speech quality across great distances was very poor. What was needed were electrical circuits that would improve reception and transmission of the speech across great distances.

Needless to say, R.A. Mac, at the insistence and prompting of Professor Mallett, eventually relented on the issue of whether or not to consider Alan Blumlein regardless of whether or not he had a Jewish-sounding surname, and decided to offer him an interview. Blumlein must have impressed Mac at the meeting as he was offered a position at IWE within a week. Moreover, Blumlein was offered the quite considerable salary of £225 per annum which, when one considers that he was just 21 years of age, and fresh out of City and Guilds, bears reflection on the impression that he must have made, the evident talent that he had, and the need that R.A. Mac and International Western Electric had for good engineers. Alan Blumlein accepted the offer, and he joined IWE on 15 September 1924.

# Chapter 2
# Telegraphy and telephony

On 15 September 1924, Alan Blumlein took up his first formal position of work at International Western Electric (IWE), a division of Bell Laboratories of America (soon to become International Standard Electric Corporation and later still Standard Telephones & Cables). His salary was to be £225 per annum, quite a reasonable sum for the time considering that Blumlein, just 21 years of age, was fresh out of City and Guilds. The industry that he had joined was about to go through a revolutionary period as telegraphy and telephony spread across Europe at an ever quickening pace with the introduction of new long-distance telephone lines and exchanges.

This was therefore, an opportune time to join a company such as IWE, which was a large company with many interests in telegraphy and telephony and was well aware of the major changes that were taking place in its industry and the fact that they would need to be acted upon. The company had secured large contracts for European and domestic telephone exchanges, which were being fought for and won by many companies desperate to be involved in this fast-growing industry.

In the United States the growth of the transcontinental telephone networks had spread much faster than they had in Europe; this had made many of the American companies working in the same field as IWE, such as Bell Laboratories (of which IWE were a part) and Atlantic Telephones and Telegraphs (AT&T), very wealthy indeed. Part of the reason why the European companies had lagged behind their American counterparts can be attributed to the fact that many of the countries in Europe had only recently been fighting each other in the First World War. Nevertheless, Europe remained a long way behind America when it came to installing long-distance and international telephone lines, large exchanges and the research and development that was destined to improve the quality of the telephone reception that was now so badly needed.

Alan Blumlein would eventually be placed within a team headed by John Collard, which had been given the task of investigating the phenomenon of interference, or perhaps more accurately, the reduction of cross-talk and interference in electrical circuits. Initially however, R.A. Mac had not quite decided where to place his new young engineer and so, on his first day at IWE, Blumlein would meet and immediately befriend Joseph B. Kaye (who was always known as 'J.B.' to friends and colleagues alike). J.B. Kaye was destined to become a lifelong friend of the Blumlein family; the two young engineers would shortly decide to share expenses by living in digs together (Kaye would eventually be best man at Alan Blumlein's wedding in 1933). Joseph Kaye, like Blumlein was a young graduate in his first year at IWE, having joined the company directly from Cambridge University where he had graduated in June 1923, and come down to London in October. At Cambridge Kaye had taken a degree in science which had consisted of various subjects including chemistry, physics and of all things geology, though by his own admission he hadn't done that well.

IWE were, at that time, looking for new engineers and having offered Kaye an interview, then realized that they didn't really seem to have an opening for him. However, R.A. Mac, who was the head of the engineering department evidently had taken a liking to him and decided that he could make an engineer out of him regardless. During the following year Kaye trained, working on all manner of projects that the engineering department of IWE had on the benches at the time, before Blumlein arrived in the September of 1924.

As Kaye himself recalled many years later, Mac was at a bit of a loss of what to do with Blumlein who, despite his evident skills which had been so useful to Professor Mallett at City and Guilds, seems to have arrived at his new position without anybody having considered what he was exactly to do. For that reason Mac decided to hand over his new charge to Roland Webb, who ran the transmission laboratory and who decided to put Blumlein and Kaye together:

'Well, they were a bit puzzled what to do with us, because they couldn't quite make out Blumlein, a little difficult to assess; so they put us together as a pair in the transmission laboratory. That was a laboratory that dealt with all the cable work, measurements, instrumentation and all that, to fortify people out in the field, the engineers dealing with all the cable units in Europe, where the company had contracts: Paris, Strasbourg, Milan, Turin, Genoa, Madrid, all that type of work. They put us underneath the control of Roland Webb. Well, we both must of had some sort of a funny reputation, because the betting was that Roland Webb would either cope with us or he would have a nervous breakdown. Where our reputations came from, I've never been quite clear, but one point that is of interest is that when R.A. Mac received the offer from Mallet, Professor Mallet, of one Blumlein, he wouldn't look at him. He had a Jewish name, and I was told at the time that the Western Electric Company in America, the Bell Labs would not engage Jews. Why, I have never known. It was not correct because in my own particular field there was Fondiller and we had celebrated three coil-loading systems from a Fondiller and Shaw Patent in Bell Labs. I think [he] was a Jew, I met him, and what a charming man. Now, we both took to one another, Blumlein and I, possibly because we seemed ... oh perhaps I should add that Mac did, under pressure from Mallet to some extent, who told Mac, if he didn't take on Blumlein, he was going to lose one of the best men they had ever had at all in City and Guilds on this type of work. Anyway, Mac took him on, but we were still a bit of a puzzle. The result was that we worked on all sorts of strange things, odd testing things, off the normal routine.'

During that first year at IWE, Blumlein and Kaye did indeed work on a manner of 'odd testing things', including the first examples of the metal permalloy which Frank Gill (later Sir Frank Gill), who was then head of the European engineering department, brought over to them in order that they might investigate its various electrical properties. It was during these tests that Kaye recalled probably the only time in their working career when he had been quicker to deduce the answer to a problem than Blumlein had:

'It was the only time that I can ever recall beating Blumlein to the punch, that's why it stuck in my memory. We were testing with a ballistic method, co- and traversal, in a toroidal coil, which we'd wound on the permalloy sample. It was permalloy tape only about four or five inches in diameter, just wound up and then round, on which we put the two inductive windings and that was a classic test.

We then reversed the counter one way, and we had a ballistic galvanometer at the far end on the other winding, so when you reverse a current the light reflector type galvanometer flashed across the screen. You then took the maximum deflection, and it would vary in current, you altered the voltage, actually across the winding, and then you reverted a current reversal and plotted a curve, which gave us the well known hysterisis type of curve.

Well, we'd finished the tests, I was the one who would normally do that work itself, and when Blumlein was with me, we were both at it and he said, "Oh good Lord, we've forgotten one thing. That this stuff has a very high permeability and it would easily saturate in the earth's field, and we haven't taken account of the earth's field", he said, "I'll go and work it out".

So, he sat down at a desk, and nearly half and hour later, he came along with his results and told me "It's perfectly all right, it doesn't affect it because its a toroid", and I said, "Well, as a matter of fact I'd found that out within ten seconds of you sitting down to start working that out". "Really?" he said, "How?" and I said, "Well, I put the coil in my hand and twisted it smartly..." the earth's field was there of course, cutting the coils and everything else, "...and it didn't deflect the galvanometer one iota, therefore it was having no effect at all". So that was it, but that was the only time in my recollection that I ever beat him, and I beat him by half an hour roughly.'

Blumlein and Kaye were given all manner of other items to test including an early highly pumped vacuum cathode ray tube and various items sent over from Bell Labs in America including an audiometer which could be used for testing the sensitivity of the human ear:

'It simply was an oscillator with an attenuator accurately calibrated for the output, and the person to be tested wore phones, and had to press a button as soon as it became inaudible. Though we had a tell-tale on it too, so we could render it inaudible at any time we wished during the course of the test, just to make sure that the victim wasn't fooling himself.

Anyhow, it worked, we took the ear characteristics of everyone we could find, ranging from the typing pool people down the line and all over the place, and we got an average ear. From that, Blumlein then did something that was absolutely typical, he calculated a network that would respond to the average ear, in other words, you put sound in and it behaved as if the average ear would and it gave you a reading out on the output end. The great feature was this, this is what he was after, all measurements in those days were done primarily by listening, a pair of headphones and so on, but you could have one of these things on a cable system which was being tested, and you could check through miles and miles of cable or, of course, through a dummy cable in a box, an artificial cable, and do measurements of that nature which gave us much more accurate results.

I don't know how far it was used, but anyway that was the principle, and of course the principle is used in this day in many fields, but the thing was he did the electrical equivalent of the sonic characteristics of the human ear, having first of all, somewhat arbitrarily determined the human ear's response from a large number of measurements. In the course of that exercise, we had to make, for the network, resistors of very low value, milli-ohms, and they had to be accurate.'

Though the work being done by both Blumlein and Kaye in the transmission laboratory may well have seemed somewhat arbitrary in its organization, it was however, to have far reaching consequences in the months and years to follow. As the various tests were carried out using the audiometer to calculate an electrical equivalent of the sonic characteristics of the human ear, Blumlein began to lay the foundations for the work which would eventually lead to his first major invention in 1926, and to the first of his 128 patents in 1927.

It is fact (though none the less one of the many curiosities about the life of Alan Blumlein), that he rarely wrote in any great detail about his work and, therefore had such a small quantity of his own work published. As has been established, he had always found academic subjects which did not interest him in the slightest, such as writing, far too time consuming, and therefore did not consider dedicating the time required to publishing his works. His entire contribution to technical literature actually amounts to just two papers for the Institute of Electrical Engineers, the first of which he co-wrote with Professor Mallett in 1924 (it was presented in 1925, and published later the same year), and the second in 1938.

He would say himself, some years later, that he was always too busy to think about writing, which is very probably the case; but the plain fact is that coupled with the secrecy that surrounded some of his later work on radar, as well as the fact that some of his work was misunderstood by his peers until quite some time after his death, very little of the enormous amount of work which can be attributed to Blumlein has been fully or widely appreciated. Despite this aversion to writing, even Alan Blumlein realized quite early on in his career, that publishing your discoveries was one path to recognition among your peers and a chance to further a career. However, it is doubtful that such fanciful notions interested him in the normal manner, as Blumlein was not an ambitious man in that way.

Strangely then, having worked with Professor Mallett on their paper entitled 'A New Measurement of High-Frequency Resistance Measurement' (which had created the first real interest in the young engineer when it had been presented to the Institute of Electrical Engineers in 1924), Blumlein now decided that perhaps this might be a worthy first article for publication. The paper had in fact won the IEE's 'Premium' award for innovation and, very probably IWE, perhaps realizing that they had a valuable prodigy on their hands, encouraged Blumlein to publish some of his work.

In 1925, the most widely read publication for the electrical engineering industry was *Wireless World*. In conjunction with Norman V. Kipping (later Sir Norman Kipping), a colleague at IWE, who actually drafted the work, they approached *Wireless World* in June 1925, with an article entitled 'Introduction to Wireless Theory'. The piece was intended to outline the potential of what electricity could do, rather than what it actually was, to an elementary-level readership, who were fascinated by the seemingly boundless possibilities of electricity and magnetism, and it was written in a very basic, easily digested manner.

*Wireless World* was interested enough in the piece and decided it could be run as a series throughout the autumn and winter of 1925. The first instalment of the article appeared in the issue of 21 October 1925, of *Wireless World* and subsequent instalments ran each week (*Wireless World* only later became a monthly periodical), until 9 December 1925, a total of seven parts (there was a break during the issue of 25 November 1925, for reasons which are unknown). These seven parts were titled weekly: 'Elementary Principles of Magnetism and Electricity', 21 October 1925; 'Power and Energy in Electrical Circuits', 28 October 1925; 'Series and Parallel Connections of Batteries and Resistance's', 4 November 1925; 'Magnetism and Electromagnetic Induction', 11 November 1925; 'Electrical Condensers and Alternating Currents', 18 November 1925; 'Inductance in Alternating Current Circuits', 2 December 1925; and finally, 'Capacity and Inductance in Alternating Current Circuits', 9 December 1925.

Compared to the later works for which Blumlein would be responsible, the article is somewhat basic. Perhaps this is to be expected from somebody who, until then, had only had to write large pieces of text for his degree, and who was after all not the most accomplished of writers at the best of times. In fact, throughout his life, Blumlein had always to work hard to dedicate himself to the time needed for actually writing down the details of his works, especially those of his later patents. He was much more likely to keep notes on scraps of paper, the backs of envelopes or any piece of paper that came to hand, and even then he regularly put these in his pockets and would forget about them; certainly he adopted this rather than a neatly kept series of diaries of his progress.

The article 'An Introduction to Wireless Theory' is written in what can only be considered rather elementary terminology, with the series covering subjects such as electromotive forces, conductors and insulators, electrical resistance and potential difference. As the article develops the subjects do become a little more detailed and Kipping and Blumlein eventually discuss potential gradient, series and parallel connections and the actions of a condenser. Blumlein's dislike of mathematical formulae is quite evident throughout and very little is used to demonstrate anything other than the most basic principles.

Only in the final article of the series, dated 9 December 1925, do Kipping and Blumlein delve into what might be considered anything more than elementary electronics when they discuss capacity and inductance in alternating current circuits. Indeed, the editor of *Wireless World* felt obliged to include a small explanation box within this particular article, pointing out that: 'In this, the concluding article in the series, the authors explain the principles underlying the tuning of wireless receivers. In previous articles... [they] have dealt with all the more elementary aspects of electrical theory, and form an excellent introduction to the technical articles appearing from time to time in the pages of this journal.' One gets the feeling that the editor is almost apologizing to the reader for having published an introductory article which has, by the final instalment, become rather more specific requiring a higher degree of knowledge.

It is debatable exactly what can be determined from this first published work by

Blumlein and Kipping. While *Wireless World* was undoubtedly widely read and could certainly be considered one of the foremost journals of the day, it has to be said that it would probably not have furthered either's career very much. Certainly the article, even though it was published at a time when Blumlein had just turned 22, can only be thought of as elementary and, considering the calibre of the work that he would be publishing in the next few years, the piece is probably written to be deliberately simplistic. One can be tempted into thinking that the article had in fact been conceived by Kipping with technical expertise added from Blumlein and indeed this may well have been the case, though no corroboration on this can be found from either Blumlein's or Kipping's notes to my knowledge.

Certainly the style of the writing, while undoubtedly edited by *Wireless World*, does not read like later Blumlein texts which are far more articulate, precise and deal with subject matters well beyond those encountered in the *Wireless World* article. Whether Kipping and Blumlein determined that the journal needed an introduction of this nature or they were commissioned to write on the subject is also open to speculation. At this distance, no records exist at *Wireless World* in its current form which can be used to shed light on the matter, though needless to say, Blumlein did continue to have his work published, he neither co-wrote with Kipping or used *Wireless World* again.

In the meantime, Blumlein and Kaye had a rather jolly time working, so it would seem, very much to a schedule of their own making, on any and every apparatus that came their way or, indeed for that matter, which they could lay their hands on. In some cases this did not necessarily belong to them as J.B. Kaye again recalled:

'We had a bridge in one of the laboratories. We were then working not in the laboratory that was in general use, but in one of our own which was used under the control of E.K. Sandeman. Anyway, Blumlein and I were by ourselves, busily working on the resistors, and we got it down to as accurate a point as we could, and this bridge was of the post office box type which had plugs to adjust the ratio arms and the value against which you were checking your resistor.

Now it also had two keys, one put the power on, you press that first, and the second on (the second press button), put the galvanometer into circuit. That was really to enable you to get somewhere near your balance point, and then you really brought it in. I can't remember whether there was a rough guide to balance or not. Anyhow, we got it down, and the laboratory voltage we used had batteries, car batteries, and the normal voltage in the communications field in those days was twenty-four volts.

He suddenly said, "Good Lord, it isn't accurate enough, we must get it really accurate", we were just scratching at copper wire to get the thing to the right value, he said, "More volts, that's what we need". So we slapped on another twelve volts, pressed the bar button, there was a loud "whooompppth", and then smoke started to emerge from this celebrated and expensive bridge.

We hastily disconnected everything, carried it out through the swing doors onto the fire escape. There was a light well at the back of Connaught House, and the other half opposite us was occupied by the Air Ministry. So, they had a vision of two very young engineers emerging through the fire escape, with a very expensive looking box that was puffing out smoke. Well, we were a bit shaken.

Anyway, we let everything cool down, and then we unscrewed the lid; well, there was a funny smell coming out of it as well as smoke by that time, and these resistors were coloured with what looked like carbon. It had a fabric-textile insulation, but luckily you see, they were the resistors against which we were balancing a resistance of milli-ohms and they were the ones that had very low resistance, which meant that they were pretty good thick wires, and we could hardly believe it when we cleaned the bridge up inside and then quietly took it elsewhere, away from observation, and measured the standards in the bridge by means of another bridge with equivalent accuracy, and found that it hadn't been damaged at all. So, I'm still wondering where that bridge is, because we put the lid on and screwed the screws down and that was it, but you know it caught us out badly that time.'

The recollections of J.B. Kaye give an insight into the kind of man that Blumlein was at this time and, perhaps more specifically, the lighter side of his nature. Alan Blumlein enjoyed practical jokes and clowning around as much as the next man did, and, on several occasions subjected a victim or two to the pranks that seem to have been carried out regularly at IWE:

'We managed to bring a little lightness into the place. Blumlein was a natural clown and, I'm told I was too. That's possibly not immediately recognized; they (IWE) were a bit puzzled over it when we arrived. You would find, for instance, with one of the laboratory oscillators, audio oscillators, that Blumlein found that he could play the opening bars of Peer Gynt on the oscillator. We always came to a gap, and we never got over that gap because he would be working on the opening bars and I was trying to cover the gap, but I didn't have three hands and couldn't quite get the frequency that was required.

We also had one or two experiments in dynamics. The laboratory in the offices were all equipped with these desk chairs that would rotate and, unlike the modern ones, these would really spin. You could adjust the height with a great bit-screw-thing at the bottom, so, we would lower it, then persuade some innocent victim to come and take part in the experiment.

The victim would be persuaded to sit in the chair, stretch his legs out forward and his arms out sideways as far as they would go, and then give him the spin. We would spin it very hard and he'd been instructed that, at a signal from us, he should draw in his arms and legs quickly. Well, when he did that, he hadn't realized what would happen of course. The speed of rotation accelerated tremendously. I forget which law of dynamics it is that covers the situation, but the speed was really quite dramatic and, in at least one case, the occupant of the chair came flying out, and landed across the lab. That was quite interesting.'

Of course there was serious work to be done as well, and the primary function of the engineering departments in which Blumlein and Kaye were a part of, was to determine which were the elements of electrical interference in telephone circuits that could be damped and to what degree. The cables worked at audio frequencies only and each cable contained many pairs of wires. These in turn had to be prevented from inducing cross-talk between one pair and another and also from picking up any interference from external sources such as the transients on the power supplies of electric railways. The need to improve the quality of communication of telephone systems was essential to the growth of national telephone networks which were becoming linked together and which, in turn, were linked to international telephone networks. The problems of extremely long wires, compatibility, overall speech quality and the delay time of the voice carried across such distances became more and more apparent the further the lines were spread.

Various departments had been set-up at IWE to investigate interference from power cables and the many and various pieces of electrical equipment used in telecommunications. The principal method for testing at that time was 'balancing', which involved arranging each pair of wires in a given cable to receive equal signals from the interference source (once it had been established of course). The capacitances then had to be measured for each wire in the cable and its near neighbours and this needed to be done for the entire length of the cable after it had been laid, but before each length of cable (usually 100 to 200 yard-long lengths), were joined up. As you can imagine, this testing process could require thousands of measurements of capacitance to be carried out in all conditions open to the elements and to within an accuracy of a few picofarads.

Perhaps it should also be pointed out that by 1925, the principal business of telecommunications technology had become very much a science, but obviously the testing procedure hindered the expansion of the telegraphy and telephony industry because of the time it took to test these cables for interference. It became apparent therefore that with the ever-growing need for international telephone exchanges the problem of line interference was one that had to be solved quickly and efficiently.

The European engineering department, headed by Frank Gill, was being run by John Collard, who had, for some time, been working on a series of projects that were designed to solve the then very serious problems associated with interference in long-distance telephone cable. Blumlein was beginning to get a bit of a reputation around IWE as a man who was determined to get a task solved, and Collard got to hear of this. He decided to take Blumlein into his team not only because he had shown an aptitude for electrical engineering and circuit design, but for problem solving as well; his reputation for being stubborn to the point where he almost took it personally if a problem could not be overcome, meant that he was just the kind of man Collard needed for the interference project.

Initially the project was based around a series of interference tests mostly conducted in the laboratory, though some were performed in the field. Blumlein spent much of the remainder of 1925, and the early part of 1926, working from the Aldwych offices in London, of the newly renamed International Standard Electric Corporation (International Western Electric had changed name in early 1926). His work, which was at that time just a small part of the many experiments that were taking place in the European engineering department, now included interference tests on all manner of electrical devices which International Standard Electric Corporation (ISEC), regularly delivered to the laboratory. This included completion tests on inter-city telephone lines which involved measuring the returned signal strength over long distances and comparing these with the known transmitted strength to determine what degree of interference had taken place.

## European trips, flying lessons and patents

In May 1926, the General Strike presented an intriguing chance for Blumlein to enjoy another favourite pastime of his, that of trains and railways. As a child he had stood for hours at a time, watching the steam trains as they passed from a vantage-point near his home, or at St John's Wood Station. Of course, during the days of the strike, it was essential to keep the railways going to supply the country with food and coal, and they were run by volunteers. Blumlein, Kaye and another colleague from ISEC, D.L.B. Lithgow, all decided to volunteer to help with the railways. The IEE had declined to have any connection with the strike because they were not allowed to take part in any political activity, but an office had been opened at Savoy Hill for engineers who might help in manning the communication networks.

So, Blumlein, Kaye and Lithgow went to notify the head of the engineering depart-ment at ISEC that they wished to volunteer for this duty, but were promptly told that as they were part of an international company, and therefore in effect they had 'no nationality', their only loyalty was to the company. It was also mentioned that they need not expect to have a job waiting for them when they came back if they insisted on going. As Kaye later explained: 'It wasn't as though it was the sack because nobody knew how the strike was going to end, it was just the point that if we went into this activity, if there were to be any redundancies, they would be ours.'

Regardless of the warning, they decided to go anyway as the Government was calling for volunteers, especially engineers, and so having been instructed to go to Euston Station, they were told that their job would be to keep the telephone exchange at Euston working. Once they had arrived, they discovered that the switchboards were not in very good condition, the cord circuits had broken down and so on and, while there were girls operating the switchboard during the day, their job entailed taking over from eight o'clock at night until eight the following morning. So, during this time they set about wherever they could replacing the cord circuits, and so Alan Blumlein, J.B. Kaye and D.L.B. Lithgow kept the Euston station telephone exchange running during the General Strike of 1926.

Luckily the switchboard was not a big one, perhaps four or five full size boards, but at times it was very busy and at one point they saw something that none of them had ever seen before and that was the complete telephone exchange with every calling light on. At the end of the strike on 12 May 1926, without wishing to sound con-ceited, J.B. Kaye said: 'that switchboard, or the switchboards in that exchange, were working at a level of efficiency, technically, that had never previously been achieved'. During the days of the strike the railway company paid them a weekly wage for what they were doing, and on 13 May, which was the day after the strike fin-ished, they were called in to make sure the boards were all right. When they went to the office they found that they were received with open arms, the railway company had told ISEC how pleased they were with what had been done, and International Standard Electric Corporation were tickled pink because they were very anxious to be on good terms with the railway company because of the prospect of getting sig-nalling equipment contracts from them. So, in the end they had been paid by the railway company, ISEC who were employing them during the day ended up paying them their full rates of pay and, to cap it all off, some of the railway companies paid a bonus to them and gave each a medal for the work that they had done.

It was around this time that all the work which Blumlein had earlier carried out with Kaye on the audiometer, and the subsequent frequency network tables which he had collected and calculated to devise an electrical equivalent of the sonic charac-teristics of the human ear, would finally come into its own. Blumlein used his system on the many different types of cables being used, and was able to check through miles and miles of it obtaining results that gave a fairly good representation of the frequencies that the human ear would pick up.

The work that had been done in late 1925 and early 1926 would become the basis for the first patent that Blumlein would file with another of his colleagues in the European engineering department, John Percy Johns. It was Johns and Blumlein who conceived the idea for 'A method for reducing mutual interference between channels (cross-talk) in long-distance telephone cable systems' (eventually to become Patent No. 291,511), and they were certain that much of the interference that was being produced was as a direct result of the three-coil inductively loaded cables that were being used at the time.

Long distance telephone systems were being developed which used a modulated carrier frequency in which the conversation would be shifted into separate frequency bands. However, these would not be available for a number of years. In 1926, the conversations were still being carried along the telephone lines at the original speech frequency. In order to equalize the signal velocity over the entire frequency band, the cables (which were formed into what was known as 'quads', that is, four wires, one pair of which form the side-circuit while the other pair form a phantom circuit) were inductively loaded by three-coil units and were the source of much of the cross-talk.

Blumlein and Johns studied these three-coil units, identifying each of the cross-talk characteristics and its source with great accuracy. They realized that one of the major reasons for the cross-talk was the capacity imbalances between the various component parts of the quaded circuit, including the wires, coils and connecting wires with the resistance and inductance unbalance of the windings, with mutual inductance between the circuits also contributing. The result of this work was the idea to rearrange the windings to concentrate the imbalances at the point of connection, and then to deal with each connection and imbalance individually. The capacities to core of the inner layers of the windings would therefore be concentrated at one side of the coil (as regards the lead-in wires from the cable). It was then a simple matter to effect a balance at one point by means of condensers or pairs of wires, the mutual capacities between wires of each pair forming the balancing capacities.

Once it had been found that these capacity unbalances were being caused in the phantom coil by the capacities between the terminal wires on the coil and the outer layers of the windings which they embraced, the terminal wires and connecting wires were brought round the coil so that the capacities between them and the outer layer of the windings were balanced. Blumlein had invented a measuring set to meet the conditions that the test engineers would have to encounter working in the field which consisted of a capacitance bridge in which the ratio arms were the windings of a mutual inductor basically a transformer. The method worked, so well in fact that it was used in its various (improved) forms until very recently, only the advent of digital/optical telephone systems making it obsolete.

International Standard Electric Corporation were naturally very pleased with this important development, which not only greatly improved the quality of the telephone systems that they were installing all over Europe, but cut down enormously the time taken to test for interference by the teams of engineers in the field, thereby saving the company enormous sums of money. Blumlein and Johns had shown that with accurate analysis of a problem it could be narrowed down and overcome, and their persistence was rewarded in September 1926 with a £250 bonus, practically a year's wages at the time. With the money, Blumlein decided to buy a motor car, a bull-nosed Morris, and this little car was to give him a great deal of joy as, once again J.B. Kaye recalls:

'Again, his experimental interests came well to the fore. He came in to the office one day, and told me, "I've discovered an interesting way of skidding. I can get round the Marble Arch quicker than anyone else. What you do is, get your wheels in the gutter, and then you accelerate. If you can control your speed..." this can only be done really on a wet day. "...you can keep it just within that periphery..." the sort of ground on which the Marble Arch was standing. "...If you overdo it, you're liable to become detached, and go sideways."

So he took me off to demonstrate it around Russell Square. It was quite fascinating, a bit nerve-racking for the passenger though. I gather that one night, a wet night, he was using this technique in Wigmore Street, and I think he was turning down one of the side streets, anyway, he did become detached. He broke away right across the road and got stuck into a paned glass window. Another event, he was coming to the office one morning down the hill from Hampstead towards Camden Town cruising at a pleasant pace, and suddenly he saw a wheel pass him, a motor car wheel. He was still carrying on normally and just thought, "somebody's lost a wheel, what fun!" This wheel went careering on, and to his horror he looked in the mirror and saw there were no other cars on the road behind him which could have possibly shed a wheel. He came to the logical conclusion that the wheel had come off his bull-nosed Morris, which it had. It shot down the road, struck the kerb, went straight up in the air and landed in somebody's front garden. Meanwhile, Blumlein, with considerable skill, he could think very quickly on all occasions, brought that car to rest by careful braking, and steering until it came to a stop and sat down on the brake drum.

Another time, it was when he had the car and we were living in the digs, we set off one morning for the office, and drove along the Embankment, when he said, "I'll have to get some petrol shortly". Suddenly a policeman stepped out in front of the car, the usual manner of those days, and signalled us to stop, exclaiming, "You have been timed over the measured furlong and you have exceeded the speed limit". Blumlein, with a broad grin said, "If I hadn't run out of petrol on that furlong I would've done it much quicker!"'

Blumlein's driving would become legendary in the years that followed. He would think nothing of explaining a complicated engineering problem to whoever was in the car with him by drawing it on the windscreen – as he was driving!

It was around this time that ISEC put Blumlein in charge of his own team which had the responsibility of developing yet more analytical processes for improving the performance of the system, and it was through this team that several other major breakthroughs were to be made. It was because they had carried out such detailed and accurate analysis that Blumlein and Johns had been able to improve the interference in the ISEC telephone system, but this had posed another problem which they could not have foreseen. In order to carry out the development it was necessary to measure very small differences in impedance, far smaller in fact than any of the laboratory test equipment which was available to them could accomplish.

To Blumlein, ever practical, the problem was simple: an accurate tool did not exist, so they would have to invent one. The existing impedance bridges of the type available to Blumlein and Johns had two major problems for the type of measurements that they wished to carry out; firstly the ratio arms were insufficiently precise; secondly, they were affected too much by stray shunt capacitances. What was required was a bridge that could measure direct capacitance between two conductors regardless of capacitances from them to earth. The original form of the bridge which he came up with, was called 'the closely coupled inductor ratio-arm bridge' (this would eventually become Blumlein's second specification, Patent No. 323,037).

At first, this looks to be a conventional AC bridge network and a deceptively simple device, but this is perhaps the beauty of the method which Blumlein came up with, for the unit is capable of measuring capacitance across both arms of the bridge with a great degree of accuracy. Blumlein's idea was that by the close inductive coupling of the two ratio arms of a bridge, you can bring the three terminals of these arms to very nearly the same potential and so, by connecting the centre terminal to earth, automatically get rid of the effects of earth-capacitance at all four corners of the bridge. Bridge networks in general have most difficulty in realizing that 'at balance' the connections are effectively at earth potential. This means that because there is no potential difference across them, any capacitance admitted across the bridge has no effect, and cannot therefore be measured.

Blumlein overcame this problem by using bifilar winding (the start of the two windings is connected together, instead of one start being joined to the other finish) making the two coils occupy substantially the same position as their iron core. In this the construction of the bridge automatically achieved a 1:1 ratio with a much higher degree of precision and constancy than any conventional resistance arm. At balance then, the close coupling now ensured that there was no potential difference across either ratio arm and therefore even small admittances across them can be measured in the presence of much larger capacitances to earth.

It was a quite brilliant design and, in the years that followed would be exploited in all manner of applications, though curiously at first ISEC did not produce large numbers of the bridge. This was probably to restrict its use by their competitors such as the Post Office who were still using the Schering bridge with a Wagner earth many years after Blumlein's patent had been first published. While the Post Office must have been aware of Blumlein's patent, they probably did not appreciate quite how important it was, possibly due to this restricted production run.

By now Blumlein and Kaye were sharing digs together in Hampstead and, as both men were very fond of dancing, they often used to go to the Chelsea Arts Ball together. On one occasion however, Kaye had stayed in and Blumlein had gone out with some other friends and had evidently not arrived home again until the very small hours of the following morning. When Kaye got up in the morning and walked down to his bedroom, banging on the door to make sure he was awake, there came a groaning sound from within followed by, 'I do feel ill, I feel terrible.' J.B. explained,

> 'Well, of course, he was suffering a bit of a hangover, although I'd never known him the worse for liquor, but I think he was just whacked out and had had a

merry evening and that was it. So I said, "Well, I think you better stay in bed", and he said he felt rather like it, and didn't feel at all like getting up. I said "Righto. Leave this to me and I'll tell your boss", who was at that time John Collard, who always seemed to me to be a sad type of individual. I went trotting along to the office and went and saw Collard, and I said to him, "Blumlein won't be coming in this morning, he's a bit off colour". "Oh dear", he said, "I'm sorry about that, what's the trouble?" I said, "Well, he looked very pale to me and I think he's suffering from a stomach upset". "Oh dear..." said Collard, "...and he's in digs?"

Now Collard, I think, was a man who had never lived in digs as far as I can make out, and therefore had great sympathy for those that had to live in digs, and, as he thought, rough-it. Anyway, so I laid it on a bit, and Collard said, "Is he having a doctor?" I said, "No, I think a little rest will do him good", and elaborated on this theme a little bit. Anyhow, I overdid it you see, and Collard was going to go round himself. Well, as I had persuaded him not to go, I'd finished saying my piece, and suddenly the door opened, and in walked Blumlein as cheerful as a lark, grinning all over his face.'

In late 1926, John Collard decided to take Blumlein on the first of a series of trips around Europe. They were going to sort out the interference problems in some of ISEC territories abroad. The company had interests in many European countries at this time and so Collard, seeing the potential in young Blumlein, included him in a team of engineers who were taken abroad to improve the telephone systems in Strasbourg and Paris in France, Madrid in Spain, Milan in Italy, and Geneva and Lausanne in Switzerland. One interesting advantage to all this foreign travel that Blumlein soon discovered was that he could indulge his passion for trains and railways. During the time that he and Collard were laying telephone cables across Switzerland, they found that the geography precluded following any kind of straight line in the valleys and the only way to lay the cable was alongside the Swiss AC railways.

These were mainly the sixteen and two third cycle, fifteenth K, Swiss railway system which were working on the same voltage in the 1920s as they do today; and we know that Alan Blumlein took a great interest in this because years later his son Simon (who incidentally also has a passion for trains and railways), came across his records. Among these were photographs of engines that Alan had taken in the mid-1920s, and were identical to pictures that Simon had taken in Switzerland when he had become interested in Swiss railways, decades after his father had died. The laying of the telephone cables alongside the railways tracks caused interference from power circuits since, on some routes such as that over the St Gothard, the power lines for the railway system ran in very close proximity to them. This caused interference associated with the starting, acceleration, deceleration and stopping of the electric locomotives themselves.

Blumlein's work in overcoming this interference was evidently very highly thought of by the Swiss engineers and, in a report from H. Melling (who was associated with some of the work being done in Switzerland), the following account was given: 'I learnt at first hand from Swiss engineers of all grades in the Cortaillod and Government organisations what they thought of this brilliant young English scientist engineer. They were almost as much impressed by his physical determination and endurance as by his technical genius and tenacity.'

Figure 2.1 In 1927, Collard (standing outside the car) and Blumlein travelled to Switzerland for ST&C to carry out coil tests on the level of interference in power lines which followed the route of the Swiss electric railways.

The 'grand tour' of Europe kept Blumlein away from the office in London for some considerable time and, while the result was yet more accumulative data for his interference tests (which in turn resulted in successfully reducing the interference in all the telephone systems that the team went to), it did mean that long periods were spent out of England. Blumlein appeared rather raw at first, Collard later recalled; it was nevertheless his first trip overseas and young Blumlein, unaccustomed to the vagaries of diplomacy, must have seemed somewhat innocent and ignorant to the worldly Collard: 'When he joined he was a difficult person to take to meetings, especially in other countries. He could be gauche and he could be very rude. He had no patience with anyone less brilliant than himself. He would work all hours himself and was quite irritated with anyone else who, for personal reasons or just for pleasure, would not join him. In time, however, he lost some of his aggressiveness and acquired a certain amount of tact.'

The picture that Collard paints of Blumlein is vastly different to that given by J.B. Kaye and other friends and colleagues later on in his career. While everyone who knew him would say that Blumlein was indeed a workaholic and subject to impatience with those he considered to be fools, it is also the case that he had infinite patience with people at times. Provided they were trying hard enough to understand the problem at hand. His time as an assistant to Professor Mallett is a good example of this and, much later at Columbia and EMI, he would spend a great deal of time with junior members of his team explaining things. Blumlein too had a lighter side to his character which had all the elements of tomfoolery and practical joking. However, it would seem that Collard was not a party to this side of his nature. It is perhaps the case that Blumlein and Collard never quite saw eye-to-eye on certain subjects. Indeed, there may well have been a little professional envy on the part of Collard. If this was the case, it never came out in the open and, indeed years later, Blumlein would call upon his former travelling partner to come and join him at EMI.

Because of his extended visits to Europe with Collard, this might well account for the fact that Blumlein's first patent was not applied for until March 1927 when he got back, though the work had been completed in mid-1926. Certainly, Blumlein was far too busy with his inventing and the application of his inventions, to concern himself with such time-consuming trivia as publishing. Thankfully, Standard Telephone & Cables (as ISEC had now finally become), had seen the potential of both the method for reducing mutual interference between channels in long-distance telephone cable systems, and the closely coupled inductor ratio-arm bridge, and it instigated the application for the first of Blumlein's eventual 128 patents.

Applied for in the name of Standard Telephones & Cables, Columbia House, 2 Aldwych, London, British Patent No. 291,511, it is titled 'Improvements relating to Loaded Telephone Circuits and particularly to the reduction of Cross-talk therein'. The specification was applied for on 1 March 1927, and accepted (completely) on 1 June 1928, listing Alan Blumlein and John Percy Johns as the inventors. During the remainder of 1927, no further patents were applied for by Blumlein or ST&C, though undoubtedly the work leading up to the solving of the problem of the cross-talk in the telephone cables had been a direct result of, or resulted in the invention of Blumlein's second patent, the Blumlein Bridge. The reason that no further patents were applied for in 1927 is probably that Blumlein and Collard had spent a large portion of the year travelling again. There simply was not enough time for Blumlein to write down the necessary details for the patent application department of ST&C to produce the specification in order that it could be applied for. Blumlein, as usual, had not yet written his work in any format capable of being published.

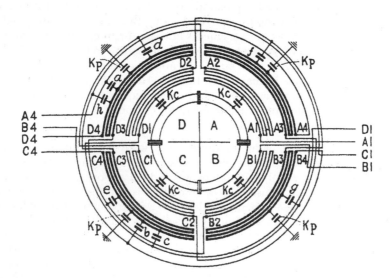

Figure 2.2 Detail from Blumlein's first patent, No. 291,511 (1927), written with John Percy Johns.

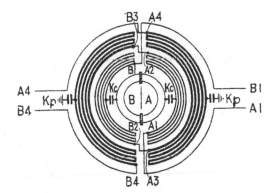

Figure 2.3
Detail from Patent No. 291,511
(1927) showing a cable
cross-section.

From 4 November 1927 to 1 December 1927, Blumlein and Collard were at Amsteg carrying out a series of tests on the Altdorf-Gothard cable, which was the main telephone link over the Alps into Italy. This was a classic test route for any telephone interference and Blumlein spent a great deal of time considering the problems which they encountered on this particular cable. Upon his return to England, he wrote a 35-page memorandum called 'Balancing to sheath on the Altdorf-Goerschenen cable', in which he pointed out that 'the object of balancing to sheath was to reduce those unbalances which produce noise in a cable'.

It was this work on balancing, coupled with the fact that they had needed to build an accurate bridge back at the laboratory in Aldwych, and very probably prompting from ST&C, that finally encouraged Blumlein to set down the principles of his second patent, No. 323,037 in the early summer of 1928. During this time, a weekly periodical, entitled *The Illustrated Official Journal (Patents)*, produced a shortened version of his first specification with J.P. Johns in their issue of 25 July 1928. The journal printed British patents only in their numerical order and only continued publication until 1931, by which time it would carry similarly shortened versions of the first six of Blumlein's patents and their associated diagrams.

It had been eighteen months since his first British patent had been applied for and, in that time having returned from his expeditions with John Collard, Blumlein finally had enough time on his hands to work at ST&C preparing his second patent; by which time his colleagues, using it in his absence, had christened the circuit the 'Blumlein Bridge'. It became one of Blumlein's most famous and useful inventions (all the more impressive when you consider it is only his second from an eventual total of 128) and, during his lifetime, Blumlein would continually use and improve upon the design, constantly being able to find new uses for the circuit.

It is easy see why this quite extraordinary breakthrough in AC bridge design was nicknamed the 'Blumlein Bridge', if only because 'the closely coupled inductor ratio-arm bridge' is a bit of a mouthful. It was, and still is, the most important advance in bridge design since Maxwell applied Wheatstone's circuit to AC in 1865. The curiosity is that, while his contemporary colleagues at ST&C (and later EMI) would continue to call it as such, the name and Blumlein's association with it has not survived into the present day. Unlike Maxwell, Wheatstone and perhaps Hay and Campbell, all of whom have their names associated with the circuits they

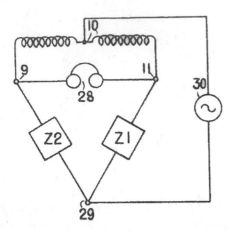

Figure 2.4
Detail from Patent No. 323,037
(1928), Blumlein's celebrated closely
coupled inductor ratio-arm bridge.

designed, Blumlein, with the exception of those who know of his work and who knew him personally, does not. This is, once again, very probably his own fault in many ways, for not publishing more of his work in his own lifetime.

Patent No. 323,037, titled 'Alternating Current Bridge Circuits', was applied for on 13 September 1928, in the name of Standard Telephones & Cables and A.D. Blumlein. It undoubtedly provided ST&C with a very valuable source of indirect revenue throughout the life span of the patent held in their name and it served to draw the attention of the electronics world to Alan Blumlein. Many years later, Blumlein, who had written very little in his lifetime about his work, as has already been established, came to the conclusion that this particular invention of his was in fact so useful, that he would write a paper to be presented before the IEE.

The document, written by Blumlein in January 1941, was unfortunately never published, ending up a series of uncoordinated papers which would probably never have come to light were it not for the attention of two of his colleagues at EMI who, long after Blumlein had died, decided that the essence of Patent No. 323,037 and Blumlein's thoughts and writings on it, should finally be published. Philip B. Vanderlyn and Henry A.M. Clark therefore produced a paper of their own in which they credit the vast majority of the work to Alan Blumlein and specifically refer to his bridge patent. They originally presented their work before the Measurements and Radio Section of the IEE on 21 January 1948, and the paper was finally read on 11 January 1949.

During 1927, Blumlein showed an ever-increasing interest in all aspects of flying. These were the days of Charles Lindbergh and Amy Johnson and so, as soon as he had enough time on his hands, Blumlein announced to J.B. Kaye in early 1928, that he had decided he wanted to learn to fly. By pure coincidence, his flying instructor, a man by the name of Matthews, happened to be the same man who had taught Amy Johnson to fly, so he was in good hands. The instruction club that he chose was at Hendon, and had been the old London Club first formed by Graham White, and now part of a social club called the London Aero Club. Evidently Alan Blumlein was a confident enough student and Matthews an able enough instructor as he managed to teach him to fly in a short time, and Blumlein was eventually granted a pilot's licence on 5 October 1928.

The location proved rather fortuitous as ST&C had laboratories in Hendon as well as the main ones at Connaught House in The Aldwych. So Blumlein, who found it convenient (or perhaps the instructor did), to have his lessons on a Thursday afternoon, had a good excuse to go there for flying lessons. However, while it was quite common practice for people to go up to Hendon during the day and come back to London later in the evening, it was not for the purpose of having private flying instruction. So, Blumlein continued his flying lessons and evidently made quite a good student, concentrating on all aspects of aeronautics at home and mastering all the physics of flight as best he could before taking to the air solo.

However, on several separate Thursday afternoons he was very nearly found out by the company, and it was left to his friend and colleague, J.B. Kaye to cover for him: 'On a Thursday afternoon, not once but possibly two or three times, the man in charge of us would say, "Oh, where's Blumlein this afternoon?" and I would say "Oh sir, he's up at Hendon", which was truthful, descriptive and we got away with it every time. Of course Blumlein was 'up', but he was well above the laboratories, not down in them.'

Blumlein had several rather interesting experiences while flying at Hendon, one of which occurred the first time he flew in a plane with slotted wings, a de Havilland Gipsy Moth. He was coming in to land and had cut the speed down, but had forgotten that slotted wings would greatly decrease his progress towards the ground. Normally if such an approach had been made the aircraft would have stalled and spun in, but the slots actually prevented him from spinning. The net result was that he floated down rather horizontally, Blumlein only realized that there was a problem when he saw games of tennis very close to him on the ground below. Being in a somewhat precarious position, he opened the throttle wide and pulled the nose up, fortunately not too far, and did a hop, and landed right in the middle of Hendon aerodrome. As he explained to J.B. later, he didn't roll at all, but just 'sat down'. The plane appeared to sit down around him, flapped its wings a bit, and he got out unscathed. The people at the London Aero Club later took the plane to pieces, because they couldn't believe it would take such a bump without over-stressing something.

He used to tell J. B. Kaye about his many theories on flying: 'These planes...' Blumlein would explain, '...they're very stable. If you take one of them high enough, theoretically you can, if you've got enough air space, let go of everything, put it in a very awkward position like a climb or something, or even upside down and just let go, just take your feet off the rudder bar and your hands off the joystick, and let it go.' 'I believe actually, he was doing tricks like that', J. B. recalled. 'Of course, he had to do them at considerable altitude. He once said to me, "You have to have sufficient time in the air to enable it to come out and fly straight". But that was the way the machine was designed in his opinion and therefore it should do it.'

'One lunch time, as we were walking down Kingsway – we very usually had lunch together – we were talking about aeroplanes. We always talked about aeroplanes or machines of some sort, or the sort of things we were doing in the lab, and having a jolly good laugh at things, but he said, "You know, one day, these aeroplanes will fly without engines as we know them. They'll be propelled", and he said, "They will be burning diesel oil and not petrol..." or whatever oil was used in those days, I don't know quite what it was; and he said that "...If you can get the right shape on a surface, and you can pour oil over it, and you can blow air hard at that surface, it will ignite. And if you can so arrange it that you have a sort of cupola in front, with the nozzle down which the air comes in, then, it will ignite inside. And the ignition of that oil combustion can blow against this sort of cover in the front, and you can deflect that back through the tail. It'll blow the flame backwards and, so blow the engine forward taking the aeroplane at the same time..." he said. "The only problem is, to get it up to the speed at which this thing would ignite." I don't think he used the word "jet", he may have done, but that was a good prediction of things to come.'

## Undersea cables, then time to move on

From September 1928, the next six months at Standard Telephones & Cables were extremely busy for Alan Blumlein as the potential of his two initial specifications were fully realized, and the work being carried out on international telephone exchanges increased. These two inventions were to lead to several others mostly as a direct result of the work that the engineering team now found themselves engaged in. The irony of the work carried out during this period is that the five patents that resulted from it, all in the name of Standard Telephones & Cables and A.D. Blumlein, were each applied for after Blumlein had in fact left the company to take up a new challenge at the Columbia Graphophone Company in March 1929.

Blumlein's third patent is the first in which he alone is credited for the work, and not with his employers, Standard Telephones & Cables. Instead, the specification gives the inventor as 'A.D. Blumlein of 20 Lynchcroft Gardens, Hampstead, London' (a modest house in Hampstead where Blumlein had taken up lodgings not long after joining ST&C and which he shared with J.B. Kaye). Patent No. 334,652 was applied for on 29 June 1929 (Blumlein's twenty-sixth birthday) and in it Blumlein returns to his bridge in considering its application with a telephone loading and phantoming coil.

Pointing out that in order to correct for, or introduce distortion in magnitude or phase in an electrical transmission system, the use may be made of networks which are inserted in the system to modify the relations between the input and output currents at various frequencies, Blumlein explains that the object of his invention was to combine the efficiency of transmission with economy in the cost of installing such a system by reducing the complexity of the networks employed. Blumlein uses what he calls a 'half-Wheatstone' network, which was derived from the original Wheatstone bridge with fifteen diagrams showing various modifications that Blumlein had devised. Much of the specification is then given over to these variations and the mathematics associated with the results obtained from the attenuation, resistances, capacities and impedance values which the modifications brought about.

By the time this specification had been fully accepted, Blumlein, of course, no longer worked for ST&C and therefore could not, in all reality, work with his bridge without infringing ST&C's interests in the patent. This accounts for one reason why he returned to his design so often. The other reason was that he never seemed to run out of uses for the little circuit and constantly found ways of improving and adapting it which, in turn, led to further specifications on the same theme.

During the last few months that Blumlein spent at ST&C, there were many lighter moments in life, some of which were recalled by J.B. Kaye, coupled with examples of how patient and considerate Blumlein could be when faced with the problem of a junior member of staff trying to solve a problem:

'When we were together at Standard Telephones & Cables, there was someone there who was very, very slow. He was one of the old brigade and he was so slow in getting any action going, it was almost unbelievable. For some reason he'd been nicknamed "Beezelbub". It's possibly just as well I can't remember his real name, anyway quite a decent bloke, but he just was terribly slow, so we had a unit called "The Beezel". But, it was such an enormous unit; well it was really a ratio of the time it took Beezelbub to do a job, and the time it took a normal bloke in the office to do it. Anyway there was the Beezel, but for all practical purposes, we all had to work in Deci-beezels.

So as I say, one of the main things I can always recall in those days was laughter, and it wasn't the brittle laughter that you have in industry in the present day, it was good hearty laughter, it was great fun. We weren't plagued either by mathematicians who could prevent young engineers and people from following a course they wished because it wouldn't work. I've seen more people damaged, I think, young men in laboratories, by these purely mathematical, theoretical types who, will try to point out how damn silly it is to do it the way the youngster is thinking. They do more damage by that than almost any other thing in a laboratory.

Now Blumlein never did that. I never heard him ridicule the youngster who was trying. I've heard him ridicule people who've really deserved it, and who should have been able to hit back if they so wished in the same terms; but, he never did anything that would hurt anybody in that way.'

Blumlein's quick thinking did not always, however, produce the desired results for himself. On several occasions he could be known to be impatient at getting the job done, and getting it done quickly. He was impatient with people who didn't want to learn, but it didn't matter how little knowledge one had, if the person wanted to increase it, and Blumlein could play a part in that, he never tired of helping them. One could be the biggest fathead in a particular problem, and Blumlein would make endless attempts to help one see the problem clearly. Effort and time would mean nothing to him, so long as he could get the solution across.

He never used bad language. Blumlein didn't swear, he didn't waste words, but he could be extremely witty at times without being cutting. J.B. Kaye recalled one case in which he had an argument of some sort, with the Institute of Electrical Engineers on passive and active networks: 'I won't name the individual with whom the argument arose, except to say that the following morning that individual rung him up to continue this discussion, and someone who was in Blumlein's office at the time heard him say to the man who'd called him on the phone, "Mr so and so, I may not know the difference between a passive network and an active network, but I do know the difference between a passive network and an active nitwit! Good morning."'

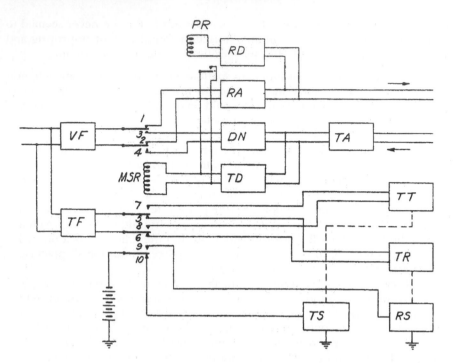

Figure 2.5 Detail from Patent No. 335,935 (1929), a simultaneous telegraphy and tele-phony system based upon time division multiplexing designed by Blumlein.

There is a fine example of Alan Blumlein's talent for putting together an entire picture in his mind, a concept of how a complete electrical system could come together, and this is seen in one of his last patents while still at Standard Telephones & Cables, worked on in late 1928. Patent specification 335,935, entitled a 'Complete Signalling System'. It was, in fact, a simultaneous telegraphy and telephony system based upon time division multiplexing and years ahead of its time.

Blumlein had observed that in a composite system in which the circuit is alterna-tively arranged for speech transmission in one or other direction only, the telegraph facilities also operate in the same direction as the telephone transmissions. If, for example, the telephony proceeds in an east – west direction, then the telegraphy also proceeds in the same direction but, because the telegraphic information only worked at frequency bands well below those used for conversations, and were much slower to transmit, there was always the possibility that the telegraphic data might be cut short in transmission.

The invention that Blumlein came up with had a series of networked filters which ensured that the telegraphic systems were set up to include a series of pulses before the start of each set of data so that, should information be cut short when trans-mission circumstances were re-established, they continued in the same manner as those previously encountered and the message was delivered intact. Blumlein called this method a 'flip-flop' device, for want of a better word (multiplexed would be today's answer of course), which he saw as a circuit which would arrange the speech transmission in alternative directions only.

Furthermore, when these speech currents were transmitted, they passed through an amplified delay network, which detected signals whose amplitude needed to be pulled up and automatically increased the amplitude of these signals as they passed along the transmission cables. The concept of time division multiplexing, with a connector matrix at each end of the transmission system, was not a new one and had been used before. What Blumlein did was to apply a series of electrical circuits to the TDM network in order to allow telegraphic and speech information to pass along the circuit using the frequencies that were not being utilized within the bandwidth of the conversation. The problem was that this form of multiplexed transmissions, with filtered layers of frequency paths, was so ahead of its time that ST&C did not fully realize its potential. This would not be the last time that Blumlein's mind would far outstrip his employers' ability to adopt or apply his work.

International telecommunications was obviously a field of great potential for ST&C and one that could be dramatically expanded if the quality of the product, i.e. telephone communication, could be improved. With the reduction in cross-talk outlined in the Blumlein/Johns patent, the path was open for ST&C to expand their undersea cable industry dramatically at a time when this was the only method available for all international telephone communications. Each cable had individually to lie under the ocean, with giant cables many miles long being reeled out from the stern of a ship as it travelled to the destination country. The problems with these undersea cables were as apparent however as with land-based overhead wires; the greater the distance, the worse the interference.

In his Patent No. 337,134, Blumlein investigated the properties of attenuating the disturbances in the earth circuit due to dissimilarity between the two legs of circuit without attenuating the telephone conversation signals travelling along the cable itself. Applied for on 19 June 1929, the specification provides for a circuit in which the natural capacities of the main and sea-earth cables co-operate with an inductance produced by two coils wound so as to have a near zero inductance for the signal current, but an effective inductance for the currents in the earth circuit. This co-operation produces a low pass filter, which attenuates currents of signal frequency to ensure that disturbances cannot pass from the loaded to the unloaded section of the cable.

In addition, condensers can be added to the circuit should DC signalling need to be used, with one of these connected in series with an inductance, or can be replaced by an impedance network designed to attenuate certain frequencies more than others. A layer of magnetic material could even be wound around the two cores of the submarine cable to insulate it from the sea and protect it from corrosion and wear.

On 21 August 1929, Standard Telephones & Cables applied for Blumlein's penultimate specification for them in which he had returned to measuring instrument circuit designs. In Patent No. 338,588, Blumlein provided a fixed external inductance coupled through a magnetic core to inductances in at least two of the arms of the bridge, while keeping the resistance of the bridge constant. At the same time, a mutual inductance was varied by current fed into the external inductance. The introduced variable mutual inductance had been effectively obtained because the intensity of the current flowing into the circuit had a constant coefficient of mutual inductance with the other circuit, one of which was traversed by an AC current of constant frequency.

The effect of this mutual inductance between two circuits was to introduce an electromotive force, which was comprised of relatively fixed inductances. This was conductively connected and fed from a common source of alternating current, with the ratio of the current to potential on the different members being altered by means of a shunted network placed across the inductance, through which the current was varied.

Various arrangements were shown in the diagrams accompanying the specification, including one where the variable resistance could be calibrated in inductance values with the measuring instrument connected, and then measured in at least two of the arms of the bridge for comparison. In another, the coupled coils of the variable mutual inductance were both external to the bridge, and then employed in shielded bridges without affecting the shielding.

Blumlein's final patent while with ST&C was also to do with submarine cables and their shielding. It had been well known that electrical disturbances at the ends of the cables where they were near or at the shoreline caused conductors which interfered with the transmission of information through the cable itself. This interference was only generated right at the end of the cable and the usual practice was to provide a screen in the form of metallic tape wound around the cable conductor to insulate them and therefore provide a sheath against extraneous electrical influences. It was also known that the shielding effect of the sea itself was quite considerable and that therefore it was of great importance to get the cable to descend as rapidly as possible. The usual method for ensuring this was the heavy armour covering which served both to protect the cable from wear and corrosion and to sink it to the sea floor as quickly as possible.

In his Patent No. 357,229, Blumlein devised a solution to the problem of screening the shore-ends of submarine cables by surrounding the entire cable with a metallic sheath which insulated the conductors, but would remain separate from the armouring of the cable and specifically did not cover the ends of the cables where they met the shoreline. This was because the armouring itself continued up along the cable, covering the ends which came ashore and was then connected through an impedance to the armouring or another earth.

The metallic shielding that ran the length of the cable under the sea and inside the armour covering was designed so that it could be graded in thickness (getting thicker as it got closer to the shore), in order to obtain the best shielding characteristics under any given conditions, and a doubly shielded repeating coil at the cable terminations ensured that the voltage was greatly reduced across the capacity unbalances.

Applied for by ST&C on 19 June 1930, the specification was in fact not fully accepted until 21 September 1931, by which time of course Alan Blumlein had left the company some two and half years earlier. It would seem that ST&C were more readily aware of the potential of Blumlein's work and the inventions therein than perhaps he was himself. The fact that Blumlein tired of telephony and needed a new challenge is perhaps understandable. The world in which he worked must have seemed to offer him few opportunities in his own particular fields of interest. We

Figure 2.6 Detail from Patent No. 357,229 (1930) showing how a submarine cable could be designed to taper towards the shore anchor-end.

know that he loved fast cars, flying and everything to do with aeroplanes, trains and railways and was a quite a marksman.

Blumlein however, also had another great interest in his life, that of music. It was to this, or rather the science of audio engineering and the recording of music that he was now to turn his attentions. As has been seen, Blumlein very probably had already been in contact with or had certainly been made aware of the work of A.H. Reeves (who would subsequently invent pulse code modulation (PCM)), and Geoffrey F. Dutton (who had investigated the properties of the electrical reproduction of sound). Bell Laboratories and IWE already held patents at that time on the then totally new system of recording sound electrically, and while they had initially intended to license it solely to RCA, Columbia persuaded IWE that it would be in their interest (and income) if other gramophone companies could use it. As IWE were essentially the R&D department of Bell Laboratories and also manufacturer of most of their products, this proposal made very good economic sense and, naturally they ended up licensing to anybody wishing to press records using their system.

Having been invented by two engineers from Bell Telephone Laboratories, Joseph P. Maxfield and H.C. Harrison, Western Electric had then incorporated their transducer design within the semi-mechanical recording system that they manufactured and which had first been introduced in 1925. As they pointed out in their definitive paper 'Methods of high quality recording and reproducing of music and speech based on telephone research' of 1926: 'The advance has been so great that the knowledge of electric systems has surpassed our previous engineering knowledge of mechanical wave transmission systems. The result is, therefore, that mechanical transmission systems can be designed more successfully if they are viewed as analogues of electric circuits. While there are mechanical analogues for nearly every form of electrical circuit imaginable, there is one particular class of electrical circuit whose study has led to ideas of the utmost value in guiding the course of the present development.'

The electrical circuits that Maxfield and Harrison were referring to were filters, and it was through the use of electrical bandpass filter theory and application that they had shown how an electromagnetic recorder and a mechanical reproducer with an enhanced frequency response, could produce recordings which were far superior than any previously possible.

This semi-mechanical recorder/cutter was then being used to produce the large majority of all the records being manufactured, and the royalty being imposed upon anybody wishing to use the system was between 0.875d (pence) and 1.25d on every record cut.

Naturally, many rival companies such as HMV, the Gramophone Company and the Columbia Graphophone Company were investigating ways of getting around the Maxfield and Harrison patents, but by early 1929, nobody had yet done so. At the Columbia Graphophone Company, Herbert Holman, one of their engineers, had managed to devise a method of bypassing the Western Electric system by putting rubber mounts at the side of a moving iron needle, but the recorded results were poor.

Isaac Shoenberg, who was the General Manager of the Columbia Graphophone Company, had been given the task, following his appointment from the Marconiphone Company in early 1929, of overcoming the patents on the Maxfield and Harrison system as it was costing Columbia a fortune. Shoenberg was looking around for the right man to tackle the problem and Blumlein, as an employee of ST&C which was, after all a subsidiary of Western Electric and Bell Laboratories, was ideally placed to have had access to, or at least knowledge of, the electrical circuit design of the Western Electric system.

It was therefore, this set of circumstances, Blumlein's growing boredom with the world of telephony, and Shoenberg's need to find somebody to overcome the Maxfield and Harrison patents on the Western Electric recording system, that led to Blumlein eventually being offered an interview with Isaac Shoenberg at the Columbia Graphophone Company in February 1929.

# Chapter 3
# The audio patents

By late 1928, Standard Telephones & Cables had certainly had their worth out of Alan Blumlein. He had applied for seven patents during his time with them, invented an entire signalling system, improved several elements of their submarine cabling system and designed various electrical circuits. Most notable of these was the AC bridge circuit of course, and all of which would earn ST&C money for many years to come. However, Alan Blumlein was restless, perhaps even bored.

There is no way of knowing now, at this distance, exactly when and why Alan Blumlein made up his mind that it was time for a change of direction and career. Suffice to say that, by the end of 1928, he felt (even if his employers did not) that he had gone as far as he considered he might within the field of telegraphy. What was needed was a new challenge, something that would stimulate his mind. As luck would have it in early 1929 such an opportunity came along. Blumlein took his chance, changed career and never looked back.

Early in February 1929, Alan Blumlein arrived at the offices of Isaac Shoenberg, general manager, the Columbia Graphophone Company in 'Petit France', London, for a job interview. Shoenberg had known about Blumlein for some time, first hearing of him during his days at Marconiphone (developers of Columbia's range of domestic receivers and gramophones) when Blumlein was making quite a name for himself as a bright young circuit designer in telegraphy and telephony circles. Though Shoenberg had heard that Blumlein had expressed an interest in furthering his career in a different field, he was not aware of his growing impatience to leave Standard Telephones & Cables and 'get on' with some new direction in his life.

At the age of 25, Blumlein had in fact decided some time before this that he needed to change and move on to new challenges. He felt that he had exhausted all of his options within the field of telephony and submarine cabling. He had even applied for several positions, which he hoped were available. These were mostly with companies that dealt with the relatively new science of recording sound. Writing in one letter, he explained 'I believe my ability lies more in a sound understanding of physical and engineering principals (sic) than in the knowledge of telephone engineering obtained by meticulous reading.'

When Shoenberg had been promoted to general manager at Columbia in January 1929, he had been given the task of finding a way, any way, of overcoming the Bell Telephone Laboratories patent held by Joseph P. Maxfield and H.C. Harrison. This was incorporated within the Western Electric semi-mechanical recording system (first introduced in 1925, the patents were Maxfield/Harrison/Bell's and Western manufactured the recorder/cutter), which imposed a royalty of between 0.875d (pence) and 1.25d on every record that was cut using it. In 1929, the combined sales of the Gramophone Company and the Columbia Graphophone Company were just over 30 million units. Every one of these had to have a royalty on it paid to a rival manufacturer, so it was imperative that a way be found as quickly as possible to overcome this.

In fact Columbia had not one, but two high-priority projects that were now the responsibility of Shoenberg. The first was the overcoming of the Western patent and the development of a moving iron recording cutter, and the second was the development of a high frequency, high quality microphone. As Shoenberg took stock of the situation, he became aware very quickly that if he did not recruit some new engineers, the tasks simply would not be completed. Of course, what he needed most of all was a top-flight engineer, someone that could tackle the entire project. The question was who?

Blumlein was not the only candidate that Shoenberg interviewed. There were in fact two to be considered; the job had been originally offered to one of Blumlein's colleagues at ST&C, E.K. Sandeman, for whom Blumlein and J.B. Kaye were working at the time. Sandeman however, while attending the interview with Shoenberg, had also been offered a more attractive position with the BBC and so turned down the opportunity of working for Columbia. He did, however, inform Shoenberg about Blumlein's interest, pointing out to him that, 'He will be the most expensive, but he will be the best.' It was almost certainly Sandeman who, returning to ST&C following the interview, told Blumlein that the position at Columbia was available and that he should apply for it.

Shoenberg, in the meantime, approached Eric Nind for information. He was one of the new young men at Columbia and had been one of the students that Blumlein

Figure 3.1
Isaac Shoenberg (later Sir) who would become Blumlein's employer, mentor and friend. (*Courtesy of EMI*)

had helped when he had been an assistant demonstrator at City and Guilds. Nind remembered Blumlein as a 'Delightful man, very human indeed, and very good at explaining anything. He didn't get exasperated if you did not understand, and would go through a point again and again. He had plenty of inexhaustible patience and a great facility for converting quite complicated mathematics into very simple circuit elements.' Shoenberg was convinced and invited Alan Blumlein to attend an interview.

At his interview, after the usual pleasantries and once Blumlein had been given the essence of the task that Shoenberg needed him for, he was asked if he thought he could do the job. Without any detailed understanding of the magnitude of the project, Blumlein told Shoenberg he was sure he could. Shoenberg then offered Blumlein the position and asked him what salary he required. Blumlein told Shoenberg that he might not want him after he had said how much he wanted. 'How much do you want then?' Shoenberg asked. Blumlein, put on the spot as he now was, named a figure that he felt certain was too much, but probably worth a try. After all, he had nothing to lose. Shoenberg made his decision there and then. 'You are engaged', was his classic response. Shoenberg then offered an annual salary somewhat in excess of what he had just asked. Shoenberg had made his point and was sure he had his man. He would not be disappointed in his choice.

Blumlein, perhaps realizing he could have asked for more than he had and still got

Figure 3.2 The Works Designs, Research and Development Building at EMI, Hayes. Blumlein's office is on the third floor, second window from the right (with one window open). His secretary, Gladys Ball, had the office next door to the left. This photograph was taken on 26 January 1934 during the construction of an experimental aerial mast on the roof. (*Courtesy of EMI*)

it none the less accepted the position, returned to Standard Telephones & Cables to hand in his notice, and say his goodbyes to the many friends he had made during the five years he had been there. It was time to move to his new challenge.

## The Columbia recording system

In March 1929, Alan Blumlein joined the Columbia Graphophone Company Limited, reporting directly to Isaac Shoenberg, the man to whom he would report for the rest of his life. Blumlein was ready to face the not inconsiderable task that was ahead of him. Columbia, however, was not the only company that had decided that there must be a way of getting around the Bell patent. HMV (the Gramophone Company), had also decided that it would invest in a team of engineers to design and build a completely new system. The HMV team was headed by A.W. Whitaker, working on a mechanical cutting system. Shoenberg however was certain that, with the team he was putting together and with Blumlein's expertise, if anybody could produce the right apparatus it was Columbia.

The task was ideal for a man like Alan Blumlein. Not only did he possess all the technical and electrical attributes to ensure that the actual day-to-day running of the project would go well, but his insatiable desire to complete the job at hand meant that much of this enthusiasm would rub off on the men working with him. Blumlein was also intensely interested in music, listening to all manner of recordings over and over again. He especially liked Beethoven and analysed the music in a mathematical way as many deep thinking men have done in the past. Beethoven's music lends itself to that form of thought. Though there is absolutely no evidence that

Figure 3.3
Herbert Edward Holman, whose initial 'H' along with Blumlein's would become one half of the name for the 'HB1A' microphone, designed by them in 1930. (*Courtesy of EMI*)

Blumlein's love of music had any bearing on his decision to accept the job at Columbia (or that the task he was employed to solve had anything to do with producing better quality music recordings), it is very probable that these considerations did cross his mind at the time.

When Blumlein arrived at the laboratory at Columbia, he found that the electric recording apparatus was rudimentary at best, though it has to be said, not uncommon for the time. It had only been a matter of a few years since recording had made the enormous leap from acoustic recordings, that is, effectively shouting into a horn driven direct-cutting system straight onto the wax record. At Columbia, Blumlein found a capacitive microphone, an amplifier based on a public address system and a moving iron cutter, similar to that which Western Electric were manufacturing.

A team of engineers had been assembled by Shoenberg already and they had made a start on the principles of a system to overcome the Bell patents, but lacked that electrical expertise that Blumlein now brought. Among his new colleagues were Herbert Edward Holman, Henry Arthur Maish Clark, Peter William Willans, and Geoffrey F. Dutton, all of whom would become engineers, scientists and friends to Blumlein for many years to come. It was almost certainly a member of this team who had drawn up the provisional designs for the system, dated 1 May 1929.

Blumlein did not at first work on the problem of the cutter. The first record of his work for Columbia appears in a memorandum dated 1 May 1929, showing him working on the several projects which had already been started by Holman and Clark. Among these was a high frequency, high quality microphone project. In June, when the complex construction work on this had been completed, Blumlein turned his attention to the designing of the coils and diaphragms for the microphone and

Figure 3.4
Henry Arthur Maish Clark who was known to his friends and colleagues alike as 'Ham' not only because of his initials, but also because he was an avid ham-radio enthusiast. (*Courtesy of EMI*)

Figure 3.5
The recording system being used by Holman and Clark at Columbia in 1928, much as it would have appeared at the time Alan Blumlein joined the company. The Columbia recording machine had a weight-driven motor and an acoustic play-back system driven by the Western Electric amplifier rack on the left.

the estimation of stray capacities. Herbert Holman did much of the mechanical design for the construction of the first prototype system which was eventually to be given the designation 'MC1A' or Master Cutter 1A (Holman, who was Blumlein's senior by eleven years, was a first rate mechanical engineer who had been educated privately and at Redhill Technical School. He been employed by Columbia in February 1924, to develop a simple form of moving coil microphone, and was there-fore ideal for a project such as this.) Recalling Blumlein's arrival at Columbia, Holman said

'In 1929, we were joined by A.D. Blumlein, a young engineer of whom I cannot speak too highly. I was closely associated with him in many projects, and his capacity for original thought, coupled with a thorough knowledge and appre-ciation of basic engineering principles considerably impressed us. Blumlein's abilities as a circuit designer here came to the fore; equalizers not only had to produce an ideal response curve, but had also to provide for wide deviations by simple strapping of terminals if such were required. This was a characteristic of A.D.B., nothing must be left if an improvement could be made. He spared no effort himself and inspired all those who worked with him.'

By August when test runs were first done, Blumlein was definitely working on the cutter project, but the results, which had been carried out using a Western con-denser microphone calibrated by the National Physical Laboratory (as a standard), determined that there was a pronounced peak at 100 Hz and a dip around 1 kHz. On 27 August, the first comparative tests were made with the prototype system rig and the Western system in wax. Blumlein and Holman noted that there did not seem to be any discernible difference between the two systems.

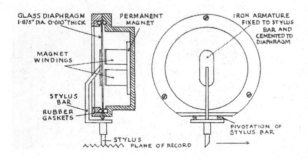

Figure 3.6
An early recording system based on a telephone earpiece. This was the type of apparatus that Holman and Clark would have been working on in 1929, just prior to Blumlein's arrival.

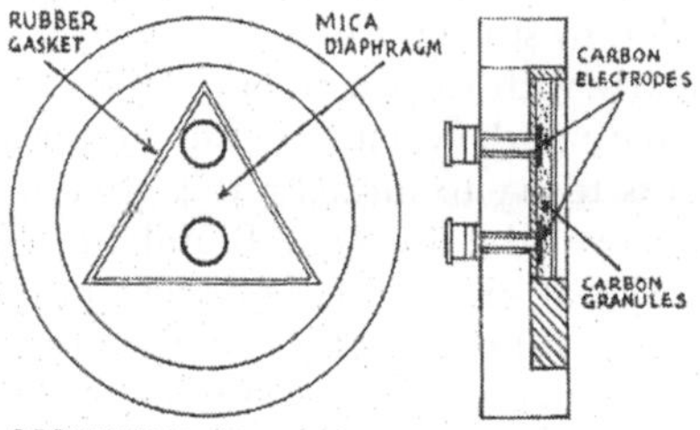

RUBBER GASKET | MICA DIAPHRAGM | CARBON ELECTRODES | CARBON GRANULES

Figure 3.7
Detail of the type of carbon
microphone that Holman was working
on in 1929.

RESISTANCE MICROPHONE WITH TRIANGULAR
MICA DIAPHRAGM                          SCALE: ½ SIZE

At the same time as Blumlein and Holman were labouring away with their prototype system in the laboratory in Hayes, Columbia had sent Eric Nind to work in Tokyo, Japan, with one of the systems which had been licensed to the Japanese record label 'Nipponophone'. These cutters were of the Western Electric type which Louis Sterling, chairman of Columbia, had been very keen to licence as soon as Western and Columbia had agreed a contract earlier in 1929. The Japanese, at Nipponophone, had been among the first to receive a system because of this. Eric Nind had been sent to Japan to oversee the installation and act as an on-site guide for the Japanese recording engineers. Nind was also encountering problems associated with cutter overload, he reported that the amplitude of the cut was not proportional to the signal, complaining that low frequency waves became 'peaky' and that Japanese music seemed particularly difficult to record.

Columbia decided that Blumlein should try to solve this problem also and a series of measurements were carried out, from which it was discovered that there was indeed considerable non-linearity, giving rise to this peaky waveform, and probably made worse with the Japanese practice of recording at very high levels. The waveform of the Japanese records (which were regularly being sent back to Columbia for comparison) was analysed by Blumlein's team under a microscope. From the Western Electric cutter in the laboratory, as well as Eric Nind's recordings made in Japan, it was discovered that, when these were compared with recordings made on the prototype system on which the team had been working, the Columbia cutter, with a sine wave input of 375 Hz (roughly the frequency of the bass resonance for the system), the device was producing 150% second harmonic and 100% third and fourth harmonics. The Western cutter was only producing 25% second harmonic and 5% third.

On 30 August, Eric Nind recorded another Japanese record to compare it with the Western system, but again once it reached Blumlein's team and was compared to their new rig in the laboratory, they had to concede that the Western system was better. In Blumlein's opinion, these problems were arising from at least two causes, the first being a reduction in the reluctance of the field magnetic circuit as the armature moved from its rest position which made the cutter more sensitive. In an extreme case, if the cutter was overloaded for example, the armature could adhere to the pole piece; indeed Blumlein had noted that this too was happening with the Western system.

Secondly, the rubber diaphragms, which were used to damp the bass resonance frequency of 375 Hz, were also causing non-linearity problems. The armature was controlled by springs which had been shaped to increase the restoring force more than linearly with displacement. This was intended to compensate for the rise in field strength and, despite several attempts by Blumlein who carried out a series of experiments to try to increase this effect, he had little success. Finally, in an attempt to gloss over the problem, Blumlein designed a circuit that would weaken the field as the signal increased. This did help during loud passages of music, but the effect could not be brought into action in less than 1/20th second and so did not help with transients.

In September 1929, frustrated with the lack of success, Blumlein took stock of the problem, and reconsidered the fundamental elements of what they were trying to do. The first decision that had to be made was the principle on which the Columbia system might be made to work. For the first four months, Blumlein had worked with the apparatus that he had found upon his arrival at the laboratory. Now he decided upon a completely new approach; they would attempt to make the recorder/cutter work using electromagnetic damping.

Blumlein, with his knowledge of electronic circuits had been aware, as had many before him, of the potential of the moving coil. Indeed, only a few years earlier in 1925, Rice and Kellogg had produced the world's first moving coil loudspeaker, where the speaker cone itself was moved in and out due to the action of an electrical current upon a moving coil. The current acting on the coil was directly representative of the amplitude of the frequencies being fed to it. If a moving coil could be used in such a manner to drive a loudspeaker, could not the same principle be used to drive a suspended cutting head (through a master recording)? Certainly nobody had done anything like it before, which Blumlein found strange; the technology was neither new nor unknown.

In order to decide whether it would work, Blumlein first had to detail the method by which the Bell patents worked. The Bell system, which Western Electric was manufacturing, used a balanced armature arrangement (for the cutting technique). This produced a satisfactory cut in the groove of the master recording from which the pressings would be made, but Blumlein had already seen the results of the armature's reciprocal action on the depth of actual cutting head. This was difficult to control and the direct attenuation of the cut was damped with a totally mechanical action. In other words, the cut made in the groove did not directly reflect the amplitude of the audio that was being recorded. Instead, a uniform cut produced what can only be described as an adequate, if rather undynamic recording, due to the main resonance of the cutting head itself.

Early in September 1929, the team gathered in the laboratory and Blumlein explained to them the basis of his decision to try electro-mechanical damping to see if they could make it work. Having decided that a moving coil transducer system would not only overcome the Bell patents, but would almost certainly produce a better recording (as the movement of the cutter would more accurately represent the audio), Blumlein now received the go ahead from Shoenberg to start engineering a recorder as this was the first critical element of any new system.

Blumlein, who had by now been placed in charge of the entire research and development team for the project, designed and then produced the circuitry, with Holman mainly responsible for producing any electro-mechanical apparatus required to construct the recorder. The records from the archives give us a reasonable account of the week-by-week, and on occasion, day-by-day activity during this period, and the problems they encountered. The first of these was the power-to-weight ratio. A coil that generated enough power to move the cutting stylus precisely enough and yet capable of moving through the wax master disc, would be so heavy as to be impractical. Blumlein came up with the idea of having a wound coil enclosed completely in a solid aluminium casing (being the most lightweight yet durable material available at the time), with a separate multi-turn signal coil rotating in a strong magnetic field from the currents induced into it from the audio. These currents in turn induced currents in the signal coil which, when passed through the driving impedance would lose power and thus provide the necessary damping action upon the cutting head.

That September, significant progress was made and Blumlein's original notes give an insight into his train of thought. Dated 11 September 1929 and entitled 'Recording Levels', the notes reveal some of the calculations that were being made. The next day the varying fields of the Western Electric system were recorded as cutting the bass by 4 dB at 100 Hz. On 24 September, they were working on the automatic field control of the Columbia cutter. Henry Clark (always known to his colleagues as 'Ham' Clark, not only because his initials were H.A.M., but also because he happened to be a leading light in the ham radio field) recalled how Blumlein's arrival and work at Columbia had changed all their lives:

'I had the good fortune to work as A.D. Blumlein's assistant and our experiments spread into the field of moving coil microphones and cutters, the main object of which was the reduction of non-linear distortion present in all the then known moving iron devices. H.E. Holman was a tower of strength in all these projects as he was responsible for most of the mechanical design. In these early days I learned that the overall performance of even such complicated devices could be calculated in advance before the device was made. Blumlein's wrath when the measured resonant frequency of a new model did not match the pre-calculated one is still a vivid memory.'

By 3 October, more test records were being cut with the auto field control now working on the Nipponophone system and, on the 18 October, Blumlein produced a large scale drawing of a representation of the 'Columbia Field System', with a series of calculations on the following pages which show the predicted figures for the probable field density. On 25 October 1929, the notes refer to the fact that a version of the Columbia recorder had arrived in Japan for Eric Nind to carry out comparison tests. Problems were also being encountered with overloads and non-linearity. The work by now had moved away from straight comparison tests with other systems such as the Western Electric cutter, to direct non-linearity analysis of the Columbia system. Originally the axis of the moving coil had been constructed horizontally and the field was provided by a single coil. However, to cut wax, it was decided that this might not be sufficiently strong to give an adequate performance at high frequencies, so Blumlein, while calculating the main characteristics of the cutter himself, gave the design of the field control to Henry Clark.

Working on the basis of a field flux density of 10,000 lines/cm$^2$, (though a lower field of 6000 lines/cm$^2$ was also considered), 'Ham' Clark did the necessary calculation and found that the power dissipation would indeed be too high for a single coil. They determined on several alternative arrangements before the eventual double-coil system with two parallel coils on a 'U' shaped core, which was tilted upwards at an angle of about 22.5° was adopted. This immediately led to the realization that there would be a significant reduction in the moment of inertia of the moving coil assembly, mainly from the reduction in the length of the stylus bar, because they had effectively changed from a horizontal to a vertical axis.

Much of the design work and construction was now carried out by Holman who had decided that the moving coil should have a hard steel bar along its axis. In this, two knife-edges were formed which rested on two hard steel plates, which in turn gave low friction and, equally important, striction. Steel wires in torsion held the knife edges onto the plates and absorbed much of the thrust. The stylus bar was then held in a slot at the lower end of the axis rod, with a cork shim of 0.003 inches thickness between the stylus bar and the coil axis with clearances around the clamping bolt-holes to avoid direct metallic contact between the two components.

In early November, the first manuscript was drawn up of the results of this work and the analysis showing how the rubber damping was holding up with the application to the balanced armature. The tests were to continue from 29 November 1929 through to 6 February 1930. There was still a lot of work to be done, though by now the team was reasonably satisfied with the results they were obtaining. The effective mass of the recorder at the stylus was calculated at 22 ounces, whereas the Western system was only 12 ounces, which meant that a totally new way of suspending the cutter would have to be found.

These matters aside, Blumlein felt by 10 January 1930 that sufficient work had been completed to write a comprehensive description of the cutter and it was this that would eventually form the basis of the patents which came later in the year. On 17 January, they completed the first run of the cutter and on 20 February 1930, Blumlein and Holman were again testing the impedance of the Western Electric cutter but now with their own system working well for comparison. Blumlein decided that the change in the axis of the moving coil from horizontal to vertical was sufficient to rename the apparatus MC1B and completed this version in the first week of March. He also introduced a much larger field coil which of course made the entire design much heavier than the MC1A with a higher centre of gravity, but this did allow the cutting head to cut a uniform groove in the wax master. By 5 March, they were engaged in simulation impedance tests and on 8 March 1930, continuing through to 11 March, when modification were made, they were concerning themselves with designing the transformer for the system. These modifications were also carried out on the Japanese system on 10 March, to give it a high frequency boost.

Blumlein's attention now turned to the amplifier. It had originally been a low-power example which, while it gave reasonable equalization and had a good smooth frequency response, had a definite spike and dip at around 3.5 kHz. It was discovered that a small resonance in the cutter was being caused by the bending of the steel shaft which was suspended at two points and which formed the axis of the moving coil. It seems that the shaft was unable to vibrate in a whirling mode. So the balance conditions for the shaft were carefully calculated, the problem being solved by replacing the steel nuts which had been used to hold the stylus arm in place with platinum ones (these being more flexible). This modification went some way toward solving the frequency spike problem in the amplifier.

In April a new push-pull, class A amplifier was devised using GEC transmitting triodes (DEM3s), running at 1000 volts on their anodes each at 250 mA current. This was known as the '500 watt' amplifier and the output was transformer-coupled via an equalizer to the cutter. This work on the coupling of the amplifier to the cutter is one of Blumlein's most impressive achievements at this time and it illustrates his incredible grasp of the complexity of the situation and its solution. High frequency boost was inserted between the driver and the power amplifier, which meant that at high frequencies the coil was being driven at a constant velocity. This, in turn, meant that the driving force required was now proportional to the frequency being applied. By inserting high frequency equalization at the input to the power stage, low frequency level was kept low in an attempt to minimize microphony, which was a major problem with triode valves at that time.

When the first trial cuts using the MC1B were made, the team found that it had a constant resonant frequency of around 250 Hz, exactly the frequency that had been predicted by Blumlein. Further tests showed that temperature runs were satisfactory and, although the performance of the system seemed to be acceptable, Blumlein was still not satisfied and continued to return to various elements of the system, constantly adjusting them until he was happy.

On 26 June 1930, Blumlein wrote to Peter Willans, the head of the laboratory, and suggested to him that they should manufacture another six cutters each with 100 watt amplifiers. He also pointed out that extensive tests still needed to be carried out involving perhaps two or three other amplifiers of differing power outputs, as well as a series of recordings made to ascertain the results of oriental music, dance bands as well as classical music. This was to ensure that the amplifier could maintain a sufficient power margin for all musical genres. He drew up a flow chart for the process of manufacture and pointed out that, although the cutter was now satisfactory, more development would need to be done on the floater. Blumlein had calculated that the power needed to drive the Columbia cutter was 2.26 times as high as that of the Western cutter at low frequencies and 22.6 times as high at high frequencies.

Though much of the electric circuitry is easy to attribute directly to Blumlein himself, as is the original idea for using electro-magnetic damping in the system, much of the mathematical calculation such as the reluctance of the magnetic circuit and the resistance and movement of inertia of the various designs of the moving coil was carried out by Clark and other members of the Columbia team. Blumlein did however work out the impedance of the wax to enable the required damping to be calculated, and it was only his and Holman's names that appeared on the patents for the new system when they were applied for in March.

## The Columbia recording system, Patent No. 350,954 and Patent No. 350,998

The new system had been built, tested and had been shown to work. Now it was presented to Shoenberg, who had been keeping a close eye on the progress that had been made by the team he had put together. The prototype MC1B would become the basis of the new Columbia recording system and, as it overcame the Western Electric system with a totally new concept of electro-mechanical damping, records cut from it would not have to pay a royalty. The system would save Columbia a fortune and at the same time earn them a considerable income. The tables had now been turned; any records that were cut on the new Columbia system would mean that a royalty would have to be paid to *them*. Shoenberg was delighted.

Patent Nos. 350,954 and 350,998 were Alan Blumlein's first for his new employer, but his eighth and ninth overall. They both date from Monday, 10 March 1930, when the applications were lodged with the Patents Office and are entitled 'Improvements in Electro-mechanical Sound Recording Devices more especially of the Moving Coil Type' (No. 350,954) and 'Improvements in Apparatus for Recording Sounds upon Wax or other Discs or Blanks' (No. 350,998). That this was just one year after he had started with Columbia gives some indication of the pace at which he could work.

There are 34 'claims' (a claim within a patent being that which the inventor wishes to be considered as entirely new compared to anything which has previously been invented), made by Blumlein and Holman in Patent No. 350,954, the outline of which is as follows:

> The Patent describes a moving coil system in which the cutting tool is con-
> nected to the drive motor by a linkage system comprising, or kinetically equiv-
> alent to, a shaft inclined or perpendicular to the wax surface. This shaft
> extends down from the motor and stylus arm assembly placed perpendicular to
> the shaft. The recorder is comprised of an electric coil mounted so as to pivot
> in a magnetic field about an axis in its own plane. The coil can then move upon
> the induction of an applied electric current by means of one or more adjacent
> electric coils. These are mounted about a common core, which forms the
> primary and secondary windings of a transformer.
>     The magnetic flux of the coil, and the same of the transformer, are arranged

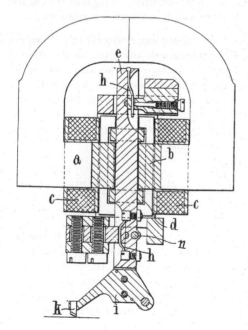

Figure 3.8
Detail from Patent No. 350,998
(1930), the Columbia Recorder/Cutter
designed by Blumlein and Holman.

so that they are at right angles to each other. The moving coil itself was to be cut from a solid block of suitable material such as aluminium. This coil would be substantially triangular in shape the sides of which would increase in width and thickness as the pivotal axis is approached. The steady magnetic field of the moving coil is achieved by providing magnetic pole pieces adjacent to the coil at right angles to the transformer core. In this way, the air gap within the moving coil assembly is reduced to a minimum, compatible with the free movement of the coil itself. The neutral position of the moving coil is maintained with a suspension system, locating the core by means of springs giving an elastic-like connection.

An additional 48 claims are made in Patent No. 350,998, again the outline of which reads as follows:

The patent describes an electro-mechanical device for the recording or reproduction of sound records in either wax or other such discs or blanks. The device is also however applicable to any other moving coil device such as a microphone which could be employed in a sound recording system, or also of a moving coil pick-up or moving coil loudspeaker. It could also be used for optical recording devices such as those employed in the cinema for sound recording and reproduction.

The moving coil element is suspended in a magnetic field and connected to an external circuit and caused to vibrate by mechanical means. This in turn generates an electromotive force (EMF) across the coil, and will cause a current to flow through any external circuit. This current, will in turn, produce a back EMF which will tend to modify the movement of the coil. If the impedance of the loading circuit is made frequency selective, the back EMF so produced will also be frequency selective. This can therefore be easily optimized to damp any natural mechanical or electrical resonance within the type of electro-mechanical system herein. In order to get the greatest damping effect from the application of this principle, three factors are necessary; first, that the impedance of the moving coil circuit should be as small as possible (for a given configuration and numbers of turns, hence the DC resistance), second, that the purely electrical impedance shall be as pure resistive as possible, and lastly that the magnetic flux density in the moving coil shall be as high as possible.

If efficient power transfer is to be obtained between a moving coil and the circuit within which it operates, it is necessary that the impedance of the operating circuit be of the same order as the electrical impedance of the moving coil to which it is connected. But, in order that the movement of the moving coil itself may be well damped at frequencies close to its resonant frequency, it is necessary that the impedance connected to the moving coil be as low as possible. The basis of the invention then, is to connect a moving coil to an operating circuit whose impedance is comparable with the electrical impedance of the moving coil, at frequencies where an efficient transfer of power is required, and thereby obtaining easily controllable electrical damping.

On 30 August 1930, the circuits for monitoring and driving the moving coil cutter were completed. Arrangements were made for waxes of 22 titles to be cut using both the Western Electric and the Columbia cutters simultaneously; the signals to both being fed from a Western CT microphone. The recordings, all of which were of popular singers, included Will Fyffe singing 'Daft Sunday', The Four Bright Sparks singing 'That Night in Vienna', and Billy Elliot singing 'Home is Heaven'. The resulting pressings were listened to by a number of people who were asked their opinion as to which they preferred, though there is sadly no information as to the listening conditions under which the tests were conducted. In the end, there was no strong preference for either, though there was a small majority in favour of the Columbia cutter.

Further trial cuts were made with reduced high tension voltage on the DEM3 valves in the power amplifier and it was on this basis that Blumlein had calculated that 100 watt amplifiers would be adequate. Somewhere along the line however something went amiss and the amplifier that was eventually used had a single DEM3 triode and ran at 250 watts. Blumlein by now was concerned about the cost of all this development and, in his notes, made comment of his desire to modify the equalizer between the amplifier and the cutter to give greater efficiency and so reduce the power needs. For a single amplifier the components had cost £92-6s-9d, the HT generator £46-19s-2d, and sundries (not listed) £13-15s-1d; a total of £153-0s-4d, quite a considerable sum in 1930, and not far off an annual wage for many of the engineers working on the project.

## An entire system evolves

Having produced the first design details of the new recording and reproduction system as early as October 1929, much of the year following was spent producing first, a prototype (February 1930), and later, production models initially for Columbia to use itself and later to license to others. It soon became obvious that the accurate control of electrical damping, coupled with electro-acoustic microphones and electromagnetic loudspeakers, could indeed produce very high-fidelity recording and reproduction sound systems.

Part of the resonance of an electro-mechanical system, such as the Columbia cutter, depended upon the elasticity of the needle and the record surface initiating a vibration, resonant with that of the effective mass of the whole device. If this was not controlled, unwanted resonance could occur, especially at the moment of inertia. Blumlein's next patent, No. 361,468 (Improvements in Sound Reproducing and Recording Devices), applied for on Friday, 12 September 1930, was designed to overcome this by taking into account the pivotation of the gramophone needle relative to the remainder of the assembly, in this case the recording apparatus as previously described.

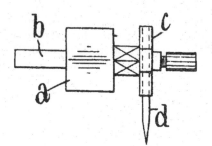

Figure 3.9
Detail from Patent No. 361,468 (1930). Co-written with Peter William Willans, the needle point is adapted to receive impulsive forces in its direction of displacement and mounted in such a way that the centre of rotation of the combined needle and armature assembly lies on the axis of rotation.

The patent, which was co-written with Peter William Willans, and initially appears to be simple in design, though it does, for the first time in a Blumlein patent, resort to mathematics. Blumlein's mathematics ability, as has been documented, only went as far as limited calculus and, while the mathematics in the patent are not overly complicated, it is likely that Peter Willans was responsible for most of it.

The invention describes the mounting of an electromagnetic pick-up on a rigid body (which Blumlein and Willans call an 'armature') which is rotated about a given axis. Attached to this is a lever, on which is mounted the needle. The needle point is at a given distance from the axis of rotation of the armature, and is adapted to receive impulsive forces in its direction of displacement. According to the invention it is desirable to mount these members in such a way that the centre of rotation of the combined needle and armature assembly lies on the axis of rotation. Assuming that the needle is fixed perpendicular to the axis, the following applies:

$I_1$ = The moment of inertia of the armature about the axis of rotation
$I_2$ = The moment of inertia of the needle about its centre of gravity
$m$ = The mass of the needle
$a$ = The distance between the point and the centre of gravity of the needle
$x$ = The distance of axis of rotation from the centre of gravity of the needle (on the side remote from the needle point (right))

The desired objective is achieved if:

$$x = \frac{I_1 + I_2}{ma}$$

In accordance with this equation, the needle can be fixed in such a position relative to the armature, that the value of $x$ corresponds to the value calculated (by this means) from the known or measurable quantities of the right hand side. The net result of the invention is a sound recording (or reproduction) device where the disposition of the various operating elements about their rotating axis is so arranged that the upper resonant frequency of the device has a maximum value. As defined here, the device comprises a rigid member adapted for rotation about a given axis, and a second member rigidly connected to the first in order to receive impulsive forces at an extremity remote from the axis of the first. In this combination, the instantaneous centre of gravity of the assembly lies on the axis of rotation. In this way, an electric pick-up in which the distance $x$ from the axis of rotation of the operating system, to the centre of gravity is calculated from the equation above.

The next patent that Blumlein applied for, No. 362,472, deals with the idea that a system such as the Columbia recorder needed to produce predictable attenuation characteristics across the entire frequency range of the disc being recorded. This attenuation had to be carried out regardless of the number of input signals being transmitted, or whether they were balanced or unbalanced lines. Only in this way was it possible to produce a uniformly recorded sound. Blumlein designed one of his most important circuits, one that would have far reaching uses then and is still used now, for the patent describes what is known today as 'the π-line attenuator'.

Applied for on 30 July 1930, Patent No. 362,472, 'Improvements in Electrical Transmission Devices', outlines an attenuation device that consists of a number of networks constituting an artificial line of any type, balanced or unbalanced, which can attenuate electric currents of all frequencies uniformly. The device would have one or more sections consisting of a network of predetermined frequency characteristics, together with sections of uniform attenuation with frequency, in order that a given frequency preference or selection can be obtained with a uniform variation of attenuation over all the frequencies.

The degree of frequency selection, or preference obtainable, could furthermore be varied at will by the operation of a moving contact switch. These sections, with their given frequency characteristics and definite impedance relationships, were to work over any of the known equalizer networks and were to have constant impedance characteristic of any wave filter provided they were sufficiently constant over the transmitting frequency to work in a system such as the Columbia recorder.

The invention was further flexible enough to allow it to work with one or more sources to feed the same load, in such a way that control over the relative attenuation and impedance relationships between the various loads could still be controlled. In this way the efficiency of transmission of an electric potential from one piece of apparatus to another by means generally referred to as an 'artificial line', can be improved and varied in a known and defined manner.

Essentially what had been invented here was a method of transmission of electrical potential where the input to the system gave a constant impedance at the generator or the load, regardless of how many or how few terminations were present. This meant that a constant volume could be controlled using a minimum of moving contacts, thereby reducing noise and effectively becoming more efficient. The system has become known as the $\pi$-line attenuator and its use as a constant impedance volume control in professional audio equipment is now commonplace.

At around this time, late summer 1930, Blumlein began the relationship that would eventually lead to his marriage. His mother, Jessie, was giving up her flat in Linden Road (just off Muswell Hill Road) and going to see her relatives in South Africa and wanted to store a rather nice Bechstein piano. Alan had, some 20 years before, been a pupil at Gothic House School and went to see Miss Chataway, the headmistress and owner of the school, to see if she would be able to store the piano at Gothic House, or at least be able to use it. Miss Chataway did indeed want a piano for the school and so asked one of her teachers, Miss Doreen Lane, who played a little, to come with her to the Blumlein residence to see the piano and to try it out. Evidently Alan Blumlein must have been at his mother's flat during the visit and was introduced to Doreen who must have made quite an impression on him.

Some years later, Doreen recalled the events of those first meetings:

'As a matter of fact I played very badly, but I used to play for the children you know, and she wanted me to go with her to see this piano. So, we went up, though we didn't see much of the piano, but had quite a nice evening, and Alan sat with me while Mrs Blumlein and Miss Chataway chatted away, and he took some pipe-cleaners, and made me three little dogs. And one was the father, the mother, and the co-respondent put behind. That was Alan's sense of humour; great sense of fun you know. And that was that. We said we would have the piano.'

A few days later, Blumlein called on Miss Chataway at Gothic House, on the basis of arranging something about the piano and, while there he asked after Doreen:

'Well, a few days after, we were having dinner that night, and Miss Chataway said, "There's Alan in his car again", so in he came, and he chatted away and said, "I've got to post a letter, will you come with me?" So I went, and when we went outside he said, "I haven't got a letter to post, I only wanted to get you out". Well that was the sort of start of it.

Well, then he said to me, what I thought he said was, "Will you come flying with me *one* day?" I was a very travelsick person, so I thought oh, I can put that off, so I said "Alright", and he said, "Right. I'll fetch you at 3 p.m." I said, "What do you mean?" and he said, "On Sunday". I said, "I didn't say I would come on Sunday, I said *one* day", "Oh no", he said, "Sunday". So I said, "Oh well, alright if it doesn't rain". So, I went back to the others and said "Pray for rain on Sunday at 3 will you, because I don't want to go up in an aeroplane in the slightest".

Anyway, he came, and we went off to Stag Lane; evidently he was a private pilot and he had his licence, and of course it was this little open Gipsy Moth, and we put on these helmets and things, and I thought "Well, I'll risk it". I suppose I was so scared of being sick in front of a strange young man. So we came down and he said, "Will we have tea here?" And I said, "No, I don't want to stay here anymore". So, we went off to a hotel in Radlett for tea, and I think that started it. Then we started going out and that's how I met him.'

Alan Blumlein was 29, and Doreen Lane 24. Their courtship would last another two and half years before they married on 22 April 1933, with J.B. Kaye as best man. Before that, Doreen would be forewarned about some of Alan's 'peculiarities':

'There was a joke amongst some of his friends, they used to call it "Blumlein-itis" or "First Class Mind". It seems that he didn't want to know anyone who didn't have a first class mind. And, some friends of his, before I married him, said to me, "Now Doreen, you're going to come up against Blumlein-itis". I said "What do you mean?" They said, "Alan won't have anything to do with anyone who hasn't got a first class mind"; and I did come up against it, but that was him you see, and it was, at times, very awkward, because at times he was unintentionally very rude to some people; he didn't seem to be able to get his brain down to their level.'

J.B. Kaye, however, was in no doubt that Doreen was the right woman for Alan Blumlein:

'I was delighted when I met Doreen because she was a character, and she had a sense of humour, very charming and I thought "Yes, this girl will be able to keep Blumlein under a sufficient degree of domestic control that will avoid any serious rifts". I thought to myself there will be a few brick-ends flying about the kitchen, but that's all part of life's rich pageant.'

The next patent that Blumlein applied for was No. 363,627, 'Improvements in or relating to Apparatus for the inter-conversion of Electrical and Mechanical Energy such as Used in Sound Recording and Reproducing Apparatus' (applied for on Friday, 12 September 1930). While the title is rather long winded, it actually refers to a very simple method for exchanging the materials used for damping the various vibrations that occur in devices such as the Columbia recorder with substitutes that would be more efficient. It was therefore a method of alleviating this phenomenon by modifying the frequency response of the apparatus by substituting the damping material, or inserting suitable material, to reduce the effect of resonance on the whole structure.

In vibratory devices such as recording and reproduction systems, it had been customary to introduce damping in the form of frictional or elastic forces imposed by a material connected with the moving system. In this way, the vibration of the moving system when oscillating (in the manner intended) would be suitably controlled. Blumlein called this the 'authorized mode' of damping. It was also possible therefore that unwanted movements such as vibrations within the moving system itself could be present (caused by inherent flexibility in the supports of the system). Blumlein called these the 'unauthorized modes' of damping.

Where damping was introduced into a system (whether the damping force was electric or electromagnetic), it would quite usually happen that these forces would be completely ineffective in suppressing the unauthorized modes. Therefore, the whole rigid structure of the device might have a resonant frequency in the working range at which the lack of damping would cause a resonance of quite undesirably large amplitudes. Worse still, these could occur from quite small forces acting upon it.

Blumlein explains it thus:

> The invention described in this patent relates to devices employed for the inter-conversion of electrical and mechanical energy, and is particularly directed at devices in which parts of the apparatus tend to vibrate at varying frequencies such as sound recording and reproducing systems. Where a damping material is introduced to alleviate the effects of vibration in an authorized mode, it does not follow that this material will have the corresponding effect on vibrations of the unauthorized mode. In fact, as is usually the case, this is what occurs. The object of the invention is to introduce a means whereby the difficulties of damping the various vibrations caused by resonant frequencies, are overcome and sufficiently suppressed so that unwanted amplitudes due to resonation do not occur.
>
> The invention consisted of a pivoted electric coil, the resonant pivotal oscillations of which are damped electromagnetically. It was designed to replace the damping methods previously used to offset the effects of vibration, such as cork, rubber compounds and other like materials. The electromagnetic damping method herein, proved sufficient to damp almost all of the oscillations that occur due to the resonant frequency of the assembly.

By the end of 1930, Alan Blumlein and Herbert Holman were tackling the question of the mounting of the cutting head assembly and the consequences that their initial inventions were having upon it. This patent, No. 368,336 was applied for on Monday, 1 December 1930.

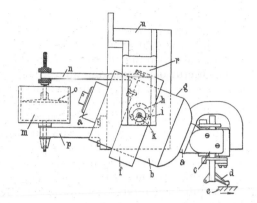

Figure 3.10
Detail from Patent No. 368,336 (1930)
showing the mounting of the moving
coil cutting head assembly.

Prior to the Columbia recorder much of the recording that had been carried out was done with the method known as lateral cutting. This allowed the cutting head to pass through the wax blank thereby cutting a groove whilst being accentuated in a lateral plane by the amplitude of the sound being recorded. The net result of this was that the head would cut the groove deeper and, naturally, there was a reciprocal effect for low volume passages causing the cut of the groove to become shallow. The lateral principle had been used (despite this effect) little changed, since the 1880s, the result of which caused a great deal of unwanted noise and was quite inefficient as a sound recording system.

Blumlein and Holman now proposed a method which was intended to allow the cutter to float in a vertical plane so that imperfections in the waxes would not cause the groove depth to vary to such a degree, but rather only from the effect of the amplitude of the sound being recorded. In so doing, their invention would not only lay the foundation for the record industry that was to follow (right up until the compact disc era of recent years), but also the stereophonic sound system that by now Blumlein must surely have been contemplating.

Patent No. 368,336, 'Improvements in and relating to the Mounting of Pivoted Apparatus, such as Electrical Sound recording Devices', relates to:

> The mounting of parts of an apparatus (having motion about a fixed axis), used with electric devices employed for recording sound on wax discs.

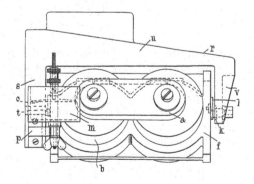

Figure 3.11
Detail from Patent No. 368,336
(1930). Similar to Figure 3.10 showing
the mounting of the moving coil
cutting head assembly but from
above.

In such a device, the recording stylus, in addition to being subjected to lateral vibratory motion, whereby the sound record is made, is usually mounted so that it may incorporate some vertical movement to accommodate for any irregularities in the wax surface, and thereby maintain a uniform depth of cut. The inertia of the apparatus in response to this vertical movement naturally has a considerable effect on the accuracy and sensitivity of the recording. One of the objects of the invention in this patent was to overcome previously encountered inertia difficulties and thereby improve the apparatus and to extend the working frequency range to which it will respond.

It consisted of a method of pivoting parts of the apparatus in such a way that the effective forces tending to move about the pivotal axis were reduced. This would also require a method of assembly for the 'floating parts' of the cutter in such a way that the effective inertia at the stylus point was reduced to a minimum. All of this was to be based around a U-shaped electromagnet on the arms of which were mounted bulky energizing windings. Between the poles of the electromagnet would be the pivoted moving coil on which the recording apparatus, stylus assembly (cutting head) would be mounted. The moving coil would in this way form a secondary winding in the actuating circuit. Being mounted thus in a magnetic field of great strength it would oscillate in a manner suitable for the cutting of a record.

The moving coil assembly is mounted in a manner whereby it oscillates in an ostensibly vertical axis (this was to allow for any unevenness of the disc surface), however, the movement of the whole device is desired about a horizontal axis. Because of the angular relationship of the electromagnetic circuit to the moving coil assembly, this is in actual fact possible, and the resulting operation is a freely moving, suspended cutting assembly, contained within the control of the electromagnetic field, able to move in both axes.

In order that the recorder may produce a uniform depth of cut (in an uneven surface of a wax blank), it is necessary that a small increase of depth of cut should produce as large an acceleration as possible of the recorder upwards, or vice versa, so as to restore the correct depths as quickly as possible. This condition is best realized when a line drawn from the sapphire (of the cutting head) to the axis of the floating assembly subtends an angle of 45° to the wax surface. If however a very uneven blank is being used, the forward movement of the sapphire will cause a change in speed. Under these conditions, it is advantageous to reduce this angle to less than 45°, in order that this forward movement for a given vertical movement (into the blank) may be reduced. It was found that angles between 30° and 50° were the most advantageous.

What Blumlein and Holman have described here is a method of cutting a blank record using the assembly they had already patented, i.e. the Columbia recorder, but having modified the cutting head assembly so that it is suspended in a magnetic field that will allow free movement of the head in the horizontal and vertical axis. In this way they could determine the optimum cutting angle for the assembly to the disc surface, and the angles subtended by the sides of the groove to the vertical, which turned out be around 45°. As time went by, this system was to become known as the '45/45 degree cutting method', and it eventually won over the 'Hill and Dale' system because the lateral movement produced a mono output, the vertical being the 'S' component, thus mono compatibility. Later, it was discovered that mono pickups could be used for binaural recordings provided their vertical compliance was good enough not to plane the groove flat.

Such was the nature of their invention that Blumlein had proposed it would be best suited to a material other than wax, and suggested that cellulose acetate be used instead as the material for the master disc. He also mentions that the cutting tip should be made from sapphire rather than the steel needles which were common-place at the time; of course this too was eventually taken up as the de-facto master-ing material for record cutting. The 45/45 degree cutting system would thereafter be used in record manufacture, almost unchanged from the date of this patent, to the present day.

## Microphones and more

Somewhere in late 1929, just as the basic construction of the MC1A recorder was being completed, Blumlein had the idea that the project for which Shoenberg had employed him had far greater potential than that which he was currently engaged in. Why not, he reasoned, continue the process to design an entire recording system having all the elements of an electromagnetically operated cutting and recording system, rather than just construct the recorder? If the recordings that were being made were of a better quality to start with, surely it would improve the overall final quality of the finished records pressed from the masters cut on the Columbia recorder.

As early as May 1929, Blumlein and Holman were engaged in conceptual designs for a high frequency microphone that they had worked on in parallel with the Columbia cutter until the end of that year. They appear to have started the work as an investigation into the effect of a diaphragm on a coil, but stopped this to con-centrate on the cutter first. By the time they returned to it, another year would have gone by. As we know, Shoenberg had already been given the task of developing a high frequency microphone and it was to this that Blumlein now turned his atten-tion.

Concentrating on the completion of the MC1B first, Blumlein decided to apply the principle of the moving coil to a pressure microphone as soon as the job was fin-ished, as he felt certain that it was equally important to have a licence-free micro-phone to pick up the sound. As soon as the Columbia recorder was ready and had been tested and approved, work on the high frequency microphone began again where it had been left at the end of 1929. This was critical to the concept of a com-plete Columbia system as, in the absence of a Columbia high frequency micro-phone, the only alternative was the Western condenser transmitter (CT) microphones that they had been using and the use of these would still require royalty payments.

The calculations on a moving coil microphone were carried out in the spring of 1930 and seemed to indicate that a permanent magnet design would have inade-quate field strength. The original intention had been to design an electromagnetic microphone which could be powered by car batteries. In fact such a design was tried using Austin Seven batteries from several cars in the car park, but these proved inad-equate. Nevertheless, Blumlein decided to go ahead with the project and had the design for a diaphragm and its surroundings ready by August 1930. The spacing between the centre and outer poles was set by a well-fitting ring of cadmium bronze of low resistance which also helped to damp the moving coil. The diaphragm itself, constructed by Holman, needed to be both lightweight and stiff. This was made from a piece of balsa wood which had been cut into a very thin layer, impregnated with celluloid, with thin sheets of shallow concave aluminium foil waxed onto each side.

The edge of the diaphragm was compressed (it had no balsa wood at this point) and at first the flat edge which had been thinned by etching was simply clamped. This however, caused the edge to wrinkle and gave the diaphragm inadequate compliance. Blumlein decided that the two ridges, which were formed between the clamp and the diaphragm, were too deep, and affected the acoustic impedance of air passages behind the diaphragm. These had been designed by 'Ham' Clark, but Blumlein redesigned them, this time with four ridges, which seemed to solve the problem. The coil was made of anodized aluminium and backed onto an aluminium former (which had a slit in it to eliminate eddy currents). This was riveted onto the diaphragm.

The grille for the microphone and the air cavity between it and the diaphragm, and the cavity behind the diaphragm were also designed by Blumlein, who was fully aware that Western had a patent for a moving coil microphone. This meant that air in this cavity could not be used for damping. Instead, Blumlein included a calculated amount of cotton wool to prevent cavity resonance. This had been determined by measurements carried out by Ivan L. Turnbull. Damping of the bass resonance was designed once again to be carried out by loading the microphone from its amplifier so that the input impedance, set by a series resonant circuit which also did most of the equalization, created the damping.

At this point, Blumlein decided that a feature of all the microphones would be that they should be interchangeable and have the same resonant frequency of 500 Hz. This meant that careful control of the diaphragm mass and the stiffness of the edge would have to be maintained, and this work, coupled with the other tests, took much of the remainder of the summer of 1930.

The first prototype microphone that Blumlein was happy with was the result of over twenty attempts at constructing the coil and the surrounding cavity. Blumlein was rarely satisfied unless something was absolutely correct in his estimation, and this particular instance is a good demonstration of how he would labour away at something regardless of time and energy expended until it was right. Curiously it was also initially called the 'MC1A', which had been the name of a prototype of the cutter. The microphone was named the MC1A because by this time the recorder was called the MC1B. In time however, to avoid confusion, the microphone was called the 'HB1A', named after 'Holman and Blumlein'.

The first HB1A microphone to be completed was received by Blumlein on 4 November 1930 and was tried the next day, though it gave rather disappointing results. The problem seemed to be that the coil was not moving freely in the gap. Refinements were made to a second prototype with the centre pole-piece being reduced by 0.01 inch, which gave some improvement, though impedance tests with a loaded diaphragm indicated that it was not moving as a true piston, but pivoting about a point near the edge. Blumlein concluded that this had arisen because the tension in the diaphragm edge was not uniform and so a screw tensioning arrangement was added, which became a permanent feature of the design and used to adjust the resonance of the diaphragm to 500 Hz. At the same time, despite Turnbull's original calculations, Blumlein decided to double the amount of cotton wool in the cavity behind the diaphragm. The specification for the second model of the microphone was now considered significantly different from the first and was thus called the HB1B and sent away for construction.

At around this time the microphones began to acquire other names; nicknames, to which many people, including Blumlein's wife-to-be Doreen, would always refer. They were called 'HBs', or 'hell's bells', which was a favourite phrase of Alan's when things didn't quite work out for him, usually followed by 'buckets of blood!' Sometimes, 'HB' took on the guise of 'hot and bothered', because Doreen thought that Alan always became so when he was frustrated, mostly with himself.

With the changes made to the original HB1A-1 (done while the HB1B was being constructed), a series of extensive tests of the motional impedance were carried out through November and into December, to enable the parameters of the equalizer for the HBs to be calculated. These tests usually had good results, but it was very time consuming and Blumlein must have become a bit disillusioned with the time it was all taking because he designed a new equalizer with the components arranged in a series of switchable banks, which was then used in yet more tests. Finally, Blumlein was satisfied that the prototype microphone and the new recorder were ready for the first tests. These were carried out on Monday, 8 December 1930.

The first recording made was of a Mr Sparks playing the piano (reference no. RWTT535/1), followed on the next day by the Van Philips Dance Band (RWTT536). There are no notes of Blumlein's or anybody else's comments, however a frequency run dated 19 December shows very good, smooth and even characteristics. There were further voice-only recording tests (RWTT549/1 and RWTT549/2) carried out that day using both the HB and CT microphones for comparison. These were repeated on 23 December (RWTT554/1 and RWTT554/2). The same day (Tuesday, 23 December 1930), Blumlein wrote down a detailed description of the HB microphone and its special features that would enable the draft specification to be drawn up (eventually Patent No. 369,063).

Yet more tests were carried out in the New Year; on 6 January 1931, The Greenings Band was recorded (curiously, the serial numbers RWTT/1 and RWTT/2 seem to have been used, but it is unlikely that these refer to the first and second ever recording sessions), using a bass 'depreciator' (which we would know today as an attenuator) of 0.7 dB at 100 Hz and 2.3 dB at 50 Hz, and a high frequency boost of 0.7 dB at 500 Hz and 2 dB at 5 kHz. The results of all these tests were noted by Blumlein and incorporated into the first production model of the HB1B microphone and its associated equalizer, which was ready by the end of January. The new microphone had the two new features which Blumlein felt sure eliminated the cause of much of the trouble in the earlier tests; the increased clearance for the moving coil, and a stretchable diaphragm which had involved increasing the size of the diaphragm itself. The profile of the inner surface of the stretching ring governed the size and shape of the cavity outside the edge of the diaphragm, and this would cause a great deal of consternation in the next few weeks during a series of tests that were carried out to determine the effect of this.

Often two or three different profiles a day were designed as Blumlein tried to get the optimum shape, the work often going on until late in the evening. It was probably now that the phrases 'hell's bells' and 'hot and bothered', began to come to the fore. In the end, eighteen profiles were made with the final design ready in late February 1931. In early March a new problem occurred; there was dissipation in the field-coil which warmed up the microphone chassis and altered the bass resonant frequency. This problem was traced to the alteration in the mass of air vibrating and it was quickly solved by changing the bass resonance setting slightly.

On 20 March, further recording tests were carried out comparing the Western CT microphone with the new HB1A-1 and the HB1B with various 'brightening' adjustments being made. By 29 March, when three discs were cut (RWTT679/1, RTWW679/2 and RTWW679/3), each using the HB1B, the brightening effect of +4 dB at 5 kHz was found to be 'most acceptable', and in May, further HB1B microphones were received from construction and tested. In this form, the microphone was to be used in EMI recording studios, by the BBC at the London Television Station, Alexandra Palace for many years to come. Curiously the BBC did not choose the HB microphone for radio work, preferring the BBC-designed Type A ribbon microphone. Perhaps this was partly due to cost, the 'Type A' being 'inexpensive' at £9-0s-0d, while the moving coil HB1B cost £40-0s-0d. There were later variants, all of which were used by the BBC for television work. The HB2, HB3 and HB4 microphones all derived from experiments with the field strength, the final design being the HB4C. Blumlein's notes on microphones end on 16 November 1933, when he made an entry for the flux measurement 'No. 82' of an HB1B microphone, though additional work was carried on (for many years) by Clark, Turnbull, Holman and others.

The company now applied for the specification for the microphone which lists Blumlein and Holman as inventors – Patent No. 369,063 – on Wednesday, 13 May 1931, as a moving coil microphone which had a rigid diaphragm suspended in an electromagnetic field. The invention consisted of an electro-acoustic device having a vibrating system comprising of a substantially rigid, piston-like, diaphragm connected at its outer edge to a flexible air-sealing surround.

> The diaphragm is composed of a light wood such as balsa covered on either or both sides by a thin plate of aluminium (or other light metal or alloy). The main resonance of the device is controlled by electromagnetic damping with an electromagnetic coil adapted to move within a magnetic field. The microphone works by the movement of sound waves impinging upon the piston-like diaphragm. In order for this to work the diaphragm should be rigid so that it is free from any resonant frequency due its own flexure, however, the diaphragm should be of sufficiently low mass that it will respond well to the high frequencies acting upon it. Part of the object of this invention was to overcome the conflicting requirements of high rigidity and low mass, this objective would be met if the diaphragm was made of a three ply construction, consisting of two thin layers of metal (such as aluminium), enclosing a centre of light wood (such as balsa).
>
> If air is permitted to pass freely from the front to the back of such a diaphragm, the response of the device at all low frequencies will be impaired due to the diaphragm receiving almost equal vibrations of sound pressure on

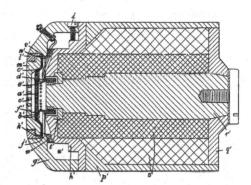

Figure 3.12
Detail from Patent No. 369,063
(1931), cross-section of the
prototype HB1A microphone.

both its front and back faces. Therefore it is necessary to provide a closed cavity (or baffle), containing or shielding the air behind the diaphragm and the sides of the air cavity.

Such a seal in a microphone cannot take the form of a frictional packing on account of the delicate nature of the movements of the diaphragm, therefore a thin elastic surround is provided between the diaphragm and the sides of the cavity. The diaphragm then forms a rigid piston, closing the mouth of the cavity, the edge of the diaphragm being sealed by a thin surround which stops the direct access of air to the cavity, but nevertheless permits the diaphragm to move. In this manner, the surround, which forms the air seal, may conveniently support the diaphragm also.

At the back of the diaphragm, within the cavity, is an attached coil which can move freely in a magnetic field provided by an electromagnet or permanent magnet. The coil serves to convert the mechanical movement of the diaphragm into electrical impulses, and may also be used for the provision of damping for the resonance of the mass of the diaphragm and its elastic constraints.

The HB1B microphone became the eventual product of this patent and it was considered so good that it immediately became the preferred choice of the newly formed EMI, and soon after HB1Bs were being sent to the BBC. Fairly soon after, EMI were manufacturing them to be sent to studios all over the country. Engineers loved it, and the recordings produced from these microphones, especially classical recordings, were the best obtained up to that point in time. The microphone was hugely successful and would be used for many years to come. Originally known as the Blumlein microphone, in later years his name was dropped from the product, but was retained in the microphone technique by which most audio engineers will have heard of him.

By September 1931, the senior management of the new EMI company had finally realized the worth of their engineering and electronics prodigy and had met to consider a special bonus for Blumlein, Herbert Holman and 'Ham' Clark for all their work on the Columbia recording system. They decided to award Blumlein the princely sum of £200 for his efforts, not an inconsiderable sum of money at the time, yet seemingly very small compared to the vast amount that the new recording system was now saving them in royalties to Western Electric and Bell Labs, and the royalties it would earn from other companies using it to cut records.

Nevertheless, typically gracious for their consideration, Blumlein wrote to his employers on 13 September 1931, thanking them:

> From: - A.D. BLUMLEIN
> 67 Earls Court Sq.
> S.W.5.

13th Sept 1931

J. Gray Esq.
Columbia Graphophone Co. Ltd.

Dear Sir,

I am writing to you to thank the company for the bonus which I have just received in connection with the new recording system. I appreciate that, that the face value of the bonus is considerably enhanced by the depressed times in which it is made, and I am therefore very grateful for it.

May I also thank the company for having given bonuses to Messrs Holman and Clark, whose very good work, I was hitherto afraid was not fully appreciated. My fears on this score, I find unfounded.

Thanking you Sir

Yours faithfully.

## Binaural sound

One day in 1931, Alan Blumlein took Doreen to the cinema and said to her during the film: 'Do you realize the sound only comes from one person?' Doreen, by her own admission, was not a technical person and so replied to him, 'Oh does it?' and he said, 'Yes. And I've got a way to make it follow the person.'

Alan Blumlein had just tried to describe his first thoughts about the system he would always call 'binaural sound', but which we have come to know better as stereophonic or stereo sound. Blumlein explained to Doreen that, if she could imagine being blind, and sitting in the cinema, she would be able to point out exactly where the person was on the screen with his system. This, of course, was what he was trying to achieve, not this 'terrible effect' where the sound comes from one side of the screen when the actor was at the other side.

In truth Blumlein had been considering binaural sound for some time before the conversation at the cinema, probably conceiving the idea somewhere between March 1931 when the HMV/Columbia merger was taking place, and 1 November 1931, when the joint research laboratories were set up. While most people were seemingly content with monophonic recordings, or single channel sound, Alan Blumlein typically, had already decided that sound recording should try as best as possible to match human hearing. As humans have two ears, and the properties of binaural hearing had been known for some time, it seemed obvious to him that recordings should also have at least two channels of sound, if not more.

Once again, he was somewhat surprised that nobody else had come up with the idea, just as with the application of electromagnetic damping to the recording system for Columbia, but seeing as nobody had, Blumlein had chosen to take on the task himself. It is true that Bell Labs at around this time had been working on a system where arrays of microphones were being fed into arrays of loudspeakers, but Blumlein's idea was fixed in his mind. As human hearing was based on two ears, so his sound theory should be based on the binaural characteristics of sound reaching those ears. Just how far into the project to complete the Columbia recording system Blumlein was at the time he decided that binaural sound was the goal, is now uncertain. Again, typical of the man, he kept very few notes on his work, usually explaining the ideas in his head on a blackboard, or the back of a scrap of paper to his colleagues, rather than write them out long-hand as he solved the problems he set himself.

Years later, during a radio interview, Philip B. Vanderlyn, who had been a young research engineer at EMI working under Blumlein and who would go on to write a paper with Henry Clark and G.F. Dutton for the Institution of Electrical Engineers on stereophonic recording derived entirely from Blumlein's work, explained how many key moments in Blumlein's train of thought had been lost this way: 'Every day he would come round the research laboratories and talk with everyone. Going back through his papers, I would have expected to find a lot of theory, thoughts and calculations. But they didn't emerge and I often wondered where the devil they were. I would have expected some sort of master plan. But the core of his work was missing.'

What is certain is that somewhere in the late summer of 1931, Blumlein began to explain to Shoenberg his idea for a totally new concept of recording sound, something which he felt sure would revolutionize the flat, lifeless recordings then being made. While Shoenberg was naturally excited that his star engineer had yet more potentially lucrative ideas, it was probably natural for Shoenberg to be more than a little cautious. After all, they had only just completed and patented the Columbia recording system. Wasn't that an entirely new and wonderful system for making recordings? Now, here was Blumlein with another idea again, proposing dramatic new ideas before the current one had even had a chance to be unveiled.

In a series of surviving hand-written notes dated 25 September 1931, Blumlein heads the page 'Binaural Speech Trials', though there is no direct evidence that any such trials were carried out at this time. Blumlein calculates the value of an element he calls 'K', which is the ability to modify the signal of the microphones to recreate the phase difference at the listener's ears. The path difference from the two loudspeakers can give sufficient phase difference when they are summed in the listener's ears, if the difference in amplitude of the signals is then correctly modified. This Blumlein does by deriving the 'sum' and 'difference' signals, the difference signal being fed through $\pi/2$ using a current-fed capacitor which also reduces the signal by 6 dB per octave (6 dB/8 ve). This new difference signal is then added to and subtracted from the sum signal to form two new signals to the loudspeakers. Blumlein called the process 'shuffling' (which we take today to have a different meaning to that which Blumlein originally uses here).

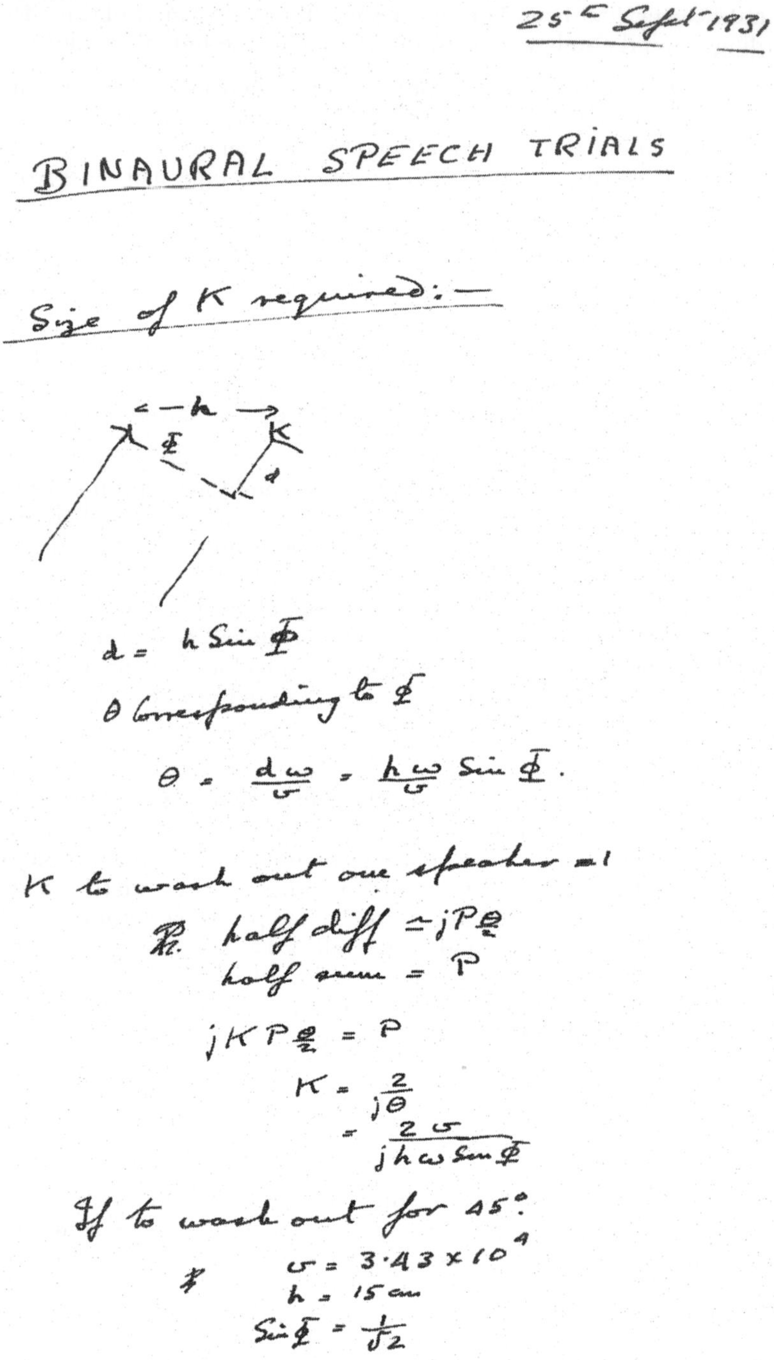

25th Sept 1931

BINAURAL SPEECH TRIALS

Size of K required:—

d = h Sin Φ

θ corresponding to Φ

$\theta = \frac{d\omega}{v} = \frac{h\omega}{v} \sin \Phi$.

K to wash out one speaker = 1

Pr. half diff $\simeq jP\frac{\theta}{2}$

half sum = P

$jKP\frac{\theta}{2} = P$

$K = \frac{2}{j\theta}$

$= \frac{2v}{jh\omega \sin \Phi}$

If to wash out for 45°:

$v = 3 \cdot 43 \times 10^4$

$h = 15$ cm

$\sin \Phi = \frac{1}{\sqrt{2}}$

Figure 3.13 Alan Blumlein's hand-written notes from the binaural speech trials dated 25 September 1931, in which the principle element 'K' is first mentioned and calculated.

Many years later, Shoenberg would admit, as would many of Blumlein's colleagues, that the idea of binaural sound was in fact so far ahead of its time, that he and the others simply could not get their heads around the concept. It was something that lived in the mind of Blumlein and was, to all intents and purposes, misunderstood by many of those around him. This did not deter Blumlein from working on the project as we now know, nor did it deter Shoenberg from encouraging him in his efforts. It is however, a little sad to think that throughout the remainder of 1931, when the binaural sound patent was finally published, probably very few of his friends and colleagues, even the engineers and scientific ones capable of understanding most of the technological advances of the time, were unaware of just how incredible and far-sighted the idea had been. They, like EMI, would have to wait another twenty-five years or more to see the fruits of Blumlein's mind ripen with the advent of stereophonic sound recordings.

Patent No. 394,325, 'Improvements in and relating to Sound-transmission, Sound-recording and Sound-reproducing Systems', was first applied for on Monday, 14 December 1931. It comprises 22 pages of outline, with 11 supporting diagrams and their legends. Quite how Blumlein intended to apply his work to a practical design we shall probably never know as many of the applications that were tried were rushed; work on the television system that EMI had committed itself to increasingly took over the research laboratories. The principle for stereophonic sound however is clearly laid out with subtle and tantalizing yet definite hints at cinematographic use as well as a multi-channel audio system beyond the specifications of this one.

Blumlein was an immensely modest man, quite often not fully aware of his own genius. However he must have had some feeling for the enormity of this work as he begins the patent by explaining (in some detail) how the human hearing system works and how this relates to the methods he was trying to apply within his 'binaural sound' system as he called it. Much of the original patent has been reproduced here, with the references to the diagrams (figures, which are reproduced alongside), which Blumlein used to outline his invention.

The invention relates to the transmission, recording and reproduction of sound, being particularly directed to systems for recording and reproducing speech, music and other sound effects especially when associated with picture effects as in talking motion pictures. The fundamental object of the invention is to provide a sound recording and reproduction system whereby a true directional impression may be conveyed to a listener, thus improving the illusion that the sound is coming from the artist or other sound source as presented to the eyes. In order to fully appreciate the physical basis of the invention the stages of its development as well as the known and established facts concerning the physical relations between sound sources, and the human ears will be briefly summarized.

Human ability to determine direction from which a sound arrives is due to binaural hearing. The brain is able to detect differences between sounds received by the two ears from the same source and is thus able to determine angular direction. This function is well known. With two microphones correctly spaced and with the two channels entirely separate, it is also known that this directional effect can also be obtained for example in a studio; but if the channels are not kept separate (for example, by replacing the headphones by two loudspeakers) the effect is largely lost.

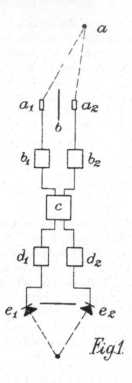

Figure 3.14
Detail from Patent No. 394,325 (1931)
Blumlein's binaural patent. Here,
showing how from a listening position
'a', two microphones '$a_1$' and '$a_2$' separated by a baffle, can have their
independent frequencies fed into a
'shuffling' network 'c' and relayed back
to the listening position via loudspeakers '$e_1$' and '$e_2$'.

The invention contemplates controlling the sound, emitted for example by loudspeakers, in such a way that the directional effect is retained.

While the operation of the ears in determining the direction of a sound source is not yet fully understood, it is fairly widely established that the main factors having effect are phase differences and intensity differences between the sounds reaching the two ears; the influence which each of these has being dependent upon the frequency of the sound being emitted.

Broadly, the invention consists of a system of controlling the intensities of sound to be (or being) emitted by a plurity of loudspeakers or similar sound sources in a suitably placed relationship to the listener. This is done in order that the listener's ears will note low frequency phase differences and high frequency intensity differences suitable for conveying to the brain the desired sense of direction of the sound origin. In other words, the direction from which the sound arrives at the microphones determines the characteristics (and more especially the intensities) of the sounds emitted by the loudspeakers in such a way as to provide this directional sensation.

Furthermore, the invention consists of a sound transmission system wherein the sound received by two or more microphones (with low frequency differences in phase of sound pressure at the microphone), is reproduced as difference in volume at the loudspeakers. Two microphones should be spaced with their axes of maximum sensitivity so directed relative to one another (this is 'shuffling' as we know it now) and to the sound source that, the relative loudness of loudspeakers which reproduce the impulses is controlled by the direction from which the sound reaches the microphones.

The invention also consists of a mechanical system where the above refer-

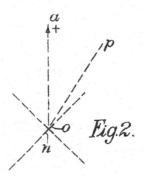

Figure 3.15
Detail from Patent No. 394,325
(1931). Two velocity microphones 'n'
and 'o' are placed with their axis perpendicular to one another and each at
an axis of 45° to the direction of the
centre of the screen and the direction
of the source of the sound.

enced two channels of sound impulses are recorded in the same groove of a
record, and in combination with the means of transmitting the same impulses
by radio telegraphy, as well as photographic recording or transmission and/or
reproduction of pictures.

For the purpose of demonstration, the sounds recorded and reproduced
may be received from a source by two pressure microphones ($a_1$ and $a_2$),
mounted on opposite sides of a baffle ($b$) which serves to provide the high frequency intensity differences at the microphones in the same way as the human
head operates on the ears.

The outputs from the two microphones are after separate amplification by
similar amplifiers ($b_1$ and $b_2$), taken to suitably arranged circuits ($c$) comprising
transformers (see *Figures 3* and *4* [Figures 3.16 and 3.17]) or bridge networks
circuits, which convert the two primary channels (called the summation and
difference channels). These are arranged so that the current flowing into the
summation channel will represent half the sum, or the mean of the current
flowing in the two original channels; while the current flowing into the difference channel will represent half the difference of the currents in the original
two channels.

Two velocity microphones ($n$ and $o$) are placed with their axis perpendicular
to one another and each at an axis of 45° to the direction of the centre of the
screen and the direction of the source of the sound. It can be seen that movement of the sound source ($a$) laterally to a position ($p$) removed from the
centre of the field, will result in the sound waves striking microphone $o$ at a
more acute angle than they strike microphone $n$, and differences in the microphone outputs will result.

The microphones are sufficiently close together to render phase differences
of the incident sound negligible, and the output amplitudes therefore differ

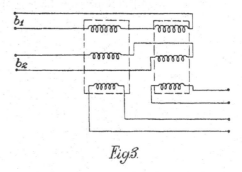

Figure 3.16
Detail from Patent No. 394,325
(1931) showing a transformer
arrangement where the input currents
from amplifiers '$b_1$' and '$b_2$' are
separately fed to windings to provide
a sum and difference output current.

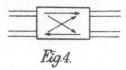

*Fig.4.*

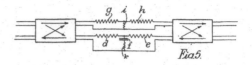

*Fig.5.*

Figure 3.17
Detail from Patent No. 394,325 (1931)
which shows a diagrammatic representation of a sum and
difference transformer.

Figure 3.18
Detail from Patent No. 394,325 (1931).
The amplification circuit for sum and differ-
ence output.

approximately proportionally to the obliquity of the incident sound. They may
therefore be amplified similarly, and supplied directly to the loudspeakers to
which they will give the correct amplitude differences for the desired directional effect, provided the relationship between the various dimensions of the
recording and reproduction layouts are correct.

*Figure 3* [Figure 3.16] shows a convenient transformer arrangement (for
*Figure 1* [Figure 3.14]), where the input currents from amplifiers $b_1$ and $b_2$ are
separately fed to two primary windings, one on each of two transformers. The
secondary winding of each transformer provides a sum or difference output
current on account of the senses in which the primary coils are wound as
shown. *Figure 4* [Figure 3.17] shows a diagrammatic representation of a sum
and difference transformer similar to that of *Figure 3* [Figure 3.16].

*Figure 5* [Figure 3.18] and *Figure 6* [Figure 3.19] represent the portion of the
circuits indicated by *c* in *Figure 1* [Figure 3.14], circuits for sum and difference
arrangement outputs, modified in order to obtain the desired sound effects.
Assuming the original currents differ in phase only, the current in the difference channel will be $\pi/2$ difference in phase from the current in the summation channel. This difference in current is passed through two resistances, *d*
and *e* in series, between which a condenser *f* forms a shunt arm. The voltage
across the condenser will be in phase with that in the summation channel. By
passing a current in the summation channel through a plain resistive attenuator network composed of a series of resistances (*g* and *h*) and a shunt resistance
(*i*), a voltage is obtained which remains in phase with the voltage across the
condenser *f* in the difference channel.

These two voltages are then combined and re-separated by a sum and difference process (such as previously adopted) to produce two final channels.
The voltage in the first channel will be sum of these voltages, and the voltage
in the second channel will be the difference between these voltages. Since

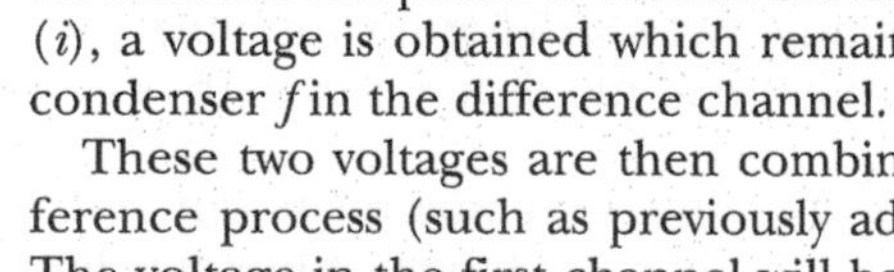

*Fig.6.*

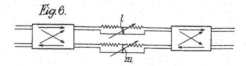

Figure 3.19
Detail from Patent No. 394,325
(1931). A similar circuit to Figure 3.18;
however, in this example the shunt
condenser and shunt
resistance are replaced by artificial
attenuators.

Figure 3.20
Detail from Patent No. 394,325
(1931) showing how an increase or
decrease of differences between
channels may be effected if no conversion of phase differences into
amplitude differences is required.
This can prove particularly useful for
the operation of more than two loudspeakers.

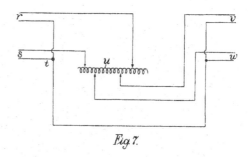

*Fig 7.*

these voltages were in phase, the two final channels will be in phase, but will differ in magnitude. By choosing the value of the shunt resistance $i$ in the summation channel and the shunt condenser $f$ in the difference channel, for a given frequency any degree of amplitude difference in the final channels can be obtained for a given phase difference in the original channels.

For higher frequencies it is not necessary to convert phase shifts into amplitude differences, but simply to reproduce amplitude differences. The shunt condenser $f$ in the difference circuit is therefore built out with a resistance $k$, whose value is substantially equal to that of resistance $i$, in which case the amplitude differences for high frequencies are passed without modification.

Where the microphones employed have a velocity type that an edge-on microphone gives, an output proportional to the obliquity of the source is desired. *Figure 6* [Figure 3.19] represents a suitable arrangement for this form where the shunt condenser $f$ and resistance $k$ are in series, and the shunt resistance $i$ is replaced by shunt resistances $l$ and $m$ which form artificial attenuators. By altering the relative attenuation in $l$ and $m$, the intensity differences in the two lines corresponding to the given obliquity of sound is controlled.

There is a simple method by which modifications for increase or decrease of differences between channels may be effected if no conversion of phase differences into amplitude differences is required. *Figure 7* [Figure 3.20] is a diagrammatic demonstration of this, which can prove particularly useful for the operation of more than two loudspeakers. If the transmission is effected in the form of two channels $r$ and $s$, of similar phase but different amplitudes, an alteration of these amplitude differences may be effected by connecting one wire of each channel $r$ and $s$ together at $t$, and connecting a choke $u$ between the other two wires of the two channels.

The outgoing channels $v$ and $w$, whose difference is to be a modification of the original difference, are connected by one wire to the common point $t$ of the original channels, and by their other wires to tappings along the choke $u$. If the differences are to be increased the tappings at which the output channels are connected lie outside the tappings to which the input channels are connected, so that the choke operates in effect as an auto-transformer amplifying the difference voltages. Similarly, for a reduction of differences, the output channels are tapped intermediately between the two input channels. This arrangement works well with a number of loudspeakers for binaural reproduction.

Each of these diagrams demonstrates a recorder assembly whereby both

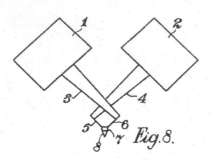

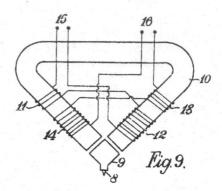

Figure 3.21
Detail from Patent No. 394,325 (1931)
demonstrating a recorder assembly
whereby both channels may be cut by
a single tool on the same groove. This
results in a recording at 45° to the
wax (or other) surface giving the sum
and difference.

Figure 3.22
Detail from Patent No. 394,325 (1931)
similar to Figure 3.21 except here the
driving force is generated from an
electromagnet '10'.

channels may be cut by a single tool on the same groove (with several variations), resulting in a recording at 45° to the wax (or other) surface giving the
sum and difference as the effective lateral and hill and dale amplitudes. A light
stylus is pulled into two directions at right angles to one another and each at
the preferred 45° angle to the surface.

In *Figure 8* [Figure 3.21] for example, *1* and *2* represent the driving elements
of two recorders normally adapted for cutting lateral-cut records. These in turn
drive *3* and *4* (drive arms), the ends of which are connected by ligaments *5* and
*6* to the end of a reed *7*. The reed carries a cutting sapphire *8* which makes the
groove in the record surface.

Movements of the recording arms *3* and *4* produce movements in the end of

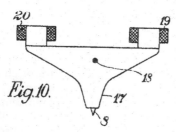

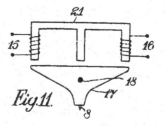

Figure 3.23
Detail from Patent No. 394,325
(1931), further example of cutting
stylus assembly.

Figure 3.24
Detail from Patent No. 394,325 (1931),
further example of cutting stylus
assembly.

the reed 7. Thus, currents in movement 1 will cause the reed 7 to move along an axis approximately 45° to the vertical, rising from left to right. Similarly currents in movement 2 will produce movement of the reed 7 in an axis at right angles to the former axis, while currents in both movements will of course result in vertical movement of the reed.

Having outlined the invention within the patent, Blumlein then went on to describe how he felt binaural recording would be, 'especially applicable to talking pictures, but not limited to such use. It may be employed in recording sound quite independently of any picture effects and in this connection (as well as when used in cinematographic work) it seems probable that the binaural effect will be found to improve the acoustic properties of recording studios, and to save any drastic acoustic treatment thereof while providing much more realistic and satisfactory records for reproduction.' Blumlein went on to conclude, 'In general, the invention is applicable, in all cases, where it is desired to give directional effects to emitted sound.'

This quite enormous piece of work, comprising as it does some 70 claims, was so far ahead of its time that again very few people (if any outside the immediate clique of engineers at EMI), understood its implications. Blumlein had been officially transferred (internally) from Columbia to EMI on 1 November 1931, when the joint research laboratories had been set-up, and the construction of the binaural system continued throughout 1932, with experiments conducted in earnest soon after, starting in early 1933. Though initially the experiments were made using poor pressure operated microphones, with omni-directional polar patterns and so gave limited results, later, when Blumlein and Holman's own microphones were used, the results became noticeably better.

Blumlein had calculated that the audio signals which would be received by a listener sitting (preferably) or standing in an off-centre position relative to a sound source,

Figure 3.25
The twin ribbon binaural microphone side elevation. It was necessary to provide an energized magnet because of the inadequacies of permanent magnet materials of the time. (*Courtesy of EMI*)

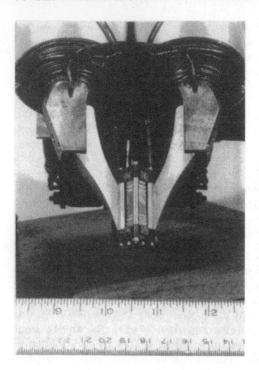

Figure 3.26
Twin ribbon binaural microphone
capsule detail. The two orthogonal
ribbons were inclined at ±45º to the
centre of the working area with
slender pole pieces to minimize the
disturbance to the acoustic field.
(*Courtesy of EMI*)

would arrive at different times at the ears. By placing two matched (identical) microphones eight inches apart (the average distance between a person's ears), to simulate a human listener, these signals could be processed in terms of phase before
applying them to a spaced pair of loudspeakers. Blumlein assumed that phase was
the only difference that needed to be calculated; in fact with the microphones as
closely spaced as he had them, he was probably correct as the path lengths of the
audio signals were so close for their attenuation to be nearly identical.

Because he was concerned about strange phase anomalies at high frequencies that
had not been investigated and understood at that time, Blumlein confined his
analysis of the response of the human ear to frequencies below 750 Hz. All that was
needed now was a series of tests at different frequencies, plotting the response of
the two microphones to build up a model of how human hearing 'works'. This
process was called 'shuffling' by Blumlein, and he could simulate this human
response through his circuit, which in turn he called a 'shuffler circuit'. Once completed, the model could be used to calculate the phase and amplitude differences
from the sound reaching each of the microphones separately and carry these signals
through to the loudspeakers. What he hoped would then be reproduced, was an
audible representation of the sum and difference of the phase and amplitude of the
sound, which we as humans detect as the perception of the directional origination
of the sound.

Though we now know that Blumlein had a firm grasp of the principles being
employed within his invention, he obviously realized that not everybody would
understand as plainly as he did exactly what it was that was being demonstrated.
Consequently he wrote an exhaustive and very descriptive report for Shoenberg and

Figure 3.27
Binaural recording head detail clearly showing how excess wax was removed from the cutting tip by means of a suction pipe. (*Courtesy of EMI*)

the senior management at EMI, hoping to explain to them just what 'binaural sound' – as he had by now christened his work – was. Unfortunately, EMI still did not fully grasp the importance of the work, and while it had no compunction in allowing Blumlein to experiment at will, the work was shelved after his death in 1942.

Blumlein continued with the binaural experiments right up to and just after his transfer to the high-definition television system, which was soon to occupy most of the engineers and scientists at EMI. He wrote an extensive report with one of the Recording Research Section engineers, Jim Castle, entitled 'Recording Research – Hill and Dale Overloading' on 20 May 1932, in which they wrote: '...the results of some tests which were made in order to check a theory of harmonic production in Hill and Dale records due to the finite size of the reproducing ball'. They concluded that: '...it would appear that the general theory and formulae (calculated in the report) for harmonic and intermodulation tone production are correct. Secondly, the Hill and Dale system as known to us is not satisfactory for levels as high as +3dB at frequencies above 3,000 cps (cycles per second), at 33 rpm [curiously, 33 rpm. Not thirty-three and a third rpm]. The fact that much higher levels than this can be tracked at middle frequencies would appear to have little advantage since scratch at these frequencies is no real trouble.'

On 4 July 1932, Blumlein wrote out a detailed document in which the binaural

Figure 3.28
Binaural recording head front detail showing the cutting tip and general construction. It was made from two Western Electric wax cutters, the left hand one moves the stylus horizontally, while the right moves it vertically. (*Courtesy of EMI*)

reproduction system was described in full. It would seem that this eighteen-page, hand-written account, served as a forerunner to a much simpler document that Blumlein would write three weeks later. In the later eight-page memorandum, written to Shoenberg on 21 July 1932, Blumlein tries to describe his binaural ideas in non-mathematical language, the main reason for which seems to be to persuade Shoenberg to sanction the forthcoming experimental programme that was being planned. Perhaps not fully grasping the depth of the system, Shoenberg none the less, possibly as a result of this memorandum, gave the go-ahead for the construction work to begin on the microphones, shuffling circuits, wax cutter and pickup. Blumlein writes, 'This memorandum describes briefly the general troubles experienced with single channel reproduction, the difficulties experienced with the more obvious binaural systems, and the proposed method of overcoming these difficulties.'

Throughout 1932, EMI was busy checking and testing any other recording/reproduction system that appeared on the horizon to see if it was better than that designed by Blumlein or, more likely, it infringed the Blumlein patents in some way. One such system was the Siemens Halske (Recording) System, which had been supplied to Carl Lindstrom in Berlin in August 1932, evaluated in a report by Blumlein to Shoenberg on 26 September 1932. Blumlein writes: 'The Siemens gear was well built and typically German in lay-out. I formed an opinion that the system would probably give a second-rate commercial product with good loud 'gramophone tone' of a rather metallic and confused type. I cannot see that the Siemens system infringes any serious patents that I know of. The microphone, though similar in appearance to the W.E. Co.'s, probably avoids their air damping patents. The amplifier H.T. is fed through a filter having both inductive and resistive series elements. The H.T. supply however, is not rectified AC. The recorder is a moving iron-type resonance device being apparently damped by the sheet rubber used to seal off the movement. Unless all frequencies below, say, 500 cps are heavily depreciated the recording must be 'dirty', in that bass notes will modulate superimposed high frequencies. The recorder employs no strikingly novel feature.' It would seem that EMI did not have much to worry about from this system.

It had also become obvious to Blumlein by this time that he, Holman and Clark needed additional engineers in order to complete all the experimental and development work that was required, and this led to a curious set of circumstances which culminated in the employment of Maurice Geoffrey Harker. One evening Blumlein came down into the car park at the side of the Research and Development building to discover that his car had a puncture. At the same time Arthur Cooper, who worked in the Works Design department came down to his car, saw that Blumlein had a puncture and proceeded to help him fix it. During the process Blumlein told Cooper that he was looking for new additions to the team and that if Cooper could think of anyone then he should let Blumlein know. As it happened, Cooper knew of Maurice Harker, then fresh out of college who had just received an honours degree in heavy electrical engineering (coincidentally the same degree that Blumlein had). Cooper gave Blumlein Harker's telephone number and the next day Blumlein called him and asked him to attend an interview on following afternoon, Thursday, 29 September 1932. Maurice Harker recalled:

'At my interview, Blumlein asked me all sorts of questions, and I told him how I'd built crystal radios and later a single valve radio set. Some time into the interview he asked me if I knew what a "Decibel" was? [The Decibel had replaced the TU (Transmission Unit) introduced by Bell in 1929.] (Well), I'd taken all the regular scientific journals of the day so I had a reasonable idea, but not an accurate one, so I said "Yes, it is the smallest degree of change in the amplitude of sound that a human ear can detect". Well, this is quite inaccurate, but accurate in a sort of way (you see), and I suppose Blumlein was reasonably satisfied with this answer, because the following morning (the Friday), he telephoned me at my parents' house in Pinner, near Watford where I was living, and told me that I was engaged and could I start the following Monday, 3 October 1932. And of course, I did; though I now had the problem of how I should get from Pinner to Hayes. (So), that weekend, I scoured the area for a suitable second-hand car and finally purchased a four-year old Austin Seven for £25 with money I had borrowed from my parents. That little car served me, my colleagues and EMI for many years to come.'

With the addition of Maurice Harker and more engineers to arrive in the months that followed, Blumlein now felt ready to commence work on the binaural experiments. The first of these took place on 8 January 1933, the work being recorded in a file by Blumlein which is headed 'Polar Curves of two W.E. C.T.'s for Binaural Work'. Using a pair of calibrated Western Electric condenser microphones (C.T.'s – condenser transmitters), spaced at about 7 inches apart a series of test were carried out to plot polar patterns. Blumlein perhaps should have used his own moving coil microphones for this experiment rather than those manufactured by Western

Figure 3.29
Maurice Geoffrey Harker from a photograph in his pilot's licence May 1939. (*Courtesy of Maurice Harker*)

Electric, but EMI were re-equipping their laboratories at the time. This meant that there would have been a shortage of EMI products and, having come from ST&C, a subsidiary of Western, Blumlein knew that these WE microphones, amplifiers and wax cutters, now all surplus to requirements, would be adequate.

On 13 February 1933, Blumlein designed a shuffler circuit, for which his hand-written notes still exist, with the main phase-shifting element now including a bridging inductance, which reduced insertion loss. He designed the low phase-shift transformers the next day and, starting on 20 February conducted tests of the microphones, circuits and the shuffler, so that by March 1933 a small team of EMI engineers were gathered together for a series of listening tests. There were still problems associated with the system, for example the 78 rpm shellac records being used for the experiments did not lend themselves that well to the subtleties we now associate with stereo sound properties. In addition, the microphones themselves introduced considerable cross-talk, which distorted and degraded the signals that were reproduced.

None the less, the method did work and on 12 July 1933, the first calibration of the recorder took place, with the bandwidth reported as being around 4 kHz. Eventually Blumlein and Holman would design and construct four binaural gramophone pickups, the earliest surviving example of which is called the 'PU3A'. It was of the moving armature type (which is surprisingly massive), and was first tested in late 1933. It still survives to this day in the EMI archives. The second pickup, the 'PU2', tested in June 1933, did not survive, while the third example, the 'PU4A', had an improved design using two coils, though it is not mentioned as being tested until January 1934. It too can be found in the EMI archives. Of the 'PU1', nothing is thought to have been constructed; it may well have been the designation for the design drawing only.

That August, Blumlein took a vacation in Cornwall staying with Doreen's family near St Ives. Maurice Harker had explained to Blumlein that he and his brother were planning a camping holiday touring Devon and Cornwall, and that they would stop off here and there along the way for a night or two.

'And Blumlein said, "Well, if you are nearby you should stop off and see us in Cornwall", to which I thanked him and, as I had met and knew Doreen Blumlein, and knew the offer to be genuine, I said that we would. As it happened, when we did arrive at Doreen's family house, Blumlein was away in Portsmouth apparently looking over a submarine. Despite this being his vacation, he was still willing to take any opportunity to extend EMI's interests, you see, and at that time we may well have been looking at the early possibilities of what would become "Sonar". He was at home next morning when we returned and the three of us went bathing in Carbis Bay, after which we were invited back for lunch which Doreen had prepared'.

By 30 August 1933, having returned from his holiday, Blumlein decided that the circuit for the binaural system could be extended to include the wax cutter, which had first been applied to the series of tests carried out in July. This circuit allowed switching to a recorder or to a pickup and included an amplifier with a 26 dB gain and variable loss pad in its output. Originally intended for a ribbon microphone, this was added into the difference channel before feeding into the shuffler evidently because Blumlein had discovered the signal in the difference channel was less than expected. Somewhere around 9 December 1933, the wax cutter was tried out for the first time with two commercial recordings being played one into each channel, one vertical, the other lateral. Blumlein reported that 'good separation', was achieved, and he judged that binaural recording was finally practicable.

When Prince Edward, the Prince of Wales (later King Edward VIII), visited Hayes in 1934, it was this recording that was demonstrated by Blumlein to him, along with a live demonstration from a suspended microphone in a room on the floor above. This was being operated by two junior lab assistants who were walking in front of it, and the sound was being fed to the loudspeakers in the room below in which the VIPs were sitting. Apparently the Prince had listened intently to Blumlein explaining how binaural worked to him, but had not understood the principle that well. During the demonstration, Prince Edward listened to the recording and then the live demonstration, and then exclaimed, 'Ah, it's moving!' He had finally understood. Maurice Harker recalled that the Prince then requested to see the microphone that was being used. This, however, posed a problem as the room in which the two lab assistants were walking and talking in front of the microphone had not been on the scheduled tour and was in fact just a bare room with two very junior (and now bored), lab assistants in it. Nevertheless, the Prince was shown to the elevator to ascend one floor. Meanwhile, Blumlein bolted up the stairs as fast as he could to warn these two young lab assistants (who, by this time, were apparently lounging on the floor), that they were about to be graced with a visit from the Prince of Wales and their future King!

On 14 December 1933, the first six 10 inch wax masters were cut using the binaural system with four more cut the following day, these have subsequently become known as the 'walking and talking' test recordings, and vary in length from 3 minutes 30 seconds up to 10 minutes and 20 seconds. Most of the binaural equipment had been placed in Room 106 which adjoined the small auditorium at EMI. This auditorium was used primarily by the company amateur theatrical group. There was a small raised stage with curtains either side and seating space for around 100 people, behind the stage curtains was a cinema screen and the auditorium was often used for the showing of movies. All around the room, the walls were lined with maple-wood which gave a rather resonant ambience and tended to create reverberal problems which would later cause Clark and Holman many problems as they struggled to perfect the microphones. Behind the curtains, which were draped along the side walls, were large metal racks on which loudspeakers had been installed by Geoffrey F. Dutton (who was mostly responsible for the experimental loudspeaker design at EMI). These could be reached by means of a gantry, which led to a narrow corridor behind the speaker boxes themselves. The first experiments carried out in the auditorium were rudimentary to say the least with Blumlein, Maurice Harker, Alfred Westlake and Frank Runcorn Trott (who was always known as 'Felix'), experimenting by walking past and around the stationary microphones as they talked and held conversations with each other. In one, Blumlein asks Felix Trott to turn the impedance down as he walks around from left to right, and then turn it up again as he moves back again. He then asks how the time is doing as the cutter was running out of wax.

Figure 3.30
Frank Runcorn Trott, always known
as 'Felix'. (*Courtesy of EMI*)

Years later, Felix Trott recalled those earliest experiments: 'Walking and talking was done in a large room known as the auditorium, looking rather like a completely empty cinema auditorium except the floor wasn't sloped, it was flat. And it was laid with hard Canadian maple, so that it made good loud, bang, bang, bang noises if you walked flat-footed round. So you had the two sorts of things to locate on: the voice, which of course had a whole lot of different noise patterns, and Blumlein's shoes going firmly down on this hard maple floor.'

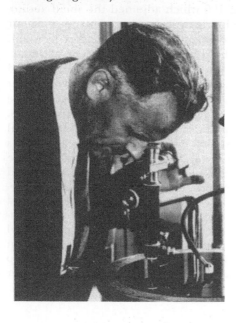

Figure 3.31
Alfred L. Westlake shown here
inspecting cut master discs. (*Courtesy of EMI*)

A progress report of 16 December remarks that the recordings made have a 'definite binaural effect'. A complete list of the recordings was drawn up for the files at EMI on 19 December 1933, and designated the tests as Nos. 5757, and 5758. In Test No. 5757, Thursday, 14 December 1933, Cut No. 1 has Blumlein, Westlake, Trott and Harker talking in the auditorium with Blumlein and Westlake in front. Cut No. 2 is as the first but with a different arrangement, now Blumlein is at the back. Cut No. 3 has all of them walking and talking in a muddle and changing positions. Cut No. 4 has them talking in turn, in pairs, more slowly but with muddled change-overs. Cut No. 5 is described as 'Heavy Shuffle', with the difference on each channel reduced by 3 dB. The people arrangement is as Cut No. 4. The final cut for Thursday, Cut No. 6, is again listed as 'Heavy Shuffle', with the difference in each channel reduced again by 3 dB, and now with Blumlein talking alone in the auditorium. It is now that he speaks to Felix Trott about the time remaining. A channel plotting is given for recordings one to five as 'A Channel = 18, B Channel = 21', and the plotting for recording six as 'A Channel = 19, B Channel = 22'.

Test session No. 5758, the following day, Friday, 15 December 1933, lists the four recordings made that day. Cut No. 1 has Blumlein only walking and talking in the auditorium. It is again given as 'Heavy Shuffle' with the difference channel as '10,000 Ω and 70 Ω'. Plotting levels were 'A =18, B =21'. Cut No. 2 has the same arrangement as Cut No. 1, but with a 'Light Shuffle', 10,000 Ω, 70 Ω and 40 Ω. Plotting levels were 'A =18, B = 21'. Cut No. 3 has Westlake, Trott, Harker and Turnbull talking in turn with all the shuffles as the previous cut and the plottings also the same. In the final cut that Friday, the auditorium arrangements were as for Cut No. 3, but the order of shuffling was changed from 'Heavy 40 Ω, 10,000 Ω and

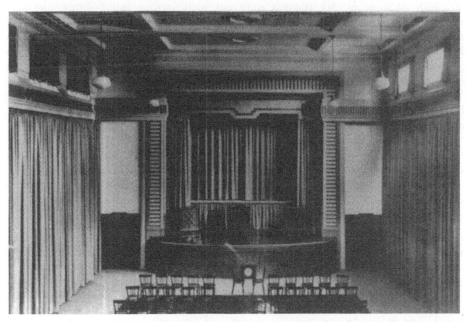

Figure 3.32 The auditorium at EMI where many of the binaural recording tests were made and some of the binaural films shot. This photograph was taken in April 1929. (*Courtesy of EMI*)

70 Ω' to 'Light 70 Ω and 40 Ω', and back to 'Heavy 40 Ω'. The plottings for the channels were 'A = 19, B = 21'.

The recordings gave a fair account of each man's position in the room as the cuttings were made and the results, which were played to Shoenberg, were sufficiently good to warrant the apparatus being taken to the then fairly new EMI recording studios in Abbey Road. The Abbey Road studios in St John's Wood (which was actually No. 3 Abbey Road), had been purchased as a private house by the Gramophone Company, for the then huge sum of £16,500 on 3 December 1929. During the next two years, EMI would spend over £100,000 converting the property into recording studios. It became part of EMI following the merger in April 1931, and was officially opened on 12 November 1931.

At Abbey Road, a complete binaural record cutting system was installed by Ivan Turnbull, Ham Clark and their assistant in the then largest room, Studio No. 1 (this is now called 'Studio 2'. It is the same room later used by The Beatles, and is often referred to as 'The Beatles Studio'). The system was used initially on Thursday, 11 January 1934, to record two cuts of Ray Noble's Dance Band, with the microphones placed approximately 45 feet distant from the musicians. The two recordings, which were given the internal numbers TT.1557-1, and TT.1557-2, were a result of plottings from 'A (lateral setting) = 18, B (hill and dale setting) = 18', for the first cut, and 'A = 24, B = 24', for the second, with the amplifiers set +6 and +6 for the first and 0 and +6 for the second. Both cuts were using 'heavy shuffling', with '6dB brightening (5000~sharp)'.

The next day, Friday, 12 January 1934, three pianos were placed converging towards a standard HB microphone with the outer pianos making an angle of approximately 60° with each other. Six records were cut with various microphone plottings recorded: Cut No. 1 (Test No. 5769-4), had the binaural microphones approximately 12 feet from the tip of each piano and the recording was noticed as too heavy, plotting 'A = 26, B = 22'. Cut No. 2 (Test No. 5769-1), was with the microphones positioned as in Cut No. 1, but the plotting was reduced to 'A = 22, B = 18'. Cut No. 3 was a repeat of Cut No. 2 but with a plotting of 'A = 22, B = 20'. Cuts No. 1, No. 2 and No. 3 were of the *Hungarian Rhapsody* by Brahms. For Cut No. 4, the binaural microphones were advanced 6 feet nearer the pianos, which placed them approximately 18 inches behind the HB microphone. The plotting is given as 'A = 18, B = 16'. Cut No. 5 had the microphones restored to their original positions i.e. as they had been for Cuts 1, 2 and 3, but with a plotting of 'A = 22, B = 20'. Both Cut Nos. 4 and 5 are of *Ride of the Valkyries* by Wagner. Finally, Cut No. 6 was made after the main sessions had finished and consisted of snippets of music from a rehearsal. The microphones were withdrawn to approximately 25 feet from the ends of the pianos and the plottings were given as 'A = 26, B = 24'.

A progress report of these recordings was drawn up on 13 January, by Turnbull in which he states that: 'Binaural gives opening out on music, but the effect is more marked for speech'. The report was circulated on 16 January to those who had been present. It had been decided, based on these two days recordings, to go ahead and try something a little more ambitious. Therefore on Friday, 19 January 1934, nine sides were cut of a rehearsal of Mozart's Symphony No. 41 '*The Jupiter*', with the London Philharmonic Orchestra, conducted by Sir Thomas Beecham. This must have been a special thrill for Blumlein as he was a great admirer of Beecham and had a comprehensive set of records of his work which he and Doreen would sit and listen to for hours at a time at home in the evenings. Felix Trott and Maurice Harker both recalled working with the great man who left little doubt who was in charge of proceedings: 'We would make sure that Sir Thomas had everything he required...' Felix Trott recounted, '...down to the fresh bottle of 'good' whisky in his dressing room, which quite often would be replaced with a second, fresh one, in the afternoon!'

Turnbull's progress report of the Beecham recordings, which was drawn up on 20 January, and circulated on 22 January 1934, states of the binaural effect, that there had been, '...solidarity, but less effective than on speech'. The report goes on to list the following details: 'All subs. cut with heavy shuffling and 6 dB sharp boost at 5,000 cps. Amplifier plotting = +6 and 0'. Cut No. 1 had the microphones placed approximately 13 feet high and approximately 13 feet from the nearest first violin. The plotting is given as 'A = 24, B = 22'. Cut No. 2 was exactly as Cut No. 1 but is of part two of the symphony. For Cut No. 3, the microphones were moved approximately six feet further back and one foot higher. The plottings remained the same. For Cut No. 4, the positions remained but the plottings changed to 'A = 26, B = 24'.

Recording No. 5 saw the microphones moved approximately 7 feet further back, i.e. approximately 25 feet from the nearest first violin. This was for part three of the symphony and the plottings stayed the same. Cut No. 6 was as Cut No. 5, but the plotting changed to 'A = 28, B = 26'. Cut No. 7 was as Cut No. 5 but again the plotting changed, this time to 'A = 30, B = 28'. Cuts No. 8 and No. 9 were of part four of the symphony with No. 8 having a plotting of 'A = 26, B = 24', and No. 9 having 'A = 28, B = 26'.

The results were judged by Blumlein and his colleagues from 'not bad' to 'marginal', and the discs remain, to this day, in the vast and complete archives of EMI at Hayes, where a record of every cut and pressing ever made was logged and kept. From the ensuing tests, carried out over three months from February to May 1934, many of the stereo microphone techniques were devised, which are still commonly used to this day by recording engineers all over the world. The musical trials were hurriedly completed, as Blumlein was eager to get on with main aim of the work, which was to perfect binaural film. It had also been some time since Blumlein had begun simultaneously working on the development of the high-definition television system, among which were some of the pioneer engineers from the original sound-to-film processes of the late 1920s. It is a testimony to the engineers working with Blumlein that, while he may not always have been present for these experiments, and

Figure 3.33
Sir Thomas Beecham (at the back)
inspecting the Columbia recording
system at Abbey Road studios in 1933.

was not directly responsible for all the microphone techniques, his colleagues applied the name of 'Blumlein' to these techniques regardless. Therefore, his name has remained in use to this day, in most cases. Indeed it is usually through 'the Blumlein stereo microphone technique', that most audio engineers first hear his name.

Some time later, on 7 February 1935, Blumlein would apply for an additional patent, No. 456,444, 'Improvements in and relating to Electrical Sound Transmission Systems', in which he outlined many of the microphone positioning techniques that had resulted from the use of the binaural system. The patent explains how, with their various outputs mixed, microphones can be used in a manner of ways including the 'A-B' microphone technique ('A' is always left, 'B' is always right. Incidentally, the Germans call this method the 'X' and 'Y' technique. Current usage of A-B and X-Y is that A-B is 'spaced' and X-Y is 'coincident'), which despite having the same initials as Alan Blumlein's name, probably refers more to the fact that he calls two such microphones 'A' and 'B' in his patent, rather than after himself. There is also mention of the method that we now call 'M and S' microphone (mid and side). Blumlein knew this method as S-M microphone, S = Sum, and M = Difference, 'M' = $\frac{1}{2}(A + B)$, and 'S' = $\frac{1}{2}(A - B)$.

## Binaural film experiments

In Patent No. 394,325, Blumlein describes the possibility of using his binaural system for adding an extra soundtrack to film, with the existing soundtrack in the normal position carrying the 'sum' signal, and an additional track carrying the 'difference'. This was done to ensure that a binaural soundtrack could be played in an unmodified cinema theatre. The job of getting this to work was given to Cecil Oswald Browne, who had worked on the original HMV mono sound-to-film system in 1928/9 which, while technically quite successful had, for commercial reasons, been shelved after only two films had been shot. Both HMV and Columbia continued thereafter to produce 16-inch pressings used for sound in many movie theatres.

By 1933, Browne, like many of the staff at Hayes was deeply engaged in the development of television. Blumlein had explained binaural sound however to an intrigued Browne in 1933, and the two of them resolved that a binaural film process could be constructed with a series of new projectors, amplifiers and shufflers. Many of these were eventually designed and made by Holman, Turnbull, Harker, Clark and their assistants, under the daily supervision of Blumlein.

Browne himself designed and constructed the 35mm film recorder and had to invent two totally new galvanometers in order to fit the two recorded tracks sufficiently close together, instead of the one that the then standard monophonic film systems were using. It was decided that variable area recording would be used instead of the so-called 'squeeze' tracks. The soundtrack itself comprised two 'unilateral' variable area soundtracks (one edge modulated), recorded side by side in approximately the same space that a monophonic soundtrack of the time would have taken up. For playback, a double cathode photocell was needed to read the sum and difference tracks, and the task of designing and constructing this was given to William F. Tedham. He had in fact developed many of the photocells for the earlier work and was also busy working on television and the development of cathode ray tubes.

Figure 3.34
Cecil Oswald Browne.
(*Courtesy of EMI*)

It is worth pointing out at this stage just how radical the project that was being undertaken must have seemed, coming as it did just a matter of a few years after the very first 'talkies' had appeared in cinemas. Talking pictures had originally been dismissed as a novelty when they first appeared in 1927, though their popularity with the cinema-going public ensured their success and quick adoption. Now Blumlein was proposing that it was possible to record stereo films before the first stereo recordings had even been released.

Browne completed the camera/recorder in early 1935 (which had a fixed focal length lens), and the other equipment, the amplifiers, shufflers and projectors were ready by early May 1935. The first test on the twin recorder had been carried out in April 1935, with a designed upper resonance of around 9 kHz, but there were subsidiary resonances at lower frequencies which needed to be damped by immersing the movement in oil. This was not very effective at first because the amount of power to be dissipated heated the oil and actually reduced its viscosity.

Figure 3.35
Philip B. Vanderlyn.
(*Courtesy of EMI*)

Castor oil was tried and, while it gave sufficient damping (the galvanometer was flat to 1 kHz with a gentle rise of 4 dB to 9 kHz), the oil proved insufficiently transparent to allow enough light through to expose the film fully. By the end of May 1935 however, a better optical system had been developed with new mirrors which worked much better. At around this time the reproducer sound head was tested and found at 8 kHz to be about 12 dB below that which it should be, most of which was attributed to the slit and preamplifier.

By June 1935, the first series of tests on the binaural film system were ready to be carried out. Although there was some frequency modulation, probably because of vibration, the major problem was a noticeable degree of noise because the squeeze track had not yet been incorporated, and the channels were unbalanced. This gave rise to a degree of overloading, which was considered a bearable evil until the problem could be solved. A movie camera was borrowed as a temporary solution, probably a left-over relic from Browne's work in 1928/9, and this was used to determine how much the binaural system would override reverberation problems in making talking movies (a big problem in the early films).

It was now that a series of logistical problems began to occur which hindered the further development of the system for a while. First, the amplifiers needed to be borrowed for several days to contribute to a major demonstration of the company's television system given at Abbey Road. Second, several members of the staff had to be found to design the sound system for the Alexandra Palace television station contract, which obviously took priority, and this held things up further. This all meant that it was not until the middle of June before the first actual binaural filming could take place and, even then, a suitable location had to be found. It is for this reason that the first films were actually taken outside in early July with the 'Trains at Hayes Station', 'Throwing Stones' and 'Fire Engines' films all being shot in the first twelve days of that month.

It was decided that the auditorium at Hayes should be used once again for the inside filming. It was 40 feet wide and had acoustics similar to those of a cinema, certainly more than any of the labs. It had also been used in March 1935 for the first loudspeaker tests of the binaural audio system. Though it had been the cause of some reverberal problems for Clark and Holman (still busy perfecting the latest versions of the HB microphone), the room provided the best option for these tests and it was equipped with both microphone types, crystal (which Holman had originally been working on before Blumlein's arrival at Columbia. This had a glass diaphragm) and velocity microphones. A 14 foot-wide set was hastily constructed; a suspended curtain on the stage area, and 16 kW of lighting rigged. Most of the recordings were to be made with the velocity microphones which had given better results in film conditions and, during the first week of July, a few test films were made without sound.

At first, it would seem that Blumlein did not have any specific plan in mind for the binaural film tests, but instead, circumstances forced him outside while the auditorium was prepared. The first of these outside films has become known as the 'Throwing Stones' film, and is actually a silent test (which could have been for any number of reasons, including the possibility that the audio tracks were out of synchronization). The silent reel, which is 103 feet in length and lasts for just 1 minute and 6 seconds, has a sequence showing Maurice Harker, Herbert Holman, Felix Trott, Albert Westlake and Philip Vanderlyn, outside the EMI research building throwing stones. There is no specific date recorded for when this filming took place, but it is reasonable to assume that it was shot in the first twelve days of July 1935.

The first of the binaural sound reels is the one which has become known as 'Trains at Hayes Station'. It is 487 feet, lasts for 5 minutes 11 seconds, and shows various steam trains arriving and departing from Hayes station. Again, there is no date for the filming, but it too would have taken place in early July. Blumlein's team (Blumlein was not present himself for the filming) took the binaural sound system and camera to an empty office building which overlooked Hayes railway station and started to film trains as they arrived and departed. The scenes are photographed from a high vantage point to give the camera as wide a view of the station as possible including the surrounding tracks. In the distance, among the various buildings and sheds which have smoke (and presumably steam) coming from them, it is clearly possible to read the sign 'Nestlé' on the roof of one building just beyond the station platform.

As the 'Trains' film progresses, it is obvious that several microphone positions are being tried out, with frequent alterations of both microphone location and type. It is very probable that they were trying to demonstrate just how distant they could achieve the effect of the binaural sound and not surprisingly this film has some of the best stereo effect recordings of all. At one point, a fast passing train is captured on film and the soundtrack clearly distorts as the microphones (which on this occasion were probably ribbon-velocity types) suffer from being totally outside the limits of their magnetic field. This would have been due to the sheer amount of sound pressure level being placed upon them.

Figure 3.36 Still image from the binaural film 'Trains at Hayes Station', July 1935.

Figure 3.37
The binaural microphone with its protective wind-shield suspended from a lamp post at the side of the railway. This lamp post is clearly visible in the right foreground of Figure 3.36. (*Courtesy of Maurice Harker*)

Once the auditorium was completed and readied, the binaural film equipment was moved inside for further tests and the next reel, which is almost certainly the first shot inside, does have a date on it, Friday 12 July 1935. It runs for 116 feet, a total of 1 minute 15 seconds, and it shows Westlake, Vanderlyn and Trott playing tricks with a short pole which they hold and then try to climb over it without removing their hands from the ends. This footage thus became known as the 'stick-trick' film.

Figure 3.38
The camera position was atop The Aeolian Company Limited Building which overlooked the railway and Hayes Station. Philip Vanderlyn is behind the camera (which is running and capturing the scenes we see in the film today) while Felix Trott has his back to Maurice Harker who took this picture. (*Courtesy of Maurice Harker*)

The sequence opens with Turnbull holding a clapperboard, which looks to be made of two pieces of wood approximately two feet long. The clapperboard system was probably needed to synchronize the sound to the film. There is a real element of comedy to the footage as Westlake hands Vanderlyn the 'stick' and then proceeds to contort himself and climb through it. Felix Trott then appears from the right wearing a hat and his white laboratory coat. Trott wants to have a go at the stick trick, so he takes off his coat while Vanderlyn, standing on the left, takes off his hat for him. 'Oh no', says Trott, 'I can't do it without my hat on', and takes it back from Vanderlyn who had placed it on his own head. Trott then tries, without much success at first, to perform the stick trick, though he does eventually almost get there. Just as Trott has almost completed the trick a voice from off-stage, which sounds like Blumlein, but could also be Turnbull, says, 'OK. Stop' and the film ends there.

The contrast of the soundtrack on the film itself is very low (determined by how opaque the dark parts on the film are relative to the transparent area of the film). The soundtrack therefore suffers from not having a simultaneous development process to the pictures in the chemical bath (this did not come about until some years later when colour development was understood). The level of modulation on the film is also quite low and these two factors combined led to very noisy reproduction when played back.

Four days later on Tuesday, 16 July 1935, two further tests were filmed, the first of which is a section of footage 115 feet long, lasting for 1 minute and 14 seconds, and

**Figure 3.39**
Still image from the binaural film 'Stick Trick'. Philip Vanderlyn is on the left holding the lab coat, Felix Trott is in the middle wearing his hat and with the stick, while Alfred Westlake, on the right, attempts to help him perform the 'trick'.

**Figure 3.40**
Another still image from 'Stick Trick' showing Westlake now on the left, Trott halfway through the 'trick' and Vanderlyn having moved over to the right, July 1935.

again shows Westlake, Vanderlyn and Trott this time talking among themselves. Following this is a second section of the reel, which is just 35 feet long. It is this section which has become known as the 'Walking and Talking' film. This short section is particularly fascinating as Blumlein himself takes part in the test; it is believed to be the only known instance of Alan Blumlein captured on film. It is also one of the few occasions where his voice is recorded (apart from the short sequence of talking with Felix Trott on the binaural recording of 14 December 1933).

The sequence opens with a rather bored-looking, white-coated Felix Trott, standing on the stage with the large drape hanging down behind him, holding the clapper-board. By this time the microphone had been improved so that it could be mounted closer to the camera and it could record sound from the whole of the stage. Felix Trott, looking directly at the camera, then 'claps' the pieces of wood together and walks off-stage to the right saying nothing. From the left appears Herbert Holman, wearing a dark suit, striding five paces to cross the stage to the right saying 'one-two-three-four-five', as he goes, but without ever once deviating from his gaze ahead to look at the camera.

Then Blumlein appears, looking directly at the camera. He is wearing a lighter, grey suit, the jacket of which he is attempting to button as he too strides across the stage from left to right, covering the area in front of the camera in six steps, and counting 'one-two-three-four-five-six'. It is just possible to hear him say 'seven' as he disappears off stage having successfully buttoned the inside button of his jacket by this time. Blumlein's voice is clear, concise and determined sounding. He has a quite definite London accent and, compared to his colleagues, sounds quite a bit louder – perhaps even a little droney.

Next comes Westlake, who is wearing a long white coat over a dark suit and looks as if he is fixing his tie just as he walks onto the stage. His voice is distinctly quieter than Blumlein as he too chants 'one-two-three-four-five-six', also glancing at the camera. Westlake is followed by the tall, thin figure of Philip Vanderlyn who appears, looking directly at the camera, wearing light grey trousers and a white shirt (which looks as if it has no tie), with his sleeves rolled up. Vanderlyn is the clearest of the speakers as he walks quite precisely across the stage counting to 'six'. Following Vanderlyn and appearing before he has reached 'five', is Ivan Turnbull, who appears some time before Vanderlyn has exited the stage to the right. Turnbull speaks rather

Figure 3.41
Still image from the binaural film 'Walking and Talking' showing Blumlein on the left walking on to the stage while Holman on the right is just exiting, July 1935.

quietly as he strolls quickly across the stage also wearing a white shirt, sleeves rolled up (this time definitely with a tie). Turnbull also only gets as far as 'five', when Felix Trott appears and walks across the stage, not really looking at the camera, but getting to quite a distinct 'seven'.

The sequence then begins again with Holman appearing for a second time, crossing again without looking at the camera, counting to 'six'. Blumlein now also appears for a second time, looking directly at the camera again, his jacket securely buttoned as he walks to a point about one third of the way across the stage. He only counts as far as 'two' however, before the audio on the film stops. Blumlein continues to walk across the stage counting in silence until he reaches 'five', by which time Westlake has appeared again following him. It is here that the film runs out. This entire sequence lasts for just over 24 seconds.

Blumlein decided to use the HB1Bs for the tests, with an additional pair of high frequency microphones as the existing pressure microphones available would be too large and cumbersome to be spaced sufficiently close enough together. The high frequency microphones were of the ribbon type, which are very delicate to work with and these particular ones each had two ribbons inside, with an absorber behind the ribbon to turn it from velocity to pressure sensitive operation. At first, these high frequency microphones gave enormous trouble to the team, failing to give either a decent frequency response or any degree of sensitivity. Henry Clark decided to apply Blumlein's velocity calculations to the two new microphones and discovered that, with a few alterations, they too could be made to work, which they did eventually after a series of configuration tests.

Following the 'Walking and Talking' film, several other tests were carried out in July 1935. There is an additional piece of footage, 117 feet long, running for 1 minute 16 seconds, which has the EMI fire engine (there was a small company fire department at Hayes) driving across a field which was just around the corner from the factory building. Though the 'Fire Engine' film is not dated, it is entirely possible that this was filmed after 16 July 1935. Pictures taken by Maurice Harker on the day clearly show that it was a bright summer's day presenting ideal lighting conditions.

The fire engine drove back and forth across the field past the binaural microphones which were placed on stands and protected from the wind by square shields. They were driven by two Austin Seven batteries which had been borrowed from two cars in the EMI car park for the afternoon. As the filming progressed, Blumlein left his office to join Vanderlyn, Trott, Turnbull and Harker to see how everything was pro-

Figure 3.42
The EMI company fire engine.
(*Courtesy of Maurice Harker*)

Figure 3.43
The binaural film camera which had been constructed by Cecil Browne from equipment he had originally used in his film sound experiments of 1928 and 1929. (*Courtesy of Maurice Harker*)

gressing. By pure coincidence, just as he strode across the field to the camera position, Maurice Harker took a photograph. This photograph remained unknown in a photo album kept by Harker untouched for over sixty years.

The outdoor binaural films were considered to be something of a success with perhaps 'Trains at Hayes Station' being the 'best' known of the results. Once back at EMI, the unused (at that time) top floor of the R&D building was converted into a makeshift laboratory where tests could be made with both the microphones and

Figure 3.44
The binaural microphone in the field for the 'fire engine' film. The two batteries clearly visible at the base of the microphone stand had been 'borrowed' from two Austin Sevens in the car park. (*Courtesy of Maurice Harker*)

Figure 3.45   Photograph taken by Maurice Harker as the filming of the 'Fire Engine' binaural film took place. Cecil Browne is standing behind the camera, Philip Vanderlyn is in the distance near the binaural microphone as the fire engine is about to pass by. Alan Blumlein can be seen on the extreme right of the picture striding towards the camera position. (*Courtesy of Maurice Harker*)

the camera. Any alterations that were necessary could then be made while the auditorium was being prepared for the next and most ambitious of the binaural films that would take place that summer of 1935.

Figure 3.46   The top floor of the R&D building at Hayes converted to a test area for the binaural camera and microphones. Cecil Browne is seated facing the equipment, Philip Vanderlyn is standing behind the rack, while Ivan Turnbull is seated to the right just behind the camera equipment. (*Courtesy of Maurice Harker*)

Figure 3.47
Still image from the binaural film known as 'Move the Orchestra' or the 'Playlet'. The lady has just sat down and the café manager is taking an order from the two people on the left, 12 July 1935.

Somewhere around the time that the fire engine film was being taken, acoustic damping was installed in the auditorium to help with the reverberation problems that had been encountered and, on Friday, 26 July 1935, Blumlein had decided to use the indoor stage again for a short film which became known as 'Move the Orchestra'. It has also become known as the 'Playlet' and is, in actual fact, two short versions of the same scene, in which several members of the EMI Amateur Theatrical Group produce a 'typical' pub scene, with people attempting to order drinks from an incompetent waiter.

The camera was placed front and centre of a 20 foot wide set, and does not move. The scene was lit with 30 kW of lighting and the binaural sound reproduction system, despite the earlier problems with filming indoors seemingly rectified, was still proving problematic. The scene opens in a small café or bar with a woman and a young boy sitting on the left. The manager shows in a young lady who asks for a quiet table and is shown to a seat on the right. She then asks the manager to send the waiter. At this point, the music begins. A band is supposedly playing out of shot and off to the right (the music, which can be plainly heard to the right, was produced by a gramophone to the right behind the draped curtain). The lady, on hearing the music, winces and complains: 'I said I wanted a quiet table, this is much too near the band. Can't you do something about it?' (she asks). 'I'm afraid not Madam. You see, the band is very popular with the rest of my clients', the manager replies in a French accent. 'Well can't you blow it up or move it or something?' she says. 'Oh Madam. I could not blow it up!' replies the astonished manager, 'I could move it.'

He then claps his hands, tells the band to move and, magically, the sound moves to the left-hand side of the shot, appeasing the customer. What had happened? Two juniors had been recruited to carry the gramophone from one side of the set to the other, behind the backdrop, from the right-hand side of the scene, where the lady had complained, to the left side as a waiter waves the 'band' across. The scene then continues when a gentleman joins the lady at her table, the waiter finally arrives and they order two 'gin and its'. The waiter then proceeds to drop everything he carries, first off to the right and then, after bringing the drinks and being paid (1/6d), he disappears off to the left and more crashing sounds are heard.

Blumlein, despite the obvious humour intended, was actually trying to demonstrate the practicality of associating sound location with visual representation of the source of the sound itself. It was, and indeed still, is an integral part of film production, where the 'soundscape' is very often backing-up the visual image. The 'Playlet' was filmed twice, once at 396 feet, lasting for 4 minutes 13 seconds, and again for 336 feet, lasting 3 minutes 33 seconds. The first take was reviewed and the sound level considered too low, and so it was raised for the second take, though it has to be said it hardly improves.

The following day, Saturday, 27 July 1935, the film was taken to Humphries' Laboratories for processing. The edited, finished version of 'Move the Orchestra' was completed by early September 1935, just when work on the binaural film project was abruptly halted, by which time Blumlein in his 'weekly reports', had considered the binaural sound to be 'fairly satisfactory technically'. In August, the squeeze track had been added and a noise reduction circuit included as well as new galvanometers with mechanical adjustment and, by mid-September, improvements were also made to both the crystal and shuffled velocity microphones. In addition, EMI installed a purpose-made cine-screen to give 50% more picture brightness. The tests were halted however, probably because of the pressing need to complete the high-definition television system first. Shoenberg, always a realist, had concluded that it was better to learn to walk before trying to run, and so binaural audio and film tests were shelved for the time being and Blumlein's talents, and those of the staff working with him, were re-directed to other tasks.

The results of the binaural film tests however, all of which were filmed on the volatile nitrate stock, are a fascinating insight into the world of Blumlein, EMI and his colleagues at the time. There are seven reels in all and they still lie in the EMI archives in their original tins (though now, thankfully, due to the unstable chemical nature of the nitrate film, they have been transferred to acetate-based stock and various forms of digital media). This needed to be done because nitrate is highly volatile, prone to degradation and eventually self-combustion. Like much of Blumlein's work, the material was to end up in the vaults of EMI, where it lay for over 50 years untouched. In 1982, and only after determined and constant persistence from a number of concerned individuals, EMI undertook to ask Norman W. Green, of the Independent Television Companies Association, if he would be willing to transfer the now delicate film to a modern acetate-based safety stock (see Author's note on p. 98). It was from this video copies could be made and eventually EMI also transferred the footage to laser disc. Alan Blumlein's binaural films were shown in public for the first time at a theatrical presentation in February 1988, and at the Audio Engineering Society 'Sound with Pictures' Symposium in May 1988. They are, despite their historical significance, still not available on video for the general public to see, though the now safe acetate-based copies and the laser disc once again reside in the EMI archives.

As for binaural sound recordings, suffice to say that it was not until 1958, some 16 years after Alan Blumlein had died and, 27 years after it had first been patented, that 'binaural' made its next appearance as 'stereo' on long playing records. By this time, unfortunately for EMI, the patents had long since expired. After the war, because of the nature of the commercial losses that had befallen so many manufacturers, the British Patent Office ruled that patents registered pre-war could be extended for an additional few years. This allowed EMI to extend Alan Blumlein's binaural sound patent, which had expired in 1947, for another five years until 1952. However, they still did not apply the material within the patent to any practical project that they were researching at the time. When the patent finally expired on

Figure 3.48
Arthur Haddy, who was in charge of
the Decca stereophonic sound experiments in the 1950s. (*Courtesy of Eric
White*)

13 December 1952, EMI had not used Blumlein's material based on binaural sound
for a project of any kind since the mid-1930s.

In 1955, Decca, one of EMI's great rivals, largely unaware of Blumlein's work some
24 years earlier, tried to patent a dual-channel sound system which they called
'stereophonic sound'. Almost at once, Decca's engineers realized that they were covering ground which Alan Blumlein had explored more than twenty years earlier.
Arthur Haddy, who was in charge of Decca's technical department at the time, thus
became aware that many of the problems that they had already been considering for
the last five years, had all been worked on by Blumlein in 1931. Despite much effort
on the part of the research team at Decca, every possible alternative to their system
had already been considered and contained within Alan Blumlein's specification
No. 394,325; Decca conceded defeat. The exercise did however, at last bring to light
the possibilities of stereophonic sound and EMI, finally realizing what they had on
their hands, began experimenting again with the system in April 1955. At Abbey
Road a series of tests were made by Sir Malcolm Sargent and, a few months later,
stereo tape recordings were being released for the first time. Eventually this led to
the first stereo long playing records appearing in 1958.

What Alan Blumlein had described to his wife Doreen in that cinema one day in
1931, was eventually to prove to be one of the most significant developments in
audio engineering of the twentieth century.


## Moving on again

By the time the last of Blumlein's eleven audio patents was being applied for, his
attention had long since been diverted to the all-out effort being made by everybody

at EMI to perfect a high-definition television system, to succeed in providing Britain with a practical public television service. It is a measure of the man however, that even during so frantic a period as that which pervaded the EMI laboratories during 1934 and 1935, that Blumlein still had the time, energy and will to improve upon aspects of the Columbia recorder (which, incidentally was known at Columbia as 'The Bacon Slicer').

In 1932, he had re-designed the cutter itself, with the new version designated the MC4. This had rubber-mounting blocks to aid suspension instead of the steel springs which the earlier version had used. In this way the mass of the moving elements and drive power were considerably reduced. This was improved upon further in the MC4A, which was capable of recording frequencies up to 11 kHz, primarily because of the lower moving mass. Then finally, another version, the MC4B, was introduced with rubber interposed in the path of the stylus. This was matched to a power amplifier, a PX25, with an output impedance capable of handling the new load on the cutter and it was this version that eventually became part of the commercial system. Several of these were installed at Abbey Road with one remaining in constant use, unaltered, until 1948.

Blumlein's patent, No. 417,718, 'Improvements in or relating to Vibratory Devices such, for example as are used in the Electrical Recording or Reproduction of Sound', was applied for on Tuesday, 7 March 1933. It was an improvement to the vibratory properties of the recording system outlined in the earlier patent, No. 350,998, which, together with Patent No. 350,954, constituted the Columbia recorder. Though the original work had been co-written with Herbert Holman in 1929/30, the improvements outlined here were apparently all Blumlein's work. This would appear to be evidence of his known trait for returning to various pieces of his work from time to time, where he would ponder over various aspects and then invariably come up with some improvement. In this patent, sixteen additional claims are made, along with several re-designed drawings to illustrate the changes.

> The patent describes a vibratory device incorporating a moveable armature which is mounted so that it can oscillate about a substantially fixed axis, and for practical reasons this armature has a slight freedom of movement in a direction normal to this axis. It had been found that in certain sound recording devices it was necessary to make the torsion compliance of the armature mounting about its axis of vibration, fifty times or more the lateral compliance. The invention therefore outlines a torsion bar, made for example of steel, wherein the stylus bar is pivoted about a point, with sufficiently high torsion compliance so that the strain does not make it fragile to the point where it would easily break. This was to be done by a means of damping of the armature movement, which is mounted on a resilient bearing, the motion of which in the direction at right angles to the longitudinal axis of the shaft subjecting the resilient bearing to compression. A form of the armature support is also noted within the patent, outlining the use of short rubber tubes for vibration support. The compliance of these rubber supports, in respect of the rotary motion of the armature about its axis, would provide an improved form of mounting for the moveable element in the recording device.

Alan Blumlein's penultimate audio patent during this period deals with the subject of the problems that were being encountered with low and very low frequencies, when two pressure microphones are arranged in order to achieve a binaural recording effect. The relationship between the microphones became ambiguous if the distance between them exceeded half the wavelength of the sound waves in question which was obviously unsatisfactory especially for mid and high frequencies of the range. The answer had been touched upon in his binaural sound patent, No. 394,325, but not really elaborated upon.

Placement of two microphones closer to each other did produce the desired results, but led to phase cancellation difference problems at the microphone outputs. While, theoretically, this did not present a major problem, it did mean that the microphones themselves and any related circuits had to be designed so that they were very sensitive to small phase changes. This in itself required the use of large amplification and transformers designed to introduce accurately controlled phase changes at low frequencies. The introduction therefore, of great amplification could lead in turn to the introduction of unwanted noise.

The main objective of Patent No. 429,022, 'Improvements in and relating to Sound-Transmission, Sound-Recording and Sound Reproducing Systems', applied for on Monday, 23 October 1933, was to overcome these difficulties. 'The impulses, picked-up by a plurity of microphones, are reproduced by a plurity of loudspeakers, thereby conveying to the listener the sense of localization of the sound source.' This was outlined further with a series of diagrams illustrating how the patent should be applied.

The patent describes a shuffling network, whereby phase differences between the outputs of microphones, in a spaced relationship designed for the picking up of sound to convey to the listener a localization impression of the sound source, are converted into differences of loudness between loudspeaker outputs. The diagram clearly shows two pairs of pressure microphones *1* and *2*, *3* and *4*. These are arranged symmetrically about the centre line or axis of the sound field; one pair, *1* and *2* being widely spaced from one another, and the other pair, *3* and *4* being closely spaced. Separate shuffling networks *5* and *6* are provided one for each pair of microphones.

From this diagram, and the explanation of networks outlined in Patent No. 394,325 it can be seen that microphone pair *3* and *4* are suitable for dealing with middle and high frequencies, but not entirely satisfactory as regards low frequencies; while the wider spaced microphones *1* and *2*, although convenient for low frequencies introduce ambiguity in the directional impression which they provide with regard to medium and high frequencies. The shuffling networks therefore are arranged in order to deal with the specific frequency tendencies associated with the relevant pair of microphones. The voltages from the shuffling networks transformed to a common impedance ($R$) are fed to two loudspeakers which are common to the two networks.

Two pairs of leads (*11* and *12*) are shown to connect each shuffling network to each loudspeaker. The pair of leads from the high frequency shuffling network (*6*) i.e. the network associated with the microphones that are closely spaced, are in each case shunted by an inductance (*9* and *10*) the value of which is determined from the equation below. The pair of leads from the low frequency shuffling network (*5*) i.e. the network associated with the microphones that are widely spaced, are in each case shunted by a condenser (7 and 8). Because the leads to each loudspeaker are connected to the outer terminals of the appropriate inductance and condenser, these remain in series as a shunt across the loudspeaker.

The condenser 7 and inductance 9 are in series with one another as a shunt across the leads 11, while condenser 8 and inductance 10 in series, form another shunt across the leads 12. The relative values of the inductances and capacities employed are determined from the following equation:

$$\sqrt{\frac{L}{C}} = R$$

L represents the value of the inductance, C the value of the capacity, and R the value of the impedance.

Blumlein went on to discuss the possibility of inserting various filter types and several alternative types of microphone, as well as to the microphone spacing itself. In essence what is being outlined here is a form of binaural recording using two microphones. Years after, this would be known as the 'Blumlein stereo microphone technique', and is still used to this day with little or no alteration necessary.

Finally, the last of Alan Blumlein's eleven, definitively 'audio' patents, is No. 429,054, 'Improvements in and relating to Sound-Transmission, Sound-Recording and Sound Reproducing Systems', applied for on Saturday, 10 February 1934. Once again, he had returned to an older theme, that of binaural sound, and had come up with some modifications and improvements. In some respects, it can be viewed as an answer to an article that had appeared in *The Journal of the Acoustical Society of America*, Volume V, October 1933. In the article, an arrangement of two microphones is described, one described as a 'velocity' microphone and the other a 'pressure gradient' microphone. It is possible that a 'pressure gradient' microphone had been modified with a damped 'backpipe' to convert it to omni (but the term 'pressure gradient' had stuck; today, a velocity and a pressure microphone type would be considered one and the same thing. Here however, they seem to refer to a directionally sensitive and a non-directionally sensitive microphone) and had been used together to form a single microphone element. Together, they had gathered sound waves from a source in order to convey to the listener an impression of binaural sound. Alan Blumlein had in fact already discussed this possibility within Patent No. 394,325, but it was obvious, and evidently essential to clarify his meaning within a separate and defined patent, hence No. 429,054. While it is highly unlikely that the article in the American journal was intended to infringe upon any potential device from the Blumlein patent, it is also possible that there had been a misinterpretation of the use of the different microphones. Specifically, the use of a velocity and a pressure gradient microphone pairing, in order to develop a new patent outside of that being lodged by EMI. Regardless of the reasons for the American article – and it should be pointed out that many American companies were at the time also developing recording devices similar to that which Blumlein and Holman had patented for Columbia – it was felt necessary that EMI and Blumlein clarify their position with regard to this specific use of a stereo microphone technique with a definitive patent and two diagrams illustrating the point.

The invention outlines a series of processes in which various arrangements of microphone types are used to convey to a listener a binaural impression of sound. More specifically, it relates to the use of a velocity microphone used in conjunction with a pressure microphone, which together represent a single piece of apparatus. In this instance the two microphones can be brought much closer together without the previous phase problems that had been encountered in Patent No. 394,325 and described at length in Patent No. 429,022. The outputs of the microphones were fed to two channels that in turn fed the loudspeakers. One channel received and transmitted the sum of the microphone outputs, while the other receives their difference. These outputs are in phase, but their relative magnitudes depend upon the direction of the sound arriving at the microphones, and these can in turn be made suitable for binaural sound by arranging them to be insensitive to sound arriving from a central position.

The two diagrams that were used to demonstrate how Blumlein intended this to work explained the detail further:

In *Figure 1* [Figure 3.49] a directionally sensitive, velocity (or strip) microphone (*a*) is placed in close proximity to a non-directionally sensitive pressure gradient microphone (*b*). These are in turn connected to a network (*c*), the outputs of which pass through leads *d* and *e* to two loudspeakers. The strip of microphone *a* is arranged along the axis of the sound field and is clearly insensitive to sounds from a sound source $S_1$, but has a maximum sensitivity to the sounds emanating from the source $S_2$.

The sensitivity of microphone *b* is substantially independent of the sound source. The leads *d* and *e* transmit respectively half the sum and half the difference of the microphone outputs, and with the sound source $S_1$ the two outputs in the leads are the same, whereas with the sound source in position $S_2$ it is the microphone outputs that are the same. The impulses in leads *d* and *e* are therefore a maximum value as well as a zero value, which will in turn provide full sound in one loudspeaker, and substantially none in the other, thus giving the correct impression of a laterally displaced sound source.

*Figure 2* shows an alternative arrangement where microphones of a similar impedance, as in the case of two velocity sensitive microphones, can be used. The two microphones are represented by *a* and *b* respectively, with the terminal of one connected to one terminal of the other. The remaining terminals of the microphones are connected to the opposite ends of the primary winding (*f*) of a transformer.

The primary winding of a second transformer (*g*) is connected between the common microphone terminals and the mid-point of the first transformer primary (*f*). In this way the output of the secondary of one transformer will consist of the sum of the microphone impulses, while the output of the secondary winding of the other transformer will consist of the difference of the microphone impulses.

Essentially, what Blumlein was saying was, that with careful choice of microphones one being directionally sensitive of course, and suitable transformer ratios in order to obtain the desired ratio of the microphone outputs to the loudspeakers, the sound impulses could be controlled to provide the binaural effect aimed at by the system. In so doing, the relative amplitudes of the sound impulses in the two chan-

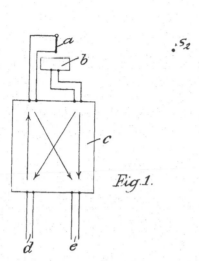

Fig.1.

Figure 3.49
Detail from Patent No. 429,054 (1934)
showing relative microphone positions
's$_1$' and 's$_2$' and a shuffling circuit.

nels could accord with the direction from which the sounds reached the microphones. It succeeded in getting the message across, and no further questions such
as those raised in *The Journal of the Acoustical Society of America* were published in his
lifetime.

## Postscript

As happened often in Alan Blumlein's life, his attention would be drawn away for
various reasons from the specific task at hand, and moved towards another subject.
This might be a subject which, at first, would seem to be totally unrelated. This
would happen on two further occasions during the period when his audio patents
were being applied for, in which he returned to the subject of telephony. In 1932,
while working on his binaural sound system, Blumlein applied for Patent No.
402,483, 'Improvements in Electric Coils'. In this specification, he provided a means
for a small and easily constructed inductance or loading coil to be constructed. It
was designed to give a small value of effective resistance for a given inductance at
telephone frequencies, but only produced a very small magnetic field, and was
therefore not liable to induction from external electric fields.

In order to achieve this, Blumlein encircled the windings of the coil with a magne-
tized circuit comprising a forward path through the coil to a point such that the line
of flux turned outwards at 45° towards the coil axis. In this way, the forward path of
reluctance was high compared to the return path reluctance for lines of force
passing far from the windings. Arrangements of the outer magnetic circuits and the
air gap spaces between them were made in order to ensure that the ratio of reluc-
tances of the forward to return paths was higher for lines far removed from the
windings than for lines closely adjacent to them. This ensured that at low frequencies, the proportion of the total magnetomotive force used to drive the flux on the

forward path of the outer magnetic circuits was much greater than the proportion used to drive the flux through the forward path of the main magnetic circuit.

It was an effective shielded loading-coil for telephone circuits. The question is, why did Blumlein suddenly, three years after he had last worked on telephone circuits, return to the subject? The answer probably lies in the subject matter which we associate him working with in June of 1932, the binaural system. Blumlein had, by this time, completed the many experiments with the Columbia recording/cutting system, which of course had an electromagnetic damping system. It was very likely this and his growing understanding of electromagnetic and magnetomotive forces, which inspired his return to telephony and an application which led to yet another patent specification being filed in his name.

Yet another example of a Blumlein patent which at first sight seems to be totally out of line with the subject he was working on at the time, was applied for on 7 March 1933, Patent No. 419,284. This is a method of removing the screening from an iron or iron-alloy core of an inductance element in order to reduce the loss of frequencies of the order of 200 kHz or more. This was due to self-capacitance of the windings of the element and the generation of eddy currents in the core.

These eddy currents in the core were produced because the resistance was not as high as that of a good radio frequency dielectric material and were compounded by the self-capacitance of the windings; the losses were therefore comparatively high. Blumlein's answer to this was to provide a radio frequency circuit having an inductance coil in which capacitive losses were reduced. He did this by making the core windings of a magnetic material in so fine a state of subdivision that the coil could be employed without any undue loss of frequencies because an electrostatic shield was effectively disposed of between the core and the windings. This patent once again demonstrated how amazingly flexible Blumlein's mind could be, as it is only marginally associated with the main topic on which he was working at the time. His colleagues would remark repeatedly on his ability to adapt trains of thought that he had to any subject and still produce an invention practically out of nothing.

## Author's note

It has been pointed out that at no time was Mr Norman Green 'engaged' by EMI for the work of transferring the Binaural films to safety stock. Mr Green gave his time entirely voluntarily having been approached by EMI for his advice. The time taken to transfer the volatile nitrate stock and its soundtracks to a modern safety stock was determined by the volume of his work as Head of Engineering at the ITCA. The Blumlein films were primarily completed in his own time, outside that of his normal work. It should also be pointed out that EMI did not pay Norman Green for his time spent on conservation of the films.

# Chapter 4
# Television

E ven before Alan Blumlein had completed his work on the Columbia recorder, EMI had turned their attention to the then very new science of television. It was, therefore, inevitable that at some stage Blumlein's services, and those of his colleagues in the research department at Hayes, should be directed towards the 'new science' and that of producing an EMI television system.

By 1932, regular experimental television programmes were being broadcast (albeit only half an hour a day), to the very few people who owned a receiver, from the London television transmitter of the BBC at Broadcasting House. These had originated from a demonstration given in April 1925, by John Logie Baird, of a mechanical transmission apparatus, which could produce crude pictures between two machines. From these rather basic beginnings, The Baird Television Company and the British Broadcasting Corporation started the world's first regular television service in 1929.

By the time EMI became interested in the possibilities of television, many other individuals and companies had been labouring to perfect a higher resolution system. With the electrical expertise that Blumlein had brought to bear on the problems of producing the Columbia recorder, and with his recently completed work on binaural sound, it was only natural that EMI considered him the most likely candidate to head the team of research engineers to produce a television system of their own.

## Origins

Television, which literally means 'seeing over distance', like many other inventions, is not so much the work of one individual but many, and, in fact, comprises many inventions all of which came together at various periods in time to produce the device we recognize today.

The earliest ideas for a system that could project electrically generated images all made use of the known properties of certain naturally occurring materials that underwent physical changes when they were introduced to light. This phenomenon is known as 'photo electricity', and without it, television would be impossible. The discovery of photo electricity can be dated from the experiments of the French scientist Alexandre Edmond Becquerel who, in 1839, at the age of just nineteen, made the first observations of the results of the electrochemical effect of light.

Edmond came from a distinguished family of scientists. His father, Antione César Becquerel, would later help to publish the results of his son's work in *Comptes*

*Rendus.* Towards the end of the nineteenth century Edmond's son, Antoine Henri Becquerel would become world famous as the discoverer of radioactivity. In 1896 Antoine Becquerel published his observations that uranium emits an invisible radiation at room temperature and could affect a photographic plate.

Edmond now observed that when two electrodes are immersed in a suitable electrolyte and illuminated by a beam of light, an electromotive force is generated between the electrodes. He had in fact discovered the basis of the first electrolytic photocell. Working under the guidance of his father, Antoine César Becquerel, Edmond presented the findings of their work to L'Academie de Sciences in Paris, in July 1839.

Then in February 1873, the British scientist Willoughby Smith wrote a letter to Latimer Clark, the Vice-President of the Society of Telegraph Engineers in London, informing him of some interesting results he had observed when experimenting with the resistance levels obtained from bars of selenium.

Selenium had been discovered in 1817 by the Swedish chemist, Jöns Jakob Berzelius, and was known to have a very high electrical resistance. It was this property which had prompted Willoughby Smith to use the material in his experiments which were conducted with the aim of finding a high-resistance substance for use with his system of testing and signalling during the submersion of long submarine cables (Willoughby Smith had been in charge of the electrical department responsible for the laying of the 1852-mile-long cable from Valentia Island, Ireland, to Heart's Content, Newfoundland, by the SS *Great Britain* between 13 July and 8 September 1866; the first really successful Atlantic submarine cable to be laid).

The letter to Latimer Clark was published in the *Journal of the Society of Telegraphic Engineers* and created quite a deal of interest and speculation at the time as to the causes, and possible uses of, the material and the observations that had been made. Within a couple of years of this, Bell had shown the world the telephone and this seems to have been the catalyst for a number of schemes that were directed towards 'seeing with electricity' rather than hearing with it. Despite the fact that these suggestions and plans appeared totally impractical at the time they were originated (and would take more than half a century before getting anywhere near becoming a reality), this does not seem to have deterred, in any way, the many who now pursued this goal.

Meanwhile, in Britain in 1878, in order to settle a debt, one of the most famous experiments in photographic history was carried out by the photographer Eadweard Muybridge assisted by John D. Isaacs. Muybridge had wagered Leland Stanford that a horse, during the gallop, raised all four feet off the ground and that he could prove it. Isaacs and he set up a battery of cameras, the shutters of which were linked to a series of strings, which crossed the path of the horse as it ran. When the horse broke the string, the cameras took a picture, each in succession, as the horse passed.

The resulting photographs proved, conclusively for the first time, that a horse does indeed raise all four hooves as it gallops. A year later, Muybridge would go on to perfect his system into the 'zoopraxiscope', which projected several rows of glass pictures onto a revolving tin disc through a small aperture. The experiment won Muybridge his bet and instant acclaim, but moreover, this experiment was to set the stage for the future development of both the motion picture camera and the television.

One of the earliest schemes put forward for the distant viewing of images by means of electricity was proposed by another Frenchman, Constantine Senlecq in 1878. The method incorporated for the first time the idea of scanning the images. Senlecq's proposal was to use: '...an ordinary camera obscura containing at the focus an unpolished glass (screen) and any system of autographic telegraphic transmission; the tracing point of the transmitter intended to traverse the surface of the unpolished glass will be formed of a small piece of selenium held by two springs acting as pincers, insulated and connected, one with a pile and the other with the line.'

Basically, Senlecq's system would have produced a reproduction of the original image on paper in the form that today we would associate as much like a facsimile. These early proposals for vision transmission, unlike modern television, did not necessarily rely on instantaneous transmission of the image. The receiver for the image was to have been a soft iron plate with paper placed upon it, above which a pencil would be suspended. The idea was that as the selenium traced across a surface on the original, that image would be composed of light and dark shaded regions. The electrical response from these regions would correspond with the degree of electrical resistance transmitted to a mechanical action, which in turn would hopefully react accordingly.

The pencil, suspended above the paper, would mechanically react to these variations in electrical impulse, and the scribe would trace the varying levels of light and dark as heavier or lighter lines on the paper. Senlecq however, was not the first Frenchman to come up with the idea of a remote facsimile transmission. In 1862, Abbé Caselli, working with an apparatus, which was a modification of the chemical recorder, used an iron stylus to mark an image on paper sensitized with cyanide of potash. This system was actually used for communication between Paris and Amiens from 1865 to 1869.

Despite the fact that Senlecq seems to have completely disregarded some of the practical problems associated with such a system, he persisted with the scheme and, in 1880, came up with a revised version which now had multiple selenium cells making up the transmitter screen.

Each cell was connected individually by means of a form of distributor (through which it was assured that each cell was only scanned once). The receiver also now consisted of multiple cells that were made up of a mosaic of fine platinum wires. Each of these wires was connected to the electro-mechanics of the distributor system from the transmitter. The method ensured that each cell in the receiver would become momentarily luminous, relative to the level of light or dark falling on the corresponding cell at the transmitter.

Again, the image that was to have been produced would have been drawn on paper with a pencil, and therefore, once again, falls into the category that we would associate as being a facsimile rather than television. All of this is somewhat immaterial however, as there is no evidence that Senlecq, who called his device a 'telectroscope', managed to either construct or demonstrate the system actually working.

There were others who proposed similar ideas based around the reactive properties of selenium, notably, George Carey of Boston who described several schemes, one of which involved a multi-cellular mosaic of selenium cells, and yet another which incorporated the elements of a practical mechanism for scanning the image, and only required a single line-wire for transmitting the signal currents. Once again however, despite producing some graphic illustrations of the proposed system, as well as publishing his work in an article in *Design and Work* in 1880, no evidence that Carey managed to demonstrate his system has survived.

In New York, John Perry and W.E. Ayrton, were concerned about reports that Alexander Graham Bell, the inventor of the telephone, had deposited with the Smithsonian Institute, a sealed document in which he proposed a means of 'seeing by telegraph'. Perry and Ayrton wrote to *Nature* on 21 April 1880, in order to diminish the claim. They too had been working on a transmitter/receiver system that was based on selenium cells with limited results, and, while they admitted in the letter that: '...the discovery of the light effect on selenium carries with it the principle of a plan for seeing by electricity', they did not think that given the complexity of the problem (based on their own experimental work), that Bell would have succeeded with his method unless it was: '...a plan of a much simpler kind than either of ours.'

## Breakthroughs

In 1881, a British scientist, Shelford Bidwell, finally made a breakthrough with selenium-based devices for transmitting a scanned image, albeit a silhouette. What is more, he lectured widely on the subject of his device and presented it to the members of the Royal Institution, the Physical Society and the British Association for the Advancement of Science.

Bidwell's apparatus (which was acquired for the nation by the Science Museum in 1908), projected a picture by means of a lens, onto the front of a small box containing a selenium cell and having a small pin-hole in the centre of the front facia. A cam mechanism caused the box to rise vertically in a linear motion, thereby scanning the image in front of the box, while a quick flyback was caused at the end of each vertical line. Progression to the next line for scanning was produced by mounting the box on a horizontal shaft with a fine screw thread.

The receiver consisted of a platinum-covered brass cylinder, mounted horizontally on a spindle similar to that of the transmitter. Around the cylinder was wrapped a piece of paper which had been soaked in a solution of potassium iodide, and resting upon this was a flexible brass arm with a platinum point attached to it. As the box scanned the image on the transmitter, the platinum point on the arm of the receiver traced out a series of closely spaced brown lines on the paper. These lines corresponded in intensity with the light and dark being picked up by the current passing through the selenium cell. Because the receiver drum rotated synchronously with the shaft of the transmitter, the rendered image on the paper was a faithful representation of the original. Despite the apparatus failing to find a practical application or commercial use, Bidwell's invention demonstrated the possibility of photo-telegraphy for the first time.

Photography as a science, was, by now, well advanced and improvements in photographic techniques were soon to culminate in the first moving picture images on film. Also in 1881, Eadweard Muybridge, had gone to Paris and met Etienne Jules Marey, a French physician and physiologist, who discussed the possibility of developing a photographic technique to capture and project images of animals in motion. The result was Muybridge's 1882 'photographic gun', which was a camera

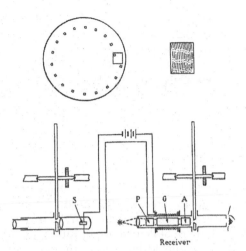

Figure 4.1
The principle of the imaging disc
invented in 1884, by Paul Nipkow.

with a cylindrical chamber, in which a dry plate revolved. With this device, Muybridge could photograph creatures such as birds in flight at a rate of around 12 frames per second. By mounting the images taken on a serial disc, he was able to show the objects in motion. In 1884, a German, Paul Nipkow, took the process of scanning one stage further with an ingenious system of light modulation. A large rotating disc with a series of holes in it near the periphery, and arranged in a spiral, would be spun in front of the object to be scanned. The area that could be scanned was quite small but could be used for example to scan small stationary objects.

The transmitter had a selenium cell, once again, to determine, through changes in electrical resistance, the light and dark areas associated with the object being scanned. The receiver was dependent upon the properties of the plane of polarization of light when a rotating piece of flint glass was situated in a magnetic field. The system did work, but poorly. This was mostly because the selenium cell only passed very small amounts of current, which Nipkow did not amplify, and thus rendered the system too insensitive for practical use. In essence though, the theory was sound and, later when Baird used much the same process (but with specially designed photo-electric cells in place of the flint glass), the results were the first true television images.

In Switzerland, in November 1889, a proposal called a 'phoroscope' was put forward by Lazare Weiller for a rotating drum or disc filled with a series of tangential mirrors around its periphery, in place of the rotating disc with holes that Nipkow had used. Each successive mirror on Weiller's device would be oriented through a small angle so that, as the drum or disc rotated, an area of the image to be scanned was done so and projected onto the selenium cell.

This system, while it also worked, presented yet another problem associated with selenium, that of time-lag. The material simply didn't respond fast enough to rapid changes in light level. This time-lag was probably the overriding factor which brought about the realization that selenium was not the ideal material to be considered if any of these devices were to be taken seriously.

In 1897, an attempt to overcome the problems associated with time-lag was put forward by a man called Jan Szczepanik. He proposed to construct a ring of selenium which would rotate at a constant speed, with a pair of oscillating mirrors that would then scan and project the image through an aperture onto a given part of the selenium ring while it rotated. In this way, the light falling on the selenium would constantly be exposing a fresh surface to the aperture. Szczepanik hoped the selenium would be capable of responding to higher frequencies, though he never built the apparatus to test his theory.

## Cathode ray tubes

Again in 1897, another of the elements that would prove essential to television, the cathode ray tube, was first produced commercially by Braun. It was not for another ten years however, until 1907, that Boris Rosing, a teacher at the Technological Institute of St Petersburg in Russia, would suggest for the first time that a cathode ray tube should be used as the receiving element in a system of remote electric vision. Rosing proposed a transmission system which had two rotating mirror-drums, much like those devised by Weiller, but placed at right angles to each other. These drums rotated at different rates, scanning and projecting as they rotated only a small part of the transmitted image at a time. The varying pulses from the transmitter were passed to the receiver where they were chased to charge a pair of deflecting plates in the Braun tube.

The fluctuating charges on these two plates caused the beam of electrons projected from the cathode to be deflected away from the aperture, though the beam itself was not sharply focused, as was common in the early tubes. In this manner, the beam, which did pass through the aperture, was only roughly proportional to the voltage of the deflecting plates and therefore to the degree of light and dark of the original object scanned. Once through the aperture, the modulated beam was caused to scan the surface of a fluorescent screen and thus produced an image as presented to the transmitter.

Rosing was hindered by the fact that he did not use any amplification within his system. No practical method of amplification would be available until after the invention of the thermionic valve. However, because of this he did not have to face any of the problems of interference as Baird would later. He did however try on several occasions to improve the quality of the image that the apparatus produced, and his experiments included a method of velocity modulation where the scanning spot contained a component which was inversely proportional to the illumination intensity.

As Rosing put it: '...the time of action of the signal upon the eye of the observer corresponds to the intensity of the light signal at the transmitting station'. While this statement is in essence true, it would seem that Rosing did not consider the fundamental that the light signal must also modulate the speed of the transmitting scanner if accurate reproduction of the original object is to be obtained at the receiver. Rosing's work received wide attention from the scientific communities of many countries including Britain, where a patent for his device was lodged later in 1907 (British Patent No. 27,570). A year earlier a Dr A. Korn had successfully demonstrated the telegraphic transmission of photographs with some degree of detail through a method of single-line spiral transmission from an image wrapped around a rotating drum. Once again, this process is more akin to facsimile than television, but the method had been improved upon by 1908, with M.E. Belin's 'télésteréographie', to the point where a 13 cm by 8 cm photograph could be transmitted over a double-wire telephone line in 22 minutes.

All this talk of 'seeing by electricity' prompted the French correspondent to *The Times* to write on 28 April 1908, that the President of the French Society of Aerial Navigation, M. Armengaud, was convinced that: '...within a year, as a consequence of the advance already made by his apparatus, we shall be watching one another across distances hundreds of miles apart'.

It must be remembered that this was a time of great scientific advance. In an age when anything seemed possible – powered-manned flight was barely five years old in 1908 – it was a bold man who voiced an opinion of disquiet. Yet that is exactly what Shelford Bidwell did when writing to *Nature* on 4 June 1908, in order, he felt, to temper the seemingly ridiculous claims made by the French which, in his opinion, did not take into account the technical difficulties involved. It was not so much that Bidwell was uncertain whether televisual transmission was possible, of that he was sure. He just wanted to balance the argument by pointing out that despite the enormous advances that had been made, the problems which lay ahead were only just beginning to be understood.

In his article, Bidwell points out that for any system of transmission of an optical image over a telegraphic wire, the photoelectric body used would need to pass over the surface of the image being scanned at least ten times per second. What Bidwell had hit upon was the fact that, in order to retain an image whose clarity could be defined as perfectly as the original object, a very large number of scans of a given unit area would need to be made.

He went on to example that: '... for an image, the definition of which would be as perfect as that presented to the eye or in a photograph, and which would be received on a screen no larger than two inches square, the number of 'elements' (units) required would be around 150,000'. This in turn would require, Bidwell estimated, a synchronized operation between the transmitting apparatus (illuminating the object) and the connection to the receiver of about a million and a half scans in every second. A quite impossible mechanical operation.

However, should it be acceptable that the definition of the object be reduced to a half-tone picture (the type of which was commonly appearing in newspapers by this time), Bidwell estimated that the number of elements could be reduced to 16,000. This however, would still entail a synchronized operation of 160,000 scans per second. As he pointed out, '...even this would be wildly impracticable, apart from other hardly less serious obstacles that would be encountered'.

As if that were not enough on its own, Bidwell then pointed out that in order to produce an apparatus which could scan a coarse-grained picture of 2 inches square, some 120 selenium cells could be employed in place of the slow and limited rotating arm/drum assembly that Korn and Belin were using. Of course, this would prove rather expensive. These 120 selenium cells would then require 120 line wires which again presented a problem of enormous cost, assuming the device was to be used over any distance.

Making rough calculations of cost based on his findings, Bidwell worked out that in order to transmit an image over a distance of 100 miles, assuming the picture to be 2 inches square, and a single scan unit to be 1/150th inch, there would be 90,000 elementary parts in the construction of such an image. There would be the selenium cells, luminosity controlling devices, and projection lenses for the receiver and conducting wires.

The selenium cells themselves would need to be mounted on an area 8 feet square, requiring the image to be projected with an achromatic lens the aperture of which would need to be at least 3 feet across. The receiving apparatus would occupy a space of around 4000 cubic feet, and the cable connections to it would have a diameter of 8 or 10 inches. Bidwell felt it could be done – for a cost of £1,250,000. He even pointed out that with the application of a three colour principle, the image received could probably even be produced in something like natural colour, but that, of course, would multiply the cost by three.

It is unlikely that the article in *Nature* was taken too seriously by most people, the costs he envisaged were so astronomical. Shelford Bidwell had not however raised the issue of cost and difficulty purely to dispel further development in television. His main aim was to shed some light on the comments of the French scientists who, unless they had 'made some quite revolutionary discoveries', were certainly not going to produce the apparatus that M. Armengaud had suggested.

One man who did take the article seriously was another British scientist, Alan Archibald Campbell-Swinton, who, like Bidwell, had also been working on the problem for some time. In reply to Bidwell's points, Campbell-Swinton wrote to *Nature* on 18 June 1908, and explained that while the 160,000 synchronized operations per second was totally impracticable, as Bidwell had stated, the problem could surely be more suitably solved if distant electric vision used the two beams of a cathode ray tube, one at the transmitting station and the other at the receiving station. This was the very first time that such a system had been suggested, and of course became the basis for the eventual television services to follow.

Campbell-Swinton envisaged the two cathode ray tubes 'to be synchronously deflected by the varying fields of two electromagnets placed at right angles to one another and energized by two alternating electric currents of widely differing frequencies, so that the moving extremities of the two beams are caused to sweep synchronously over the whole of the required surfaces within one-tenth of a second necessary to take advantage of visual persistence.'

He went on: 'Indeed, so far as far as the receiving apparatus is concerned, the moving kathode (sic) beam has only to be arranged to impinge on a sufficiently fluorescent screen, and given suitable variations in intensity, to obtain the desired result. The real difficulties lie in devising an efficient transmitter which, under the influence of light and shade, shall produce the necessary alterations in the intensity of the kathode beam of the receiver, and further in making this transmitter sufficiently rapid in its action to respond to the 160,000 variations per second that are necessary as a minimum. Possibly no electric phenomenon at present will provide what is required in this respect, but should something suitable be discovered, distant electric vision will, I think, come within the region of possibility.'

Three years later at a presidential address in front of the Röntgen Society, on

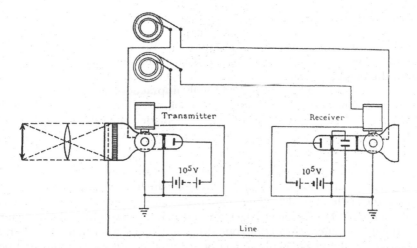

Figure 4.2 The first electronic television principle as put forward by Alan Archibald Campbell-Swinton during his address to the Röntgen Society in 1911.

7 November 1911, Campbell-Swinton elaborated on his original ideas wishing to point out that his ideas were just that, ideas only. He had never intended to actually construct the apparatus he had suggested partly due to the mechanical limitations of the time, but mostly due to the amount of experimentation he imagined would be required, not to mention the cost involved. In fact, Campbell-Swinton proposed that any serious development of an all-electric system should be taken up by a large industrial research laboratory.

Some years later, in April 1924, Campbell-Swinton wrote again of his ideas, this time in *Wireless World and Radio Review*. His article, 'The Possibilities of Television', went some way further to explaining his original concept at a time when Baird was just getting his mechanical television system to work, and the idea of an all-electric system seemed to have been forgotten. Later still, in October 1926 he wrote to *Nature* once more pointing out why he had not constructed an electric system of his own: 'I actually tried some not very successful experiments in the matter of getting an electrical effect from the combined action of light and cathode rays incident upon a selenium-coated surface.'

Campbell-Swinton's prophetic view of the future is all the more startling when one considers that at the time he wrote the original letter to *Nature* in 1908, radio transmission was but a few years old, Marconi was still working out such problems as radio communications techniques, and devices such as amplifiers had yet to be built. Couple this with the primitive photoelectric cells that were being used and the basic vacuum tubes available, and his words are all the more remarkable in their accuracy.

Campbell-Swinton died on 19 February 1930, and so, unfortunately, did not survive to see his dream of 'distant electric vision' come to pass. His work however, the suggestions he put forward and the ideas he outlined, all constitute the basis of the electric television system which would follow just a few short years later.

## The first practical television systems

From the time of Campbell-Swinton's address to the Röntgen Society in 1911, to the mid-1920s, progress in television development was painfully slow. As he had rightly foreseen, the degree of experimentation and development that was needed made any development through an individual totally impractical. That is not to say that progress was not made. During this time a number of schemes were put forward and tried but, most of the major improvements were made primarily due to the invention of the multi-stage amplifier and the thermionic valve, two essential requirements for television progress.

In America, in June 1925, Charles Francis Jenkins became one of the first to demonstrate a practical television device using a system which employed two bevel-edged glass discs, the angle of the bevelling changing continuously around the circumference of each disc. The bevelled edges formed prisms that deflected the beam of light as the discs rotated. By rotating one disc many times faster than the other, the entire surface of the image could in fact be scanned in this way. Jenkins had been working on the problems of practical television for many years and, as early as March 1922, had applied for a patent based on the prismatic glass discs for the transmission of pictures by wireless.

There were, however, problems with this method. As the discs were glass, they could only be rotated at a finite speed, which in turn meant that the system could only be used for low-definition picture generation. Also, because the system was limited to use of transmitted light as opposed to reflected light, Jenkins' prismatic disc device could transmit motion pictures faithfully, but could only generate a shadowgraph of a stationary object.

At around the same time as Jenkins' work was being done in the United States, a young Scotsman by the name of John Logie Baird was preparing to demonstrate a system which he had devised, at the London departmental store of Messrs. Selfridges & Company Limited. The apparatus which was described as being for 'wireless television', was assembled in the store in late March 1925; once operated, it produced at the receiver a blurred image of a simple form such as a cross or triangle or even letters, painted white on a black card, held up before the transmitter.

Baird was to become the name most widely associated with the invention of the television system, an honour which he hardly deserved as, by his own admission, his work was based almost entirely on that of the many others who had gone before him. Yet, it is Baird who today is remembered ostensibly as the inventor of the device. What had Baird done that none of his predecessors had?

Born in 1888, in Helensburgh, Scotland, Baird had attended the Royal Technical College and Glasgow University where he was attracted to experimental work. In 1923, he decided to devote his entire time to the development of television and busied himself studying every facet of the work that had gone before him. It occurred to him that much of the experimentation that had been done by his predecessors, especially that of Nipkow with his rotating disc with holes at the circumference, was in essence sound, but lacked the benefits of modern equipment. As Baird had access to much of the latest equipment at the time, he set about designing and then constructing a television system he felt sure would work.

In early 1924, Baird had perfected his apparatus to the point where he felt he could

Figure 4.3
John Logie Baird, often incorrectly
given the credit for 'inventing' television. (*Courtesy of APTS*)

demonstrate it for the first time; in order to do so, he arranged for a series of experiments to be conducted in front of an invited press audience. The device consisted of a lens mounted in front of a rotating disc in which a series of thirty holes were cut around the circumference in a spiral pattern. This disc, very much like that which Nipkow had invented in 1884, rotated at 600 rpm and scanned simple objects, such as a cross, which was placed in front of the lens and illuminated from behind with an arc light. Behind the perforated disc was a second larger serrated disc, which also rotated, only this time at 2000 rpm.

The arc light illuminated the cross and cast its shadow on to the lens, through which the light from the image passed. It then fell onto the perforated disc, with the light from the image passing in succession through the holes as they spun. Once beyond this, the light fell upon the serrated edge of the second rotating disc. The light flashes were interrupted by the edge of the serrated disc at a very high frequency before they finally reached a selenium photoelectric cell placed at the rear of the second rotating disc.

Baird had realized that even though the selenium had an instantaneous response to light causing it to begin to change its resistance, the old problem of time lag would be fatal to the passing of any image over a selenium cell without being taken account of. This chemical inertia could however, be overcome by the rotating serrated disc, the slots of which make the actual resistance of the cell at any given instant of no consequence, it is the pulsations which are transmitted. Writing in *The Wireless World and Radio Review* in May 1924, Baird, explaining his process, described it in the following manner: 'To make the matter clearer, it might be said that light was turned into sound. Loud for the high lights, low for the darker areas and complete silence for the darkness.'

The picture was reconstituted at the receiver by means of a synchronously-driven scanning disc between the eye of the observer and a neon lamp (the neon light had been invented by Georges Claude in 1911), the glow from which was made to alter

Figure 4.4 The basis of the Baird mechanical system from an article of 1925. This method was demonstrated at Selfridges in late March that year.

in intensity according to the variations of the current in the selenium photo-cell. Continuity of vision made the whole image appear simultaneously, though the received picture measured barely 1.5 inches by 2 inches and was a dull pinkish colour. As you can imagine, with a low-definition of 30 lines and a repetition frequency of five frames per second (that is five individual pictures on the screen each second), this led to a pronounced flicker.

Crude as the Baird system was when he began his demonstrations at Selfridges (an account of which appeared in *Nature* on 4 April 1925), the public were amazed by the apparatus. On 26 January 1926, Baird, by now working with his partner Sydney A. Moseley, had improved the device sufficiently, with addition of a gas-filled potassium photo-electric cell (to replace the selenium one), to demonstrate his system before the members of the Royal Institution. This was the first time that images of moving human faces were shown not as mere black and white outlines, but having tone gradations of light and shade, an achievement not previously considered possible with this form of device. Certain that the breakthrough had at last been made, Baird pressed the Post Office for a transmitting licence that was granted. He now embarked on a massive campaign for money from a handful of private backers and official support for the system.

By May 1927, Baird had developed the television to the point where he could transmit pictures along standard telephone lines from Glasgow to London and, by February 1928, he had repeated the process but this time from London to New York. As if to prove the versatility of the apparatus he even transmitted pictures to the ocean liner *Bavaria* in the mid Atlantic. By 1929, the Baird Television Company were ready to consider the basis of an arrangement with the BBC for the first experimental television transmissions, though Baird's 'bull-in-a-china-shop' tactics did not find favour with the BBC general manager, J.C.W. Reith, who took a distinct disliking to him. The transmissions, it was proposed, would take place from the London station, for half-hour periods, five days a week, with a definition of 30 lines and an improved repetition frequency of 12.5 frames per second which reduced the flicker. Arrangements were made for Baird to use the 2kW transmitter at the BBC's London station, which was situated in Oxford Street, though this was soon altered to a more appropriate location at the BBC's transmitter at Brookman's Park.

Baird had attempted to produce the first colour television pictures at this time, and even dabbled with the idea of a stereoscopic television system, both of which were shown to the British Association in Glasgow in August 1928. In July 1930, during a music-hall programme at the Coliseum in London, a large multi-cellular lamp screen was erected for the presentation of television pictures, made up of 2100 flash-lamp bulbs. A commutator with 2100 fixed segments and a single brush rotating at frame rate was constructed with each segment connected to a lamp. The intensity of each of the lamps was then driven individually and directly from the current arriving at the receiver.

Even as Baird was demonstrating his mechanical 30-line system for the first time, in America events were taking place that would affect the future of electric television. In January 1927, a young experimenter, Philo T. Farnsworth, had persuaded a group of backers in San Francisco that he could develop an electric television system based on the principles of a camera tube later to be called an 'image dissector'. Farnsworth set up the Television Laboratories Company in Green Street, San Francisco and set about developing his camera tube which had a light-sensitive plate formed of a fine mesh covered with a light sensitive medium such as sodium, rubidium or potassium.

An electric shutter was formed by a metallic plate, in which there was an aperture. Between the shutter and the light-sensitive plate, there were four plates to cause the electric image to be passed systematically in front of the aperture to analyse or dissect the scene to be transmitted. Farnsworth's tube was to be of a high vacuum type to permit a high potential without ionization. The system was planned to have a straight-line scanning method of 500-lines at 10 cycles per second vertically, with which he hoped he could light uniformly the entire image to be scanned.

## The iconoscope

At the research laboratories of the Westinghouse Electric & Manufacturing Company in East Pittsburgh, in early 1925, Vladimir K. Zworykin began working to develop electric methods of scanning having studied and established the limited possibilities of a mechanical system. Like Campbell-Swinton before him, Zworykin understood that what was needed was a charge storage system using a photo-electric mosaic of some kind, and had proposed a deflected beam of electrons from a cathode ray tube as the image scanning device.

When he applied for his first patent in July 1925, his design was very much like that of the proposal that had originally been put forward by Campbell-Swinton in 1911, though there is some evidence to suggest that Zworykin had not heard of his work nor had he seen any of his proposals. Nevertheless, Zworykin's 'iconoscope', as he called it, worked on the principle that the sensitivity of the device should be approximately equal to that of a photographic film operating at the speed of a motion picture camera, with the same optical system. This was a bold move for the time as it represented a high-definition system far beyond anything that had previously been achieved or proposed other than Campbell-Swinton's work.

The iconoscope consisted of an electron gun producing a fine beam of electrons and a photosensitive mosaic enclosed in a highly evacuated glass envelope. External to this assembly was a magnetic deflecting yoke, with its driving circuit, which would deflect the electron beam as it left the mosaic. An amplifier would then raise the output signal, which, as it left the cathode would be very small and insufficient, before it would be sent to the transmitter. The electron gun produced a very narrow beam of perhaps a few thousandths of an inch across on the photoelectric mosaic. This beam would be produced from two groups of electron lenses formed by coaxial cylindrical electrodes. At the first lens group was the thermionic cathode and control grid, this concentrated the raw beam of electrons into a narrow bundle which Zworykin called the 'crossover'. The second group of lenses then focused the crossover beam upon the mosaic, bunching it yet further into a narrow spot in the process.

The electron beam from the gun was now deflected vertically and horizontally by varying the magnetic fields in the deflecting yoke. This enabled the beam to scan over the surface of the mosaic in a series of parallel lines. A timing mechanism was incorporated so that in the course of one frame the entire surface of the mosaic would be scanned by the beam.

The mosaic itself was the electrical equivalent of the retina, and consisted of a small, thin layer of mica or other dielectric. It was coated on one side with a conducting film known as the 'signal plate', and on the other by huge numbers of minute glob-

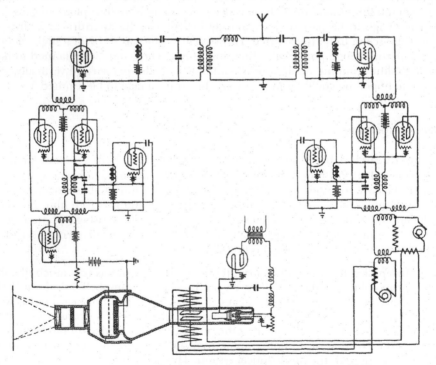

Figure 4.5 Vladimir K. Zworykin's electronic television patent from 1928.

ules of silver known as 'elements'. Each of the elements was isolated from one another (their active portions), the metal backing plate and the separate mica forming minute capacitors. The elements were photosensitized with oxygen and caesium with each one acting as a condenser with the conducting signal plate.

When light fell on the mosaic, these elements would emit photoelectrons, which charged the condensers of which they formed a part. In other words, the action of the light upon the elements caused them to acquire a positive charge, which in turn would trap an equal, yet negative, charge on the signal plate. As the scanning beam from the electron gun passed over each element in turn, it would return it to equilibrium potential, thereby releasing the stored charge. This altering of state, the release of the charge, induced a corresponding change in the signal plate and thus a current was created which was caused to flow in the signal lead.

The mechanism by which the elements were returned to equilibrium in the iconoscope corresponded with beam electrons hitting the elements with a velocity equivalent to about 1000 volts. This caused the elements to emit secondary electrons in excess of the beam electrons. If an element was at a negative potential with respect to the electrode (which acted as a collector of the secondary electrons), more electrons left the element than were supplied by the beam, resulting in the elements becoming increasingly positive.

This positive charging increased, until the elements reached such a potential that most of the secondary electrons (which only had emission velocities of a few volts), were turned back to the element. This meant that the current leaving the element was equal to that arriving at it from the beam, the element potential then being known as its equilibrium potential. This equilibrium potential of the element represented, for a typical silvered photo-electric cell of the type being used by Zworykin, about 1 or 2 volts positive with respect to the secondary emissions collector.

If the element was more positive in its charge than this, the secondary emission current leaving was less than that of the bombarding current. In this case, the elements accumulated a negative charge until they established equilibrium. These methods of reaching equilibrium had two important consequences on the results that were achieved from the iconoscope.

First, because not all of the returning secondary electrons reached the elements from which they came, they ended up being deposited all over the surface of the mosaic. As a result of this, some signal was lost due to a masking effect of the redistributed electrons, which can often result in a spurious effect known as 'black-spot' or 'shading' arising from the non-uniformity of the distribution.

Second, the mean potential of the elements was relatively high with respect to the electrode, with the result that its photo-emission was inhibited to some extent. This reduced the efficiency of the mosaic as a photo-emitter, which was obviously undesirable, but it also led to a characteristic where the rate of increase of the signal actually decreases with increasing light. The iconoscope tube therefore did not become saturated or block at very high light levels.

Zworykin conceived that his iconoscope could be used as the element of an electric camera that might be used to scan images and transmit them to a receiver, which would in turn reproduce the scanned image using a second cathode ray tube. What is all the more remarkable about Zworykin's concept is that when he addressed the scientific world in a paper explaining his device in 1933, he went on to state that he could easily imagine it would possible to achieve a scanning system that would work at a resolution of 500 lines or more. While Baird undoubtedly took mechanical television from bare concept to the boundaries of reality, Zworykin brought electric television to brink of reality. Without the iconoscope, television as we know it could never have been developed.

While Zworykin was working on perfecting the iconoscope in Pittsburgh, two other Americans, Dr Herbert E. Ives, and Dr Frank Gray, this time working from the Bell Telephone Laboratories of American Telephone and Telegraph (AT&T), had been working on another low-definition television system that was used to transmit images and sound from Washington DC to New York on 7 April 1927. At the same time, there was also a wireless version being demonstrated from Whippany, New Jersey, where a programme of amateur vaudeville was taking place, and being televised back to New York, 22 miles away.

Ives had developed a scanner similar to that of Baird's with a rapidly moving spot of light scanning an object, the signal was then synchronized with the mechanism of the receiver of which two types were used. The first receiver produced a small image of around 2 inches by 2.5 inches, only really suitable for viewing by one person. The second receiver however, produced an image by means of a multi-element neon lamp approximately 2 feet by 2.5 feet, which could be seen by an audience of some considerable size.

The system that Ives and the Bell Labs had come up with had a scanning resolution of 50-lines at a repetition frequency of 18 frames per second, somewhat superior to Baird's. It had been intended that the smaller receiver version could be used by individuals who wanted to telephone from New York to Washington for example, and who wanted to see who they were talking with. In this manner, Ives inadvertently invented a form of the videophone some sixty years before it became commercially available.

During the demonstration, the key-line speech was given by Herbert Hoover, then the Secretary of Commerce. People reported that the pictures reproduced were 'Excellent Daguerreotypes, which have come to life and started to talk'. The detail of the face of Hoover appeared in clear-cut black lines against a shiny gold background, though the detail on the larger of the two screens was significantly poorer with many features blurred by the enlargement process.

The larger-screen version of the system was intended to work in conjunction with a public address system and was in fact used in Washington to relay images of various other speakers, and then the singers and entertainers from Whippany. It may have been slightly poor in reproduction quality, but this large screen television was the first system to have the sound relayed with the pictures. A dual-cast with radio station 3XN broadcast the pictures on 191 meters, the sync signals on 1600 meters and the voice information on 207 meters.

Back in Britain, the BBC had decided not to broadcast the sound with their 'experimental' television service until April 1930, though it should be pointed out that these earliest transmissions were conducted solely by the Baird Television Company. The BBC had no share in the technical aspects of television equipment or in the programme material that was being broadcast at the time. Instead, they decided to wait and watch television progress for a while, leaving the Baird Television Company to market all commercially available television receivers. These first became available from February 1930 at the then considerable price of 25 guineas.

## The kinescope

In America, Dr Zworykin had applied for a patent on 16 November 1929 on behalf of Westinghouse for a vacuum tube, in which he claimed that it was substantially impossible up to that point to focus the cathode ray to a well defined spot while remaining in complete control of the intensity of the ray.

The development which Zworykin now put forward stated that he proposed a CRT improved in the following ways: it should be capable of operating under acceleration potentials of thousands of volts, it should not require continuous pumping, it should have electrodes that were simple in construction and small in volume, the focus of the beam should be totally unaffected by the control potentials, and it should have a spot that is well defined at all points of deflection on the screen. It should also have an acceleration potential so large in respect of the intensity controls that the deflection of the beam is not influenced by the control potentials and should be the same for all intensities and, finally, the tube in which the electrons are accelerated should be a two-stage assembly, the first stage for low potential, the second for high potential. Deflection of the electrons should therefore also occur in two stages.

The tube that Vladimir Zworykin had designed was to contain no gas; instead it should be of a very high vacuum. It also had a second accelerating anode in the form of a coating of metal on the inside of the large portion of the tube itself. The deflection devices themselves, one magnetic for horizontal deflection, and the other electrostatic for vertical deflection, were mounted external to the tube. Zworykin also suggested that the screen could have a long persistence phosphor in order to decrease the number of frames per second, with a cooling fan or cooling 'jacket' if the beam needed to remain in one position for a sustained period.

The patent was granted, and a couple of days later Zworykin delivered a paper in which he described his new picture tube, which he called a 'kinescope', before the Institute of Radio Engineers at Rochester, New York. The kinescope was a significant step forward in the development of an all-electric television system as it was discovered that three sets of signals, picture, horizontal scanning frequency and framing impulse, could now be combined into one channel with the device.

This gave the kinescope several advantages which had previously not been possible: the picture now appeared green rather than the red (which had been due to the neon light source), the picture was visible to a large number of people rather than just a few peering into a reflected mirror image, there were no moving parts which could easily breakdown or wear out, the framing of the picture was automatic (as the synchronization frequency was applied directly to the deflection coil and the framing impulse with the picture signal applied to the control electrode of the kinescope). Perhaps most important of all, the picture was brilliant enough to be seen in a moderately lit room. It could easily be argued that the device changed the future of television for all time for without the kinescope, television may well have continued to have been developed along the basis of an all or semi-mechanical system for many years to come. As it was, Vladimir Zworykin had produced the key that would unlock the potential of television as we know it to this day.

## Getting serious about television

In January 1929, the Radio Corporation of America purchased the Victor Talking Machine Company of Camden, New Jersey for $154 million. With the deal came a series of vast manufacturing plants, the Victor Talking Machine Company of Japan and agreements with the Gramophone Company Limited of Great Britain, with both companies' well organized systems of distributors and dealers. The Gramophone Company Ltd had earlier made an agreement with the Marconi Company to acquire the business rights to the Marconiphone Company (for the last fifteen years of which, Isaac Shoenberg had been the leading light, working in the field of home entertainment, this of course included television). When, in March 1929, David Sarnoff, executive vice-president of RCA, became a director of the Gramophone Company Limited, it meant that RCA began to exchange patents with their new acquisitions with regard to television development.

Sarnoff had absolutely no interest in the talking machine that Victor produced; his interests lay elsewhere within the merger. RCA, General Electric and Westinghouse agreed to consolidate their interests in development, though Sarnoff pressed the directors of RCA to build a separate manufacturing plant for RCA at a cost of $25 million, to which the staff and facilities of the other two companies could be moved. This of course included Zworykin and his staff. Sarnoff needed Zworykin.

There now began a short period where all the major players in the television development game began to exchange their patents, staff and ideas as mergers and agreements happened thick and fast. Marconi Wireless Telegraph Company now began to receive and exchange patent applications (which are earlier than finally accepted patents, thereafter made available in the public domain) with RCA and registering them in Marconi's name. The General Electric Company were assigning their patents to the British Thomson-Houston Company, while Westinghouse Electric kept many of the patents they held under their own name or that of Metropolitan-Vickers.

In the same month, March of 1929, Baird Television Limited had tried to get the BBC to grant them further facilities to broadcast programmes from their own studios. Baird and Moseley gave another demonstration of the system to the Reith at the BBC and to the Postmaster General in London but, once again, the poor quality of the mechanical system employed, produced grave doubts in the minds of those responsible for the broadcasting of the pictures. The Postmaster General stated that: 'While the (Baird) system is a noteworthy scientific achievement, it has not reached a sufficiently advanced stage to warrant its occupying a place in the broadcasting of programmes'. The BBC did relent a little; Reith, despite his personal dislike of Baird (and his tactics), allowed him to use the facilities of their studios outside the hours of broadcasting currently agreed, but they drew the line there, refusing his demands for a studio of his own.

Incensed by this and obviously having little luck with either the BBC or the Post Office, Baird threatened to take his system abroad to use the facilities of any nation which could offer his work the kind of dedication he felt that it merited. As it happened, hearing of his disappointment, he was almost immediately offered the facilities of the German broadcasting authorities in Berlin by Dr Hans Bredow who, with two assistants, had witnessed the Baird demonstrations in London, in December 1928. Baird was offered the opportunity to bring his equipment to Germany, to have use of the German broadcasting facilities and more importantly use of the Witzleben transmitter in Berlin, which stood at 350 feet, and could reach quite a large potential audience.

Baird agreed to come to Germany and to divulge much of the expertise he had achieved in mechanical television systems to the German broadcasting authorities; they would even form a new company expressly for the purpose. Fernseh A.G. was formed in Berlin on 11 June 1929, from an equal partnership between Baird Television Limited, Zeiss Ikon, Robert Bosch and Loewe Radio; the company was registered on 3 July 1929.

Baird left England for Germany with all his equipment and staff, and set up new research facilities in Berlin explaining that: 'Other broadcasting authorities are more interested in my television transmission than the BBC are!' Things did not go quite as smoothly as Baird might have liked however. His television system did broadcast from the Berlin transmitter from 15 May 1929, to 13 June 1929, but he was sharing the facility at the time with the Hungarian, Dionys von Mihály (who had patented a system of recording sound onto film in 1919, and by 1929, was also very involved in television development). It would seem that Baird and Mihály did not seem to enjoy this relationship and so while the Hungarian stayed and continued development of his 'tele-cinema' system, Baird prepared to return, with his equipment, to England.

Upon his return the Post Office, with some reluctance it has to be said, allowed Baird to use their facilities at BBC radio station 2LO with transmission signals coming from the Baird laboratories in Long Acre, near Covent Garden. The Post Office relented; probably due more to the adverse publicity that their attitude to Baird's experiments had generated once he had left for Germany, than any great hope they may have held for the successful improvement of his system. Nevertheless, still using his 30-line, 12.5 frame per second mechanical system, Baird was back in England and back in business, though he would maintain his working relationship with Fernseh until the Nazis placed all television research under military control in 1935.

Meanwhile, in America on 27 June 1929, the Bell Laboratories gave a demonstration of colour television in which a new type of photoelectric cell, using sodium (rubidium and caesium had also been considered and tried, but sodium was easier to work with), rather than potassium which had been used up to that point. Sodium it seemed responded much better into the deep red spectrum and Bell Labs constructed a simultaneous colour system with the three primary colours displayed at the same time. This was a major advance on the Baird system, which had only displayed the three primary colours in sequence in his earlier colour demonstration.

The Bell Labs system had three sets of photoelectric cells, each covered with a set of coloured filters: yellow-green for green, orange-red for red and a greenish-blue for blue. A bank of photoelectric cells, fourteen with red filters, eight with green filters and two with blue filters, was used according to the relative sensitivity of the cells. The receiver worked with a mirror arrangement, which had three lights, two argon (one with a blue filter, one with a green filter), and one neon (for the red).

A standard 16-inch, 50-hole, 18.5 frames per second scanning, rotating disc was used and, it was claimed, that when the system was operated and adjusted accordingly, a picture with natural colour was obtained. Bell announced that the pictures, which were viewed through a pair of semitransparent mirrors set at an angle of 45°, were 'Quite striking in appearance, in spite of the rather low brightness and small size characteristic of the present stage of development.'

By now several other companies in America were very interested in television and its future potential. It is probably fair to say that even though many of the developments taking place in the United States at this time were mainly concentrating on improving the large number of mechanical television systems that were appearing (rather than the direct development of a practical all-electric system), the American attitude towards television was far more enthusiastic than that of the British authorities.

On 4 October 1929, Sarnoff proposed the formation a new entity, the RCA Victor Company, to merge the talents of the three sets of engineers and managers that were currently then developing separate systems for RCA, General Electric and Westinghouse. Sarnoff had gone to Zworykin earlier in 1929 at around the time of the original buyout of Victor, and asked him how much it would cost to produce a workable television system.

Zworykin had replied that he felt it could be done in two years, and would cost $100,000, informing Sarnoff that the basic elements were already operative (the kinescope had just been developed in the Westinghouse lab), but would still require extensive and expensive testing and refining. Sarnoff was convinced and saw to it that Westinghouse gave Zworykin more funds, staff and equipment, in order to produce such a system. The RCA Victor Company was formed, with RCA owning 50% of the stock, General Electric 30% and the remaining 20% owned by Westinghouse. Sarnoff however, had not bargained for the onset of the great depression, which would begin following the Wall Street Crash just a few days later.

Sarnoff was another among these visionaries, and when, on 3 January 1930, he finally became president of RCA, it seemed that television was about to come of age. Sarnoff should have been in a position to see to it that American scientific expertise should achieve this goal first, however the depression was to have a profound effect on television development in America and Europe.

While the Americans became more and more serious about the potential of television, and subsequently wheeled and dealed themselves into positions of power and influence over the development of the system, in Britain, the emergence of details of Zworykin's developments finally began to stir into action some of the British companies on the fringes of television development.

Zworykin's kinescope details had filtered through to Britain from the various mergers and arrangements that had come about in America, and through publications such as *Radio-Craft* and *Radio News*, though the British publication, *Television*, and the German equivalent, *Fernsehen*, both inaccurately reported Zworykin's breakthrough. So, while Baird persisted with his mechanical system (along with a few others), the more far-sighted of the television developers now began to see that the future lay in an all-electric system, once it could be perfected. This included Louis Sterling, who was director of the Columbia Graphophone Company Limited.

He now spoke with Isaac Shoenberg, who was then general manager of Columbia (Shoenberg had been with the Marconi Wireless Telegraph Company for fifteen years before the merger with Columbia in January 1929, and had assumed the role of general manager at Columbia soon after), and who was just about to have his first major success through Alan Blumlein. Shoenberg had employed Blumlein in the spring of 1929, specifically to overcome the patent problem on the Western Electric recording system, the royalty on which was a halfpenny on every record cut. Sterling now offered Shoenberg the job of manager of patents at Columbia, a task which he felt sure he would blossom in, dealing as it did with a host of projects that were under way at the time. Shoenberg did not accept immediately, went away and thought it through for a while, but finally agreed to accept the post in December 1929.

As part of their arrangement with the BBC, Baird Television used the Brookman's Park transmitter in London, in January 1930, and began to double the number of broadcasts to two half-hours per week. By March, two wavelengths had been assigned to Baird specifically for television broadcasts so that sound and vision could be transmitted together, sound on 261.3 meters and vision on 356.3 meters. Transmissions began on 1 April 1930, April Fool's Day.

At HMV the Gramophone Company, advanced research had been led by A.W. Whitaker who had among the many projects on the bench at the time, a mechanical recording system which it was hoped would also get around the patent which had been granted to Western Electric (the same project that Shoenberg had employed Blumlein to solve). Mechanical television was another project which had interested them for some time.

Once the agreements with their American partners were in place, Whitaker found himself, at the invitation of David Sarnoff, travelling to Camden, New Jersey to witness the work being done by General Electric, Westinghouse and RCA. On 2 April 1930, Whitaker was given a demonstration of the Zworykin system of CRT tube and receiver and he reported that 'The high voltage CRT presented a 5 inch image and was so bright that it could be viewed in a brightly lighted room.' Whitaker returned to England with the firm conviction that mechanical television was already dead in the water, that the Zworykin system he had witnessed would undoubtedly out-perform any mechanical system under development at the time, and that all HMV work should be re-directed to this end.

With the first effects of the recession starting to give shareholders all over the world the jitters, it became known to RCA Victor that the Television Laboratories of Philo Farnsworth in San Francisco, may be for sale. They dispatched Vladimir Zworykin to see Farnsworth, and despite the fact that several patent applications made by Farnsworth had been turned down on the basis that they interfered with Zworykin's, the two men were delighted to meet with each other. Zworykin arrived on 16 April, and Farnsworth set about showing him everything that he had been working on to that point, realizing his financial predicament, and that Zworykin could be a much needed source of future financing.

As matters developed, Farnsworth still had the time and finances to continue to develop his all-electric television system, which he described in detail for the first time in November 1930. The system would consist of the image dissector tube which converted the light from the scanned image into electrical impulses, a wide-band amplifier flat to 600 kilocycles, the synchronization of two alternating currents of a sawtooth form which also turned off the beam of electrons on its return path, and a picture tube (which was essentially a modified Braun oscillograph), which Farnsworth had called an oscillite. This was magnetically focused and deflected with the horizontal frequency around 3000 cycles and the vertical frequency of 15 cycles, thus producing a 200-line image on the picture tube.

For his part, Zworykin was evidently very impressed with everything he saw from Farnsworth's results, in particular the image dissector tube, remarking that he wished he had invented it himself. This is hardly surprising. By now the image dissector had been perfected by Farnsworth, and was in fact quite superior to the camera tubes, which Zworykin had been working on himself at the RCA laboratories. He did however, feel that his kinescope picture receiving tube was far superior to Farnsworth's 'oscillite', which was about to be patented (the application was made on 14 June 1930). Zworykin returned to his RCA laboratory, where he was now in charge of group television development. The influence of what he had seen from Philo Farnsworth must have inspired him, as he re-doubled his efforts to produce a viable device, and promptly filed, in May and July, for his first patent on a camera tube since 1925.

RCA now decided to keep much of the Zworykin system under development secret, while Farnsworth had much of the details of his developments published in England, Germany and the United States. Publications such as *Wireless World*, *Fernsehen* and *Radio News* were the usual vehicles for such information, and it can be generally accepted that they were widely read by just about everybody involved at the time in television development. Bell Laboratories had earlier in the year demonstrated a 72-line mechanical system which was far superior to anything that Baird had yet produced, but most authors were writing that the Farnsworth system had produced images which were significantly better than anything the Bell Labs system was capable of.

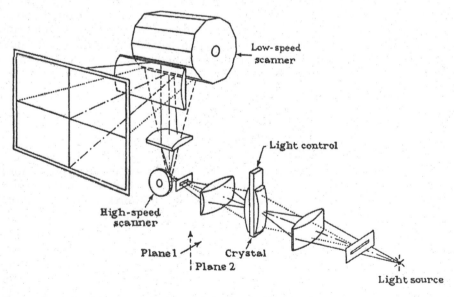

Figure 4.6 Detail of the Scophony mirror-drum scanning system.

Back in Britain, Baird Television had continued with the improvement of their system, which by now was being used for regular experimental broadcasts of programmes including plays and dramas, news and general interest items, etc. In fact the first television play, 'The man with a flower in his mouth', had been performed from their Long Acre studios on 14 July 1930. At around this time a British company called Scophony Limited had been formed by Solomon Sagall, who had been interested in the work of Dionys von Mihály and others (which he had witnessed in 1929). Though Scophony concerned themselves with the production of improved optical-mechanical television systems, their stockholders included many companies who would later play important roles in the development of all-electric television systems, such as Ferranti Electric, Gaumont-British and Oscar Deutsch of the Odeon (Oscar Deutsch Entertains Our Nation), cinema chain.

January 1931 saw the Imperial College of Science, South Kensington play host to the Physical and Optical Society's twenty-first Annual Exhibition, at which the research laboratory of the Gramophone Company, headed by George Edward Condliffe, gave a demonstration of the state of television development to that time. It was not common knowledge then, that the Gramophone Company were associated with RCA Victor, from whom they had been receiving patents since the merger in March 1929, and at which time they began television research and development work. Despite the RCA patents however, the demonstration television system that was shown at Imperial was almost entirely a development of the Gramophone Company engineers.

The demonstration of the equipment, which had a much higher definition than anything previously demonstrated due to the combining of five separate channels

Figure 4.7
George Edward Condliffe.
(*Courtesy of EMI*)

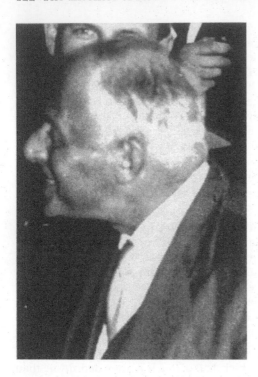

Figure 4.8
Edward Cecil Cork. (*Courtesy of EMI*)

each using mechanical scanning, was given under the direction of Cecil Oswald Browne. Browne had perfected it with R.B. Morgan, John Hardwick and W.D. Wright (who had taught at Imperial College, and who worked part-time for the Gramophone Company. Wright had been employed by Westinghouse in East Pittsburgh, working with Zworykin from March 1929 to March 1930, where he had access to all of Zworykin's early camera tube and kinescope work, but had returned to England for personal reasons.)

Cecil Browne gave his demonstration of the Gramophone Company system right under the noses of Baird Television, who naturally attended the show themselves. They were apparently totally unaware of the rival system, and naturally somewhat shocked at what they saw, having, as they did, a total monopoly on television research and development, not to mention transmissions, in Britain at that time. Baird Television immediately filed a lawsuit against the Gramophone Company, objecting to the use of the word 'televisor', which it seems Browne's system had used in some small part. The law suit came to nothing, but it made Baird Television only too acutely aware of the fact that they were no longer alone, either in Britain or the rest of the world, and that a practical, all-electric television system was not far off.

In fact, though they could not have known it at the time, the Baird mechanical system's days were already numbered. On 13 February 1931, the General Electric Company reported from America that it had successfully transmitted (low definition) television pictures from Schenectady, New York to Leipzig, Germany, and that the images had been recorded on 35-mm motion picture film. GEC reported the images '...as good, if not better' than those of the television image itself, and that they intended to use them for newsreel display.

At HMV the Gramophone Company, the head of the advanced development division, G.E. Condliffe, had, for some time, been investigating the potential of television for that company also and, on 18 March 1931, just weeks before the merger with Columbia to form EMI, Condliffe wrote a memorandum to Stanley Preston in the patents division at HMV regarding the latest Scophony specifications. Condliffe wrote: 'Returned herewith description of the Scophony System, and two Patents (No. 328,286 & No. 333,548). The information supplied by this company has been examined very carefully, and in our opinion it is very doubtful if they have anything of value to offer. In their type-written description, they show a sound appreciation of the technical problems evolved, but their claims to invention of great importance are based on very vague statements, which indicate that the desired results cannot be obtained by the methods proposed.' Clearly, the 'race' to develop television was becoming more intense.

It was the merger of these two companies, the Gramophone Company with the Columbia Graphophone Company Limited to form Electric and Musical Industries, EMI, a new holding company, which took place on 21 April 1931, that was the single most important event as far as the story of Alan Blumlein is concerned. Alfred Clark, chairman of HMV became chairman of EMI, and Louis Sterling, chairman of Columbia became managing director.

As the depression began to take hold, and spread worldwide, the merger of these two competing record companies into a joint venture was seen as an ideal way to benefit from the enormous experience that both had to offer. The new company would have at its disposal a wealth of scientific talent the like of which had not been seen before in Britain. From Columbia came Isaac Shoenberg, Alan Blumlein, Peter William Willans, Edward Cecil Cork, Herbert Edward Holman, Henry Arthur Maish Clark, Eric Arthur Nind (who had been a student that Blumlein had helped while both were at City and Guilds and while Blumlein was working as an assistant demonstrator with Professor Edward Mallett), and others. From the Gramophone Company came Cecil Oswald Browne, W.D. Wright, George Edward Condliffe, John (Jack) Hardwick and William F. Tedham.

David Sarnoff, who had sat on the board of directors of the Gramophone Company since 1929 (when RCA Victor had acquired a controlling interest in the company, since taking over the Victor Talking Machine Company), now took his place on the EMI board of directors. In April 1931, EMI were ready to tackle television.

# Chapter 5
# EMI and the Television Commission

Electric and Musical Industries, EMI, a new holding company, was formed on 21 April 1931, by merging The Gramophone Company with The Columbia Graphophone Company Limited. Alan Blumlein and all of his colleagues at Columbia now became a part of a vast scientific and engineering team at the new company whose interests were enormous at that time. They included audio recording as well as television which, through their partnership with RCA Victor in America, was about to dominate the next eight years of research.

## Co-operation

In April 1931, RCA began to send their new partners, EMI, several of Zworykin's new kinescopes in order for W.D. Wright to attempt to build a receiver that could pick up the Baird television signals (being broadcast from Long Acre and Brookman's Park). Although the results were only minimal, Cecil Browne reported that the new tubes gave a bright image, some 6 inches square, with very good illumination. EMI now set about their first task as a company, to find a way of perfecting an all-electric television system based upon the Zworykin/RCA kinescope. It was decided that the format for the EMI system should be based roughly upon that which Browne had demonstrated in January 1931, at Imperial College, namely 120-lines at 24 frames per second using a mirror drum scanner. William Tedham was put in charge of the project, as he had been working on photoelectric cells for the Gramophone Company. He was assisted by J.W. Strange and H. Neal, with H.G.M. Spratt in charge of the receiver design.

In September, EMI sent Captain A.D.G. West to visit the RCA Zworykin laboratories in Camden, New Jersey in order to assess the progress that had been made. West had been chief research engineer for the BBC in London, and so he was well qualified to report upon his return that: 'Television is on the verge of being a commercial proposition. RCA intend to erect a transmitter on top of a New York skyscraper, and if all is well to market their receiving apparatus in the autumn of 1932'. The skyscraper West mentioned was of course the newly completed Empire State Building, tallest structure in the world at the time. The results he observed were of a picture 6 inches square with a quality similar to that of an ordinary cinema if viewed from the back of the theatre. RCA Victor intended to sell the self-contained receiver cabinet, when ready, for a sum of about $470 (£100).

West made his report to the directors of EMI on 28 October 1931. They reacted by immediately authorizing the purchase of a complete RCA Victor transmitting system at a cost of some £13,000 ($50,000), with a £2000 installation charge. On 2 December 1931, West sent a second telegram from Camden in which the price of the transmitting system was decreased slightly but only as a result of the purchase order being enlarged greatly. Broken down in dollars it read: 'Television transmitter $28,600; Two sets television transmitter tubes (RF output) $4,810; Sound transmitter $13,000; Two sets sound transmitter tubes (RF output) $2,860; Sound studio Equipment $5,850; Motion picture & studio scanner combined $23,620; Two kinescope tubes $260; Two sets photoelectric cells for studio pick-up $2,080; Television sight and sound receiver $2,652; Grand Total $83,732 (£21,770).'

However, the purchase was vehemently opposed by Isaac Shoenberg who had been made, as promised, head of patents for EMI. Shoenberg's reasons were two-fold. First, he was reluctant to rely too heavily on RCA, believing, even at this early stage, that with the wealth of talent available to EMI they could produce a better system of their own given time. Shoenberg had received reports drawn up by Condliffe and Browne dated 6 November, and 12 November 1931, respectively, in which the 'Programme of work for advanced development of Television' had been assessed. Condliffe had divided the work into three categories: television, photoelectric cells and hill and dale recording and reproduction, each of which was further broken down into: progress in hand, staff that had been allocated to the work and the expected developments in these fields.

Secondly, there was the problem of Marconi Wireless and Telegraph who, as a part of the other agreements made before EMI came into existence, were also in receipt of all of RCA's patents in England. This concerned Shoenberg as it had been rumoured that Marconi were interested in joining forces with Baird Television. After all, they had both produced mechanical television systems at that time, and were in the process of developing that system further. As Shoenberg had been an employee of Marconi, he knew of their expertise in television transmitters, and undoubtedly saw this as an opportunity to bring them into the all-electric television development picture, hopefully as partners. Besides, the development of the cathode ray receiver could not proceed without a source of test signals, and a radio link of some kind was necessary.

Condliffe made enquiries as to whether the Marconi Company had such a transmitter available that could be modified for television use to suit the needs of the EMI R&D teams. Marconi had a 9.74 meter, low-power SWB-type Tx used for experimental transmissions between Sardinia and Rome, that was based at Chelmsford, and informed Condliffe, who in turn reported to Shoenberg on 14 December 1931, that it was 'available to be moved to Hayes immediately' if necessary, and it could easily be modified for the 6 or 8 meter wavelength transmissions that EMI were working on. EMI and Marconi agreed upon a deal, and the transmitter was delivered to EMI headquarters on 25 January 1932. It was the beginning of a relationship that would lead to the production of the EMI-Marconi High Definition Television System over the next few years.

Shoenberg was certain that a high-definition system could be developed by the team of scientists and engineers at EMI. He would write in later years:

In deciding the basic features of our system, we frequently had to make a choice between a comparatively easy path, leading to a mediocre result, and a more difficult one which, if successful, held the promise of better things. When we started work in 1931, the mechanically scanned receiver was the only type available, and that was under intensive development. Believing that this development could never lead to a standard of definition which would be accepted for a satisfactory public service, we decided to turn our backs on the mechanical receiver, and to put our effort into electronic scanning.

In the meantime, John Logie Baird had visited the research laboratories of General Electric and Bell Telephone where he had been shown both companies' latest developments in all-electric television. On 25 October 1931, Baird reported that he had not been very impressed with what he had witnessed, remarking that he saw 'no hope for television by means of cathode ray tubes'. Preoccupied with the development of his mechanical system, Baird's words would come back to haunt him over and over again in the next few years. It is very likely that his lack of foresight during this visit set Baird Television back for years, only finally accepting defeat for the low definition system, and adopting the all-electric high definition one, in late 1933. By then however, the Baird Television Company had been sold and any advantage they might have gained had long since been lost.

EMI applied for their first television patent on 19 November 1931, it was for an improved method of forming a fluorescent screen on the large end of the CRT, without the use of binders that created problems with the high vacuum in the tube. The method put forward conducted the electrons from the screen itself, back to the anode in order to prevent accumulation of electrons on its surface. This patent was of immense value to EMI as it represented the point where the original RCA know-

Figure 5.1
Hans Gerhard Lubszynski. (*Courtesy of EMI*)

Figure 5.2
Leonard F. Broadway.
(*Courtesy of EMI*)

Figure 5.3
William Spencer Percival.
(*Courtesy of EMI*)

ledge, which had been given to them as part of the merger, had been improved upon to a point where EMI could manufacture their own tubes for television use. RCA now began to furnish EMI with many of their patents in order to assist them with the development of a practical high definition television system.

The RCA patents were being sent to EMI and to the offices of William Tedham, who was totally dedicated to his task of developing the all-electric television system. Tedham had vast knowledge of all the aspects of television development to that point, low and high definition, and especially the work of Campbell Swinton. He had been busy since the spring working on the production of a high vacuum CRT that could be electrostatically focused. His work had been assisted by Alan Blumlein and Michael Bowman-Manifold, who were developing much of the electric circuitry that would be required, such as amplifiers, scanning circuits and high voltage supplies, and it is this research that saw Blumlein brought into the television picture for the first time.

Though still ostensibly working on his binaural system and its associated components (the patent for which, No. 394,325 was applied for on 14 December 1931), it is very likely that Blumlein had, with his typical interest in all aspects of electrical engineering, been closely following the work of Tedham for some months. Blumlein had shown and explained to Cecil Browne his binaural system already, and both Browne and Blumlein thought it could have applications in film and television. In an atmosphere such as that which prevailed at Hayes in those first months of EMI, and with

Figure 5.4
Eric Lawrence Casling White, from a
photograph taken in 1935. (*Courtesy
of Eric White*)

so much groundbreaking work going on, it is inconceivable that department heads, and especially Shoenberg, would not have informed Blumlein of where his help and advice might be needed in the future. William Tedham had been working closely with Cecil Browne at this time, on the building of the silver–silver oxide caesium photoelectric cells, that were needed for the mirror drum system (similar to that which Browne had originally demonstrated), working at 120-lines.

Shoenberg now began a recruiting programme of people with academic research experience who could be of use to EMI. Among these would eventually be included Joseph Lade Pawsey, Hans Gerhard Lubszynski, Jack E. Cairns, Leonard F. Broadway, Eric Lawrence Casling White, William Spencer Percival, Frank Blythen, Bernard Randolph Greenhead, Leonard Klatzow, Ivan L. Turnbull (who had of course been with Alan Blumlein at City and Guilds when he started there in 1921), W. Turk, Sidney Rodda and F.H. Nicoll. By the time he would be finished the formidable team at EMI consisted of 32 graduates (nine with PhDs), 32 laboratory assistants (of first year BSc standard), 33 instrument and toolmakers, glass blowers (for vacuum tubes, etc.) and mechanics, 8 draughtsmen, and 9 female assistants. They were all placed under the charge of G.E. Condliffe who was research manager.

On 1 January 1932, EMI employed the services of a young research student from Cavendish, James Dwyer McGee, who had been working under such men as Lord Ernest Rutherford and Sir James Chadwick, to assist Browne and Tedham in their field. McGee, who was an expert in the field of hard cathode ray tube development, would go on to become another pivotal member of the EMI team that would perfect the high-definition television system over the next few years. When he had been informed of the position at EMI, McGee had inquired of his supervisor, Chadwick, if he should take the job. 'Well, McGee', Chadwick replied, 'You had better take this offer, since jobs are scarce. I don't think this television business will ever come to much – but it will keep you going until we can get you a proper job.'

One of the primary hurdles that needed to be overcome was that which concerned the DC component of television signals, the importance of which had been known about for some time. The DC component represents the mean brightness of the whole picture, and can vary considerably, e.g. from a night street scene to a sun-lit countryside. AC coupled amplifiers lose it, with bad effects both at the radio transmitter and at the receiver display. To avoid overloading the transmitter on a signal from a mainly black picture with a few highlights, and at a the same time to be capable of dealing with a mainly white picture with a few black parts, without the DC component present and without cutting off the sync signals, the allowable signal excursion from tips of syncs to white is only about 55% of what it can be with the DC component present. This analysis refers to 'positive modulation' where the carrier level increases with brightness. Similar limitations apply to the display, unless the viewer is constantly adjusting the black level.

The DC component is that part of a current waveform which is direct current and with certain types of waveforms this can be simulated. This is done by adding and subtracting sinusoidal (AC) waveforms of differing values to which a direct current is added, thereby achieving a 'rectified' waveform. This is sometimes known as a 'smoothing circuit' as the application of a DC current to the sinusoidal AC waveforms has the effect of calming and smoothing their motion. In effect, a frequency cancellation process has taken place.

Figure 5.5
James Dwyer McGee, from a photograph taken at the inaugural Shoenberg Memorial Lecture in 1972. (*Courtesy of EMI*)

Figure 5.6
Jack E. Cairns. (*Courtesy of EMI*)

In a television system, the modulated frequency associated with the transmitted picture meant that at the receiver, the viewer would constantly be adjusting the brightness or black level of the screen. If a way could be found of reinserting the DC component, this would no longer be necessary. Peter Willans came up with the solution. In his patent (of 13 April 1933), Willans showed how the DC component could be reinserted provided the datum signal was either a maximum or minimum signal, the basic principle of DC reinsertion being that, provided care is taken to include the datum signals which would (were the DC component retained), be at a constant level, the DC component may be lost, and then subsequently recovered by bringing the datum signals to a fixed level.

In other words, if a peak rectifier were to be used, when the datum signals tended to rise, the peak rectifier charges its associated blocking capacitor, and counteracts this rise. If the datum signals tend to fall, the charge in this capacitor leaks away through its associated leak resistance, and this fall is counteracted. In practical terms, if the picture signals are positive and the sync signals negative, black will have an intermediate level. Willans' patent overcame this and while it was not perfect it did work; the problem of wild fluctuation in the brightness and contrast of the transmitted pictures had been solved, the DC component could be transmitted with the picture information. This had a two-fold effect, first, the transmitter operated much more efficiently because the signals no longer drifted about as the picture changed and, second, the transmitter could be more fully modulated.

Though Shoenberg had always intended to achieve the ultimate goal of an all-electric television system, EMI, at first, did not turn their backs fully on mechanical elements of the television system. After all, mechanical television had developed by this point to a degree whereby quite reasonable pictures could be transmitted and received as Browne had demonstrated at Imperial College in January 1931. In mid-1932, Browne, Morgan, Hardwick and Wright had achieved transmissions from a film processing system they had developed up to 120-lines, though the 25 frames per second scanning rate still produced a disturbing flicker on the poorly illuminated image. Browne had presented a paper on his 'Multi-Channel Television' system, before the Wireless Section of the Institution of Electronics Engineers on 5 January 1932. Intriguingly one of the panel, before whom the paper was discussed, was Robert Alexander Watson-Watt, who would, a few years hence, go on to invent radio direction finding. The Browne system most probably represented the high water mark for the mechanical television system, for though EMI had now long since decided to pursue a higher-definition system, the last main protagonist for the mechanical television method, John Logie Baird, had finally succumbed to the pressure.

On the same day in January 1932, as Browne presented his paper to the IEE, Baird Television Limited, seriously crippled by lack of finance at a time when the depression was at its depth, had been sold to Isadore Ostrer of Gaumont-British Films, following a voluntary liquidation. Baird's remarks following his trip to America coupled with his vocal scepticism of televised pictures on short-wave, despite all the evidence to the contrary had left his backers in a very precarious position. In order to save the total financial collapse of Baird Television, the only option had been to sell out.

This meant that Baird Television could continue to work, unhindered by constant financial pressure, though he no longer retained control of the company he had

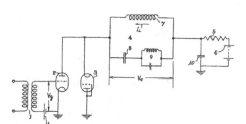

Figure 5.7
Detail from Patent No. 400,976
(1932), Blumlein's first television
patent, 'Resonant Return Scanning'.

founded. On 1 June 1932, Baird Television took a mobile unit to the races at Epsom and positioned cameras in the Grandstand and on the winning post for 'The Derby'. The transmitted picture feeds were given to the BBC who relayed them on a wavelength of 261 meters for the small, but enthusiastic home consumer market, while another transmission was relayed to the Metropole Theatre, where audiences could watch the races live, quite a novelty in 1932. Baird Television claimed some success for the day's work, tests having shown that pictures were received (albeit of dubious quality), some 120 miles away. Not long after this publicity stunt, Baird was offered better studio facilities by the BBC at last. Given a room in the sub-basement of Broadcasting House, at the top of Regent Street, the installation of the apparatus began in July and the first transmissions took place on 22 August 1932. The year before, in America, the other serious contender to RCA Victor (and especially to Vladimir Zworykin), Philo Farnsworth, had also reluctantly had to sell out. Farnsworth, like Baird needed financial backing to continue his work and so during the week when David Sarnoff finally decided to go to San Francisco to buy Television Laboratories, Farnsworth was in Philadelphia selling his company Philco.

In April 1932, Alan Blumlein filed his first EMI patent specifically aimed at solutions for television, it was for 'resonant return scanning' (British Patent No. 400,976, 'Improvements relating to oscillatory electric circuits such as may be used, for example, in connection with cathode ray tubes'), or in other words an oscillatory electric circuit which can be used for a time base in a CRT for the purpose of deflecting the beam of electrons. The patent describes a relatively simple scanning circuit which is suitable for use with hard valves. (A 'hard valve' is one where the amount of gas in the tube is so low that it has substantially no effect on the operating characteristics of the valve. The current is carried by electrons not ions. A 'soft

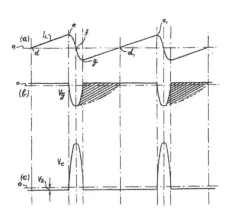

Figure 5.8
Detail from Patent No. 400,976 (1932)
showing the resulting
waveform.

valve' is one in which the valve has virtually two conditions, conductive or non-conductive, due to the presence of gas in the tube. This form is therefore unstable.)

Blumlein does however describe in the patent how the circuit might be used with a Thyratron valve. (A 'Thyratron' or gas-filled triode, is another form of valve in which the gas in the tube is sufficient to entirely affect the characteristics of the valve. When the gas is ionized, due to the anode to cathode potential, the anode current suddenly changes to a high value.) It had been common practice up to that point to charge a capacitor slowly through resistance and then to discharge it very quickly through a Thyratron (gas-filled triode, usually using mercury vapour). The capacitor voltage was applied directly to the deflection plates of the cathode ray tube.

With Blumlein's method, a current is produced for magnetic deflection, allowing the valve to produce a conducting current, which grows linearly through the coil, until a sync pulse cuts it off. The circuit then performs a half-cycle of oscillation at a relatively high frequency, forming a fly-back. Excess oscillation is controlled by the conducting of a second valve in the circuit, which limits the rate at which the current flowing through the coil, in a now negative cycle, can decrease.

On 19 July 1932, another of the essential pieces of the high definition television system jigsaw puzzle was put into place by Randall C. Ballard of RCA. Ballard applied for a patent for what he called an 'intermeshed' scanning television system. Essentially, Ballard had come up with a practical way of reducing the flicker associated with television pictures whilst at the same time allowing for the increasing of the number of scanning lines without increasing the width of the frequency channel. We now call the system he came up with 'interlaced' television, and it is the basis of all television pictures to this day.

Even though Zworykin's kinescope and newly produced iconoscope were capable of producing bright, clear pictures at the receiver, at the desired (full-motion) rate of 24 frames per second, there was still a distinct flicker due to the combination of high brightness with the relatively low frame rate. The idea of using interlaced pictures and increasing the frame rate was new for television, though it had been known for some time that motion pictures could have flicker eliminated by means of a multi-bladed shutter (which effectively gave a frame rate as high as 72-frames per second). This was not possible with a television system however.

Ballard came up with idea of using a Nipkow disc perforated with an odd number of holes (for this patent he chose 81 holes), which meant that an odd number of lines was produced giving two sets of frames. These were coincided or meshed with one another in the following manner: as the beam swept line 38.5, an impulse caused the vertical sweep generator to move the beam back to the top of the frame. It would arrive there in the middle of line 40, causing the next line to intermesh between line 1 and line 2. This would continue until line 79, at which point another impulse would cause the beam to return to the top of the frame again, this time arriving at the start of the frame at line 1. Then the process repeated itself.

Ballard's patent was passed on to EMI and it quickly became adopted by all the research and development departments of both EMI and RCA. However, despite the system being quite ingenious, several problems began to be noticed. The first was noticed by RCA who were working independently of EMI using a horizontal scanning method. It was observed that even when the standard 24 frames per second scanning rate was raised to 48 frames per second, this was not fast enough to eliminate flicker on the received picture. The 60-cycle power standard used in the United States also caused a noticeable ripple effect in the picture, though this was not so much of a problem in Britain (where the 50-cycle standard had little noticeable effect).

EMI had also tried to use the interlaced system at the end of 1932, but with far greater problems than those encountered by RCA in America. EMI were using a vertical scanning mirror drum at the time, and all of their experiments proved to be total failures, the picture appearing as nothing more than a series of lines moving across the screen at right angles to the scanning lines. It was decided that the resolution of 120 lines which had been chosen when the companies had merged (as that had been the target most had been working to at that time), was insufficient and as a result was raised to 180 lines (non-interlaced) at the end of 1932.

The obvious answer to the problem was to eliminate the mechanical element in the process altogether, that is the mirror-drum, and substitute an all-electric television camera. It was to this end that Zworykin had been working for some time, in fact ever since his visit to see Philo Farnsworth at Television Laboratories where he had been shown the image dissector. While Farnsworth had made his achievements public through such bodies as the Television Society, which was the accepted practice, Zworykin's work at Camden had been kept very secret by Sarnoff who did not wish for RCA's rivals to gain any advantage from the work being carried out there.

Whether this secrecy extended to the research team in England headed by Tedham has never been sufficiently established. Certainly many patents passed between RCA Victor and EMI during 1931 and 1932 supposedly including the work on the iconoscope, yet Tedham and McGee always denied that they had full knowledge of Zworykin's work. Regardless, they continued with their research throughout 1932, on a way of producing a practical electronic television camera which would eventually lead to the Emitron camera.

On 25 August 1932, Tedham and McGee applied for a patent for 'Improvements in or relating to Cathode Ray Tubes and the like' (British Patent No. 406,353); they had made a breakthrough. The EMI men had come up with a proposal for the depositing of a substance through a wire mesh in order to obtain a great number of insulated photoelectric elements. A photoelectric ring was formed around the mosaic and connected in a manner whereby it could be used to test the sensitivity of the tube as it worked. Tedham and McGee wrote:

> Television systems are already known in which at the transmitting station, scanning is effected with the aid of a cathode ray tube. In such systems, an image of the object to be transmitted is optically projected upon a photo-electrically active anode of a cathode ray tube, the anode being scanned by the cathode beam. The anode may consist of a glass sheet coated on one side with a 'mosaic' of small, insulated, elemental areas of oxidized silver having a coating of photoelectric material such as caesium.

EMI decided that they should go ahead and commission the building of such a tube to see what effects they could ascertain from it. There was however, a major problem that was written into the agreement between EMI and RCA. It seems that RCA had forbade EMI from working in the field of communications, as all electronic camera work was to be done specifically in Camden, in the laboratory of Dr Zworykin. On

22 August 1932, Isaac Shoenberg became director of research at EMI, while Captain West had decided to leave EMI to become chief sound engineer at Ealing Film Studios, leaving G.E. Condliffe to become head of the Advanced Development Division. Tedham and McGee knew that in attempting to build their tube they would be going strictly against company policy; yet they were faced with the dilemma of whether or not to tell Shoenberg and Condliffe, who after all had championed them all along. In the end they decided not to tell them, instead they would construct the tube on their own, telling only the few who 'needed to know', and wait to see what results they obtained.

The tube that Tedham and McGee had patented was a very complex piece of engineering and required a spherical bulb of Perspex glass about 7.5 inches in diameter with a long thin neck. In order to manufacture the device, they would have to produce it themselves with the facilities available within the EMI laboratories. A polished flat glass window was sealed on to the inside of the bulb so that an undistorted optical image of the scene to be transmitted could be focused on a mosaic by means of a series of lenses.

The photoelectric mosaic was formed on the surface of a sheet of mica only about 0.001 inches thick and roughly 4 inches by 5 inches in dimension. Though the first tube that Tedham and McGee constructed was crude by later standards, they realized that this sheet of mica needed to be carefully chosen for uniformity of surface, it needed to be, as best as possible, free of any blemishes. The opposite side of the mosaic had a layer of conducting, highly reflective metallic 'liquid silver' applied to it by means of a stencil mesh placed above the target mosaic with a density of 50 meshes to the inch.

Once the silvering process had been accomplished a scanning electron gun was inserted into the thin neck of the tube with the beam aimed at the target mosaic and fixed in position. This electron gun consisted of an indirectly heated oxide-coated cathode placed immediately behind an aperture in a modulating electrode. Two anodes extended forward along the neck of the tube away from the cathode towards the Perspex glass bulb. The first anode extended from the modulator, away from the cathode in the middle of the tube. The second anode was formed by coating the inside of the narrow neck and a portion of the Perspex glass bulb of the tube with a layer of silver.

Caesium vapour was now admitted to the tube and the entire apparatus was baked for some time at 180°C to 'cure' it. Once this had been done and the tube cooled, all the gas was pumped out and when highly evacuated the tube was sealed off. Tedham and McGee had to go and borrow a signal amplifier from Browne for the purposes of testing their device, however once a few minor adjustments had been made, an image appeared. McGee later explained that 'as if by magic a picture appeared'. It would seem that both men were so excited by their results that neither thought to take a photograph of the apparatus or the picture image that they had seen.

Very quickly the image began to fade as the gas from the electron gun which had built up inside the tube became too much for the photo-cathode. The experiment had been a success from a certain perspective. There was still the matter of who they could tell of what they had done, and how they had done it. Both men decided it was better if the tube be packed up and put away with no further tubes built until the time was right. That time came in late 1933, when Zworykin's iconoscope finally became public knowledge. Tedham and McGee could then explain what they had done, by which time of course they had the blessing of EMI to go on to build as many experimental tubes as they wished. Though no actual corroboration exists to back up the work that was carried out, McGee insisted that neither he nor Tedham had any knowledge prior to carrying out their work, of the work being carried out by Zworykin and RCA.

Further patent applications now followed. On 12 October, W.D. Wright lodged an application for a patent to overcome the keystone shape distortion of a cathode ray tube in which he planned to project an image on the side of the anode facing the cathode ray by means of a simple optical arrangement. This keystone shape occurs because the electron gun has to be set at an angle to the mosaic to allow the optical system to be normal to it. A series of lenses and mirrors were tilted in a manner which produced an image that was larger at the top than at the bottom, thereby correcting the keystone distortion. RCA too had men working on the same problem and Richard Campbell, Arthur Vance and Alva Bedford had also applied for patents to overcome the distortion problem, though their solutions had differed from Wright's. Later still, Michael Bowman-Manifold invented a better solution by suitably correcting both line and field scanning waves.

EMI now felt confident enough about their system to offer it to the BBC on 29 November 1932, for a demonstration which would hopefully show them the benefits of the EMI film transmitter and cathode ray tube receiver alternative to the Baird Television system. The BBC accepted EMI's offer, and arranged for the demonstration to take place on 6 December 1932, a fact which was publicized. The Baird Company, hearing of the demonstration, was infuriated by the possibility of a relationship between the BBC and EMI/RCA and made their objects known to Isaac Shoenberg. He, in turn, was greatly offended by Baird's objections and ordered all further publicity stopped. The demonstration of a '200-line television system' (it was actually 130-lines, although the EMI standard at the time had recently be raised to 180-lines), took place on the appointed day.

The BBC were impressed with the EMI system, seemingly coming from nowhere, and from a company which was, to all intents and purposes, just over a year and a half old. They invited EMI to move their equipment into BBC headquarters at Broadcasting House, at the top of Regent Street, which EMI accepted, planning to do so in early 1933. Thus began a long, bitter rivalry between Baird Television and EMI that would not be resolved for another five years.

## Rivalry

In early 1933, EMI proposed to the Post Office that they should be allowed to go ahead with a television service of their own, which could utilize the existing BBC ultra-short-wave transmitter (with a few minor adjustments), and with equipment supplied by the HMV division of the company. They explained that, given the go ahead from the Post Office and the BBC, they could have production of receivers completed by the autumn of that year (which was to say the very least a bit on the optimistic side). EMI had only one proviso, they would not consider working in any way with Baird Television.

The Post Office, disturbed by the growing rivalry between the two companies, were none the less curious enough to see how close EMI were to achieving this goal. They sent a team of engineers to Hayes, in February 1933, to see for themselves the latest developments that EMI had come up with, and perhaps whether or not they could deliver on their proposal. The Post Office engineers reported, somewhat surprised, that the images EMI were transmitting were very good, and that the overall impression gave an easy to follow picture with limited flicker. Despite this however, the Post Office decided that EMI should not be allowed to begin a television service until Baird Television had also been given the opportunity to further demonstrate their latest developments.

Baird Television, now quite vocal in its disapproval of the EMI system (and evidently upset by the progress that had been made by them in such a short period of time), now stated that the EMI proposal violated the agreement between Baird Television and the BBC. They explained further that 'dealing with EMI would be dealing a heavy blow at British industry and directly assisting an American concern', that EMI were in fact, nothing more than a subsidiary of an American company, namely RCA Victor. It was this fervent nationalism that was to become the Baird argument against EMI for the next three years, stating on numerous occasions that the Baird Television system was 'all-British', while EMI were nothing more than a shop window for RCA. The Post Office however were not convinced, and suggested that both companies be given the opportunity of demonstrating short-wave television as soon as the necessary arrangements could be made.

The dual demonstrations took place on the afternoon of Tuesday, 18 April 1933, just four days before Alan Blumlein and Doreen Lane got married. The EMI system transmitted from Hayes, while the Baird system transmitted from their offices in Long Acre. After the demonstrations had been compared, the engineers of the Post Office were in total agreement that the EMI system was far superior to that of the Baird system. Yet despite this, they still did not grant EMI use of Broadcasting House as they considered it might prejudice Baird's case. EMI, naturally very disappointed at not getting the go ahead for a television service of their own, resolved to return with an even more superior system, while Baird, faced with the undeniable fact that electronic television had been better, finally accepted defeat for his mechanical system. John Logie Baird now changed his mind about cathode ray tubes and decided the only way to beat EMI would be at their own game. The question was, had he left it too late?

The Blumleins' wedding took place at St John's, the Parish Church of Penzance, Cornwall on 22 April 1933. Alan Blumlein however, was working in the office at Hayes until quite late on the evening of 20 April 1933, as Maurice Harker recalled:

'We all knew that he had to get down to Cornwall for his wedding you see, and if he didn't leave that night well, he wouldn't make it. So, at about 7 p.m. I came into his office and there he was hard at work on some point of interest to him and I pointed out that he had better get in his car and get under way, or he would be late getting to Cornwall. He was so wrapped up in the work that we were doing, that he had completely lost all track of the time.'

Blumlein's friend and colleague from Standard Telephones & Cables, J.B. Kaye, was of course best man, and also a witness on their marriage certificate, as was Willie

Herbert Lane, Doreen's father. Kaye recalled the event some years later remembering it as one of the few occasions when he actually saw Blumlein nervous:

'We went to Moss Brothers, and I said, "Here's the bridegroom and here's the best man. We want all the necessary fancy dress", and we fitted it at great speed and skill, looking terribly smart, and then I think Blumlein went off to Cornwall, and I followed the night before the wedding. I think I travelled all night. I got there, and they had a room in the hotel for me, and so Blumlein and I, we both got changed there, and then went down in the lift in our clobber, feeling frightfully self conscious, and anyway, the whole ceremony passed off, according to plan.

I then discovered, to my horror, that the wedding breakfast was being held in a very large hall, which was full of people, and that I had to make a speech to toast the health, I think of the bridesmaids. It was the worst speech I have ever made in my life because I didn't know what an earth I was to say. It's terribly difficult to make speeches of that nature, it's something I've avoided ever since, so it was my first and last appearance of that sort at a wedding! And it really was a very successful married life I think I can say that without any contradiction from anybody. Very, very happy. And Doreen, with her sense of humour, and fun, recognized she was coping with someone who was not the average cabbage that you may get in industry, and that he was effervescent, and she gave him all the latitude he required for his studies.'

When he married Doreen, Alan Blumlein had been living at No. 32, Woodville Road, Ealing, but with his new status it was time for a move. Towards the end of the summer of 1933, the Blumleins moved into a small house at No. 7 Courtfield Gardens, West Ealing, where they would live for the next two and a half years.

During the spring of 1933, the EMI engineers had lodged many patents, which would prove vital to the development of their high definition system. Shoenberg by now had come to the conclusion that Blumlein should work on the television system as soon as he had concluded his work on the binaural system which had yielded so many revolutionary ideas. Isaac Shoenberg was one of very few men who, like Blumlein, actually understood the system; and yet he realized that it was so far ahead of its time, that it could only be developed slowly, over the coming years. Blumlein, Shoenberg decided, could be put to much better use designing new circuitry for the far more important television system, and re-designing existing circuitry that could be improved. In the early summer of 1933, Alan Blumlein, who had almost certainly been made aware of the goings on of the television development team, now became a part of it.

Peter Willans had filed his patent for reinserting the DC component in April and shortly afterwards left the team. It was at about this time that Eric Lawrence Casling White joined. In May, McGee applied for a patent for a two-sided camera tube which used short aluminium wires with a thin insulating coating of aluminium oxide covered by a conducting film of silver. The emissions from the screen were collected by an earthed grid with a collector set behind the screen to collect the secondary emissions, helped by an amplifier.

Blumlein filed his Patent No. 421,546 (Improvements in and relating to the supply of electrical energy to varying loads, for example to thermionic valve apparatus), on

16 June 1933, which essentially compensated a power supply with reactive elements, making it appear to load as pure resistance. As Blumlein describes it himself,

> It is common practice to supply electrical energy to thermionic valves apparatus from a generator through a filter circuit. Such a filter circuit may comprise one or more condensers arranged in series and one or more condensers arranged in parallel. The electrical energy is usually taken from the terminals of a condenser of the filter, and the size of this condenser is usually made such that it offers negligible impedance to the lowest frequency that the amplifier is to handle.
>
> For some purposes, however, for example in television, apparatus is required to operate at frequencies extending effectively to zero, and however large the condenser is made, it has been found that the performance of the apparatus is adversely affected by the variation in regulation of the generator, together with its filter, at different load current frequencies. It is the object of (this) invention to enable electrical energy to be delivered from a source associated with a reactive impedance, to a load which varies at frequencies down to effectively zero, without the supply voltage varying with the frequency.

It was, in effect, an impedance network again, and Blumlein, drawing upon his years of experience with telegraphic applications and the circuits he had designed for telegraphy, now applied a similar thinking.

More important still, on 11 July 1933, Blumlein, Browne and Hardwick applied for a patent that improved the work that had originally been carried out by Willans on DC restoration. There had been limitations in Willans' original patent and though only three months old, Blumlein, Browne and Hardwick had studied the circuitry and come up with a superior system (British Patent No. 422,914). As described in the patent itself,

> In television or the like apparatus, having means for the reinsertion of direct and low frequency components into electrical signal variations, comprising picture signal, recurrent maxima or minima and synchronizing signals in the form of current trains of oscillation, in which the reinserting means cause the signal variations to take place about a zero line coincident with or separated by a predetermined fixed amount from the peaks of the said maxima or minima, there are provided means for preventing the said trains of oscillation from affecting the reinserting means.
>
> Thus, the reinserting means may be located in an auxiliary channel branched from the main channel, through which the signal variations are transmitted, and a filter circuit, or the like, may be provided in the auxiliary channel to remove the trains of oscillation from the signals fed to the reinserting device. The latter develops corrective signals, which are fed into the main channel, the corrected signal variations in this main channel still containing the trains of oscillation which are required for synchronizing purposes.

It seems that the DC component reinsertion question fascinated Blumlein and it was a problem that he would return to again and again, with several variations leading to further improvements.

That month RCA sent Zworykin to London to visit the laboratories of EMI in order to help Shoenberg set up facilities to build English iconoscopes under the direction of McGee. EMI had finally modified their agreement with RCA with regard to communications development, and were now able to manufacture as many camera tubes as they wished. Zworykin had finally publicly announced his developments with the iconoscope in a paper presented on 26 June 1933. In the paper, he outlined a brief history of television, related how mechanical scanners could not get enough light to produce an image outdoors, and how the iconoscope, by means of charge storage in individual cells released by a scanning cathode beam was the only way forward.

In the spring, Tedham fell ill, and so McGee took over the direction of this part of the EMI development programme. With Zworykin's iconoscope achievements now finally public knowledge, McGee could proceed with his camera tube development in the open. During the illness, which lasted for several months, Tedham was temporarily replaced by Leonard Broadway who, like McGee, was another of Shoenberg's 1932 academic recruits, though this seems to have suited Tedham who was primarily a chemist rather than a vacuum physicist.

McGee and Tedham, despite their earlier work on camera tubes (still a closely kept secret at this time), always had great difficulty in understanding the actual mechanism by which the picture signals were derived from the pick-up tube. In the months that had passed, McGee had studied the problem and come to the correct conclusion that secondary emission phenomena played a vital role in the signal generation. The early tubes that Zworykin had developed and which were now at the disposal of EMI, were operated with a scanning beam of comparatively high velocity which in turn gave rise to a large amount of secondary emissions from the mosaic elements as the beam fell upon them.

Under these circumstances, the potential of the elements became unstable and practically anything could happen, the process was not under control. If an element rose in potential, it lost more secondary emissions than primary electrons received from the scanning beam. Alternatively, if the potential of the elements fell, it would lose fewer secondary electrons than primary ones supplied by the beam. The release of secondary electron emission then was governed by the energy of the scanning beam. As this was the case, or so it seemed to McGee, these same elements should react to, and behave according to, the voltage applied on the tube. As the effect of secondary emissions on the tube was to produce a 'shading' effect on the picture, the control of secondary electron emission became a vital factor in producing an image of a high definition.

This shading typically took the form of unwanted superimposed signals (and thus brightness) increasing from top to bottom and from left to right (or vice versa). Looked at on a waveform monitor, the signals were sawtooth-like, and christened 'tilt'. There was usually a noticeable curvature of the sawtooth shape christened 'bend'. Both EMI and RCA provided sawtooth generators of adjustable amplitude to cancel the unwanted components. While RCA took the route of an adjustable Fourier series of harmonics to cancel and 'bend', EMI took the more pragmatic course of using successive integrals of the sawtooth and in practice only needed parabolic components. This eventually became Patent No. 462,110. Each camera channel in the final equipment had to have its own 'tilt' and 'bend' controls, adjusted by hand (and often to taste) when there were significant changes of picture content.

On 8 August, William Percival, Cecil Browne and Eric White applied for Patent No. 425,220, which formed a frame pulse around the television picture, which prevented 'hooking' at the top of the image. It was not the first time that the synchronization information had been sent with the picture; ever since experiments with CRT displays had started it had been common practice to send line sync pulses in the period allowed for electronic line scan fly-back. The problem was how to send frame sync information with the picture. In early 1933, a rounded frame sync pulse was in use at EMI, superimposed on the line sync train by means of a low-pass filter. Experiments were being done to determine the optimum amplitude of frame pulses relative to the line ones, bearing in mind that only a limited amplitude for the two together was available. No ratio was satisfactory; either the frame sync was too small, or the reduced amplitude of the line pulses in order to accommodate the frame pulses was inadequate and led to imperfect line synchronization during and after the frame pulse, the effect called 'hooking' at the top of the picture.

The idea of a 'slotted' frame pulse, giving a continuous train of equal amplitude front edges, which was all that was needed for line synchronization, and with a few line pulses broadened to serve for frame synchronization, is given in Patent No. 425,220. Various receiver circuits were devised to provide frame sync utilizing low-pass filters, and not all by EMI. The scheme that was eventually to be adopted by RCA, especially for interlacing, is more complicated in order to allow the use of a low-pass filter if desired.

In an effort to counter the advances that had been made by EMI, the director of Baird Television now employed the services of Captain A.D.G. West from Ealing Film Studios where he had been working since leaving EMI in August 1932. Baird wanted West not only for his knowledge of the EMI early development work which included the kinescope and Zworykin's first iconoscopes, but to set up and run as technical director of Baird Television, a cathode ray tube development programme. Captain West accepted the position and as one of his first duties, wrote to Noel Ashbridge explaining that Baird Television were working on a 120-line cathode ray tube television system which, by early September they felt ready to show publicly for the first time. At this time, Baird Television also took out a long lease on the south tower of the Crystal Palace where they would transmit picture information on 6.05 meters and sound on 6.20 meters.

With the life span of the 30-line transmissions (which had continued since 1929), now obviously outdated, the Post Office announced that they would cease on 31 March 1934. The Baird 'Televisor', the receiver that picked up these transmissions, had originally been sold in kit form. With Baird's interests in Germany at Fernseh, a total of around 20,000 Televisors had been sold in Britain and throughout Europe, all of which would soon become obsolete. The British Post Office however, had finally decided that the only way to settle the matter of which system was the best, would be with a series of tests.

On 18 September 1933, Blumlein and Henry Clark applied for Patent No. 425,553 which specified by name the use of a negative feedback loop in an amplifier, to reduce the output impedance of pentode valves. This circuit is especially useful as it improves upon a theory which had been known since January 1932, when Nyquist of Bell Labs had outlined the general conditions under which an amplifier becomes stable when feedback is applied. Negative feedback is usually attributed to H.S. Black who outlined the method in January 1934, though the Blumlein and

Clark patent however, pre-dates Black by some four months with a method that combines both current and voltage feedback. Black had used an identical formula to the one which Blumlein and Clark came up with, but had only calculated the linearity of the voltage gain.

Blumlein and Clark were fully aware that negative feedback would improve the linearity of an amplifier, but they also wanted to maintain a defined output impedance over the specified frequency range being used. This however, presented the problem of current feedback increasing with the output impedance, and as independent control of the gain and output impedance of the amplifier could not be achieved if either of the feedback methods were applied on their own, they came up with a formula and designed a basic circuit that could overcome the problem.

EMI sent W.D. Wright to Camden in November 1933, to visit Zworykin's laboratories to look at the progress that had been made with interlaced picture framing. Wright had been amazed to find that the engineers there had perfected the system to the point where 243-line interlaced pictures at 24 frames per second were being transmitted. Determined to find out where EMI had been going wrong, all previous experiments at Hayes ending in failure, Wright discovered that Zworykin was using the horizontal scanning method as opposed to EMI's vertical one.

He returned to England armed with his new knowledge, and EMI altered their scanning method accordingly to a horizontal one, whereby 243-line interlaced pictures could be shown. They altered the process but a little; the camera tube used was their own and by April 1934, they had the system working well at 243-lines, but at 25 frames per second rather than 24 (this was done in order to lock the field rate of 50 frames per second to the mains. This meant that less care was necessary in receiver design to eliminate mains pick-up affecting the picture). Later in November, Prime Minister James Ramsay McDonald visited the Hayes site to see for himself how far television development in Britain had progressed. He was, by all accounts, quite impressed.

On 29 December 1933, Blumlein applied for Patent No. 432,485, in which he described a method for producing a multi-gun cathode ray tube, which would bring the electrons from the cathode to a spot focus on the screen. The control electrodes were arranged in front of the cathode and at right angles to it, with each insulated from the other and having a separate input terminal. The position of the cathode's focus spot would correspond to only one control electrode, and remain completely independent of the effect of the other control electrodes on the beam. The cathode would therefore produce an independent focus spot for each of the electrodes which, by means of a delay circuit, could be caused to follow in sequence the first spot across the screen.

In this way, each focus spot would respond much more accurately to the controlling impulses upon it, which in turn determined light and shade in the picture signal. As each phase of the controlling electrode would be different from the adjacent electrodes, by applying a delay circuit to the output of the controlling electrodes, the phase of the impulses could be determined and the resultant picture image would become clearer as light and shade were identified.

This must have been the most fascinating time to work at a place like EMI for somebody like Alan Blumlein. His insatiable appetite for all things electrical not only helped the company, but also furthered many a project to which he had no real connection at all. This all came about when, on occasions, he got himself into quite a bit

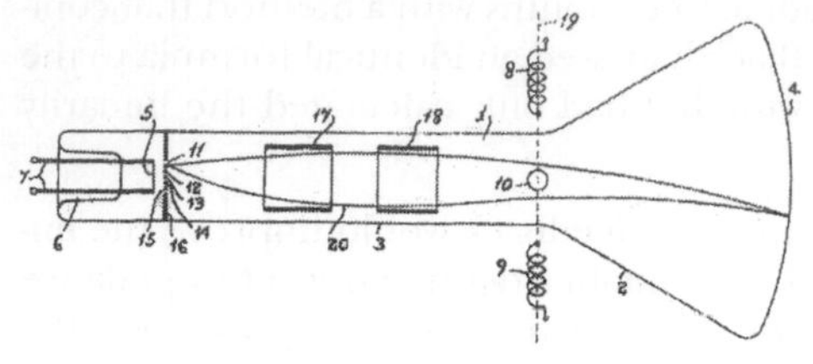

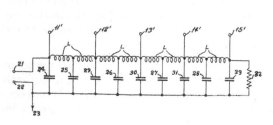

Figure 5.9
Detail from Patent No. 432,485 (1933), a multi-gun cathode ray tube, bringing the electrons from the cathode to a spot focus on the screen.

of trouble. Blumlein had a habit of passing someone's office and looking at the job on which they were working. As often as not, if they were stuck, he would simply write the answer they had been looking for, which appeared to be perfectly straightforward to him, on the back of a scrap of paper or an old envelope, and leave it nearby.

There were times however, when his natural curiosity got the better of him, as James McGee was to find out. Apparently, McGee had been working with some apparatus, which he had set-up in the laboratory where he and his group were working, and then left to go home for the weekend. When he returned to Hayes the following Monday, it had mysteriously vanished. After a brief search, McGee tracked down the missing apparatus only to find it with Blumlein. It would seem that he had been passing McGee's office, noticed the apparatus set up there, and promptly decided to try it out. Becoming totally engrossed in the equipment, Blumlein had disassembled it, taken it to his office and continued to work on it there.

Naturally somewhat annoyed by this, McGee returned the set-up to his laboratory and filed an official complaint against Blumlein, who, horrified that he might have upset a fellow worker, apologized profusely, explaining quite truthfully that he was blissfully innocent of any misdoing. He explained that he had only been curious as to the workings of the equipment, and intended only to help in some way. McGee relented of course, and seeing the innocence in Blumlein's actions revoked the complaint remarking, '...the most important aspect of the incident was really Blumlein's generous reaction when he realized that he had upset some of the group and myself'. The two men would go on to become firm friends and work very profitably together for EMI, for several years to come.

This was typical of Blumlein however. He would often help in ways which seemed perfectly innocent to him, but which may have offended the recipient gravely without Blumlein even knowing. He was however a gentleman, and a sport, and would always, upon realizing that he had offended, apologize, usually with a short note (on which the formula for solving the problem that the person in question had

Figure 5.10
Detail from Patent No. 432,485 (1933). By applying a delay circuit to the output of the controlling electrodes, the phase of the impulses could be determined and the resultant picture image would become clearer as light and shade were identified.

been working on was written), all of which would be placed next to a small animal of some description made out of bent pipe-cleaners on their desk.

On yet another occasion, recalled by W. Turk who had worked within Hans Gerhard Lubszynski's group, Blumlein had once again been working late and seen a light on in an office and curiosity got the better of him:

'Lubszynski and I had on test a complicated vacuum tube aimed at verifying some particular parameter and had assembled around it the then usual array of accumulators and HT batteries to feed its many electrodes via wander plugs and multicoloured wires. We had tried for several days to get the result we wanted and knew was possible, but the meters just wouldn't behave. The day in question had been particularly frustrating. Lubszynski and I were still fighting the circuitry at 10 p.m. in the dark when [there was] a knock at the door. It was Blumlein, who apologized for the interruption, said he'd seen a glimmer of light under our door, was worried in case something had been 'left on', drew on his pipe and asked what we were doing.

Lubszynski explained briefly and Blumlein said, "Why not put that lead here instead of there?" Lubszynski did, and everything clicked in place! We were mad! Here was a circuit engineer who knew nothing of tubes (in the heat of the moment we forgot his work with Tedham on CRTs), coming along in the dead of night, with no previous knowledge of the work we were doing, summing it up in a few seconds, and solving the problem which had been taxing us for several days.'

Not everybody however, was quite as appreciative of Blumlein's interruptions, as William Percival recalled:

'Blumlein came in and started giving his opinion and so on. When he'd gone, the section leader said "Oh, what's he doing butting-in to our work?" and it wasn't long before people realized that it might be a good thing that he did butt-in to so many facets of the work. He never seemed to be outsider as it were, butting-in on something.

This ability to cross barriers between scientific philosophies, and between people. Whoever he was talking to he was part of that set-up. Every day he would come round and see what was going on, and talk as man to man to the engineers and boys on the job, such as myself. He wasn't at all stand-offish, he would talk to anybody, no matter how junior; and it was very good for your education – it was certainly good for mine, I think a good proportion of what I know about engineering I learned from Blumlein.'

Philip Vanderlyn also recalled how Blumlein, despite being chief engineer would always help a junior engineer whenever he had the chance. 'He would keep in daily touch with all the work in the laboratories, and he would talk with the most junior of us as equals without ever making us conscious of our lack of status. This was good for our ego and even better for our technical education.'

By now, the development of the CRT camera had been continued by McGee and Broadway to the point where they felt they could explain their earlier experiments to Shoenberg. Naturally concerned that they had infringed upon the relationship with RCA (as it stood at that time), Shoenberg was enough of a realist and a visionary to overlook the matter, and concentrate on the job at hand, improving the iconoscope. By January of 1934, the development team felt confident enough to show the results of their work to EMI chairman, Alfred Clark; at a demonstration attended by representatives of the BBC, the first presentable pictures from an EMI camera tube were shown, with all parties commenting on how good the pictures were.

Not to be outdone, and infuriated that EMI seemed to be developing a rather wholesome relationship with the BBC, Baird Television also arranged a demonstration for the representatives of the BBC. It took place on 12 March 1934, and was, unfortunately for Baird, patently not as good as the EMI one had been; though of course they did not know that at the time. The company organized a series of demonstrations throughout the month of March including chairman of Baird Television, Sir Harry Greer, addressing the stockholders by means of ultra short-wave television, transmitted from Crystal Palace to receivers at Film House, Wardour Street. A few days later, 40 members of parliament also received a demonstration, which for many of them would almost certainly have been the first television pictures they would have seen.

Despite all their enthusiasm for their system, Baird Television must have been devastated by the news which they received from the BBC announcing that they must make arrangements to terminate the experimental broadcasts after 31 March 1934. As if to make matters worse, over the next few months EMI's scientists, with Alan Blumlein most prominent, filed a series of patents, which solved many of the problems which had dogged the further development of the CRT camera-based television systems. On 5 February 1934, Alan Blumlein applied for Patent No. 434,876, which solved the problem of getting enough black level signal from the camera. By cutting off the electron beam during playback, Blumlein demonstrated how the problem of transients in general illumination on the mosaic could be solved.

Because of the importance of the invention, much of the text of the patent has been produced here to outline Blumlein's thinking behind the problem. He describes the invention thus:

> When an image is cast upon the screen, each element of the mosaic emits electrons in proportion to the brightness of its illumination and these electrons pass to the anode. When the cathode ray passes over an element during the scanning operation, the element is charged negatively with respect to the anode and when the rays leave the element the potential of the element rises due to the emission of photo-electrons to the anode. The charge given to each element is therefore dependent upon the amount of photo-electric discharge that has taken place since the last scanning of the element, and consequently upon the brightness of the illumination of the element.
>
> So long as the average illumination of the mosaic screen remains constant, the average current flowing to the signal plate is zero. When the cathode scans an element which had been brightly illuminated, the potential plate is made negative because of the arrival thereupon of charging electrons which are greater in number than the sum of the secondary electrons emitted owing to the bombardment by the electrons of the cathode ray beam and the total photo-electric electrons emitted.
>
> When the cathode ray strikes an element which has not been illuminated

since the last scan, no charge is given up by the ray, the number of secondary electrons emitted being equal to the number of electrons arriving; but at this instant the potential of the signal plate will be positive owing to the steady photo-electric discharge occurring from all the elements. Thus the signal generated across the resistance will represent true differences in light and shade between successively scanned elements but will not give any indication of the average illumination.

In order that average illumination may be transmitted, the DC component of the signal may be established at some convenient point, for example in the manner set forth in co-pending Application No. 11204/33 (This was Willans' patent which was eventually given the number 422,906 when accepted.) According to the method described in this prior application, the signal is caused at intervals to assume maximum or minimum values corresponding to some fixed absolute light intensity and the DC component is inserted at any desired point with reference to these recurring maxima or minima.

These recurrent maxima or minima may be established in the case of a cathode ray system such as that above described, by cutting off the cathode ray at the end of each scanning stroke so that there is no bombardment of the mosaic screen during the return stroke, that is to say, whilst the ray is being returned to one side of the screen after scanning one strip of the image. Thus, during the return stroke, no charge is delivered to the mosaic. The signal generated will not necessarily be zero, but will depart from zero by an amount depending upon the total steady photo-electric current arising from the electrons leaving the mosaic. The signal, however, corresponds to that generated during a scanning stroke, when scanning an un-illuminated element and therefore corresponds to full black. Consequently, by shutting off the cathode ray beam during each return stroke, recurrent minima (or maxima, depending upon the sense chosen), of value corresponding to full black, are obtained, and these minima may be used in the manner described in the co-pending application above referred to, re-establish the average picture intensity.

In the above discussion, it has been assumed that the brightness of the image remained constant. If however, the average brightness should change suddenly, the operation of the tube, during the first scanning cycle after the change, will be quite different from that described above. A sudden change of image brightness will cause a sudden change in the photo-electric current from the mosaic, and this change will cause a momentary additional current in the signal plate circuit. The additional current will be opposite in sense to that associated with the change when considering changes from element to element with the average brightness constant. For example, if the image be assumed to be quite dark, then the whole mosaic will remain fully charged and no signals will be transmitted during the scanning, the conditions during the forward (or scanning) stroke being the same as those during the return stroke when the ray is cut off.

If the mosaic now be suddenly illuminated, there will flow from the signal plate a photo-electric current tending to make the signal plate positive by virtue of the resistance in series with it. The signal so generated would normally be associated with a black portion of the screen and therefore this new signal, generated by a sudden increase in average brightness, is in the reverse direction to that normally associated with brighter illumination. In effect the signal represents 'blacker than black'.

Blumlein, in his patent, describes how these transient signals can cause quite dangerous conditions in the mosaic resulting in the sudden transmission of a full black or full white picture. The transients can also be associated with the overloading of the amplifiers or receiver, none of which is very desirable. Blumlein's solution was to connect the signal plate or anode to an auxiliary photoelectric cell. This cell was connected to a variable tapping point on the resistance, with its cathode connected through a suitable bias battery (or other source of electromotive force), to earth. The auxiliary cell was mounted in a blackened box, through which the scanned image was passed by means of an aperture. By adjusting the resistance control, the relative amplitude sensitivity of the light or dark falling upon the mosaic could be controlled during the scanning process.

Despite the importance of the principles shown in this patent, Eric White recalled that 'it was not found necessary to use it in the Alexandra Palace cameras. A compromise was used in which it was assumed that the darkest part of the picture was true black, by providing a DC restoration circuit working on the blackest part of the picture signal (after excluding the spurious signals occurring during the scanning beam cut-off period).'

On 17 April, Blumlein, this time working with Cecil Browne, filed Patent No. 436,734, which solved the problem of coupling an amplifier to the photoelectric cells on the mosaic without causing a wideband dropping of DC output. This was necessary because the oscillation to be amplified may contain components of very low frequencies, including substantially zero frequencies, as the patents referring to the reinsertion of the DC component had already established.

The difficulty with this was that the amplifier needed to be capable of amplifying equally all frequencies within the range being used down to substantially zero frequency. The currents generated in the picture cell were of a very small magnitude and therefore very high magnification of the picture signal was needed. Blumlein and Browne solved the problem by coupling the control grid of the amplifier valve conductively to the anode of the photoelectric cell, and the cathode of the valve to earth through a suitable grid bias.

The value of the resistance was made so high that the voltage developed across it at low frequencies, for any given change in illumination of the photoelectric cell, was much greater than at high frequencies. The high resistance allowed the low frequencies to be passed on efficiently, but the stray capacities associated with the photoelectric cell caused attenuation of the higher frequencies. In effect the signal-to-noise ratio of low frequencies could be increased, thereby making the amplifier more responsive across a wider range of frequencies such as those being employed in television picture transmission.

## The Television Commission

The rivalry between EMI and the Baird Television Company had now grown to the intensity where something positive needed to be done to determine once and for all which of the two systems was the better. The Postmaster General, Sir Kingsley Wood, proposed that a joint committee, led by the BBC and the General Post Office, should be convened in order to report on the relative merits of the rival systems to the Government. A discussion was raised in the House of Commons on

14 May 1934, with the announcement that the commission would be set up with Lord Selsdon (Sir William Mitchell-Thomson), chairing it. Selected representatives from the British Broadcasting Corporation, the General Post Office as well as scientific advisors (who could be considered knowledgeable enough to pass comment, but who were not directly involved with either EMI, Baird, Cossor or Scophony), would make up the remainder of the committee. Baird, Scophony and EMI agreed immediately to the proposal, naturally believing in the merits of their respective systems. The British Government Television Committee therefore came into being on 16 May, meeting for the first time on 7 June 1934.

During the next seven months the committee members set about interviewing anybody and everybody who they felt might have some influence on a British television system. Notification that a television committee was being assembled was made to the press on 29 May 1934 and again on 11 June 1934, with the clear indication that the committee members were prepared to receive evidence on the subject of television from any interested society, firm or individual during this time. The notifications appeared in the entire major daily papers as well as periodicals such as *Wireless World, Popular Wireless* and *Wireless Constructor*, and by all accounts the response was far greater than had originally been expected.

The committee was obliged to listen to all comers as they had, after all, invited any interested parties to take part. In the end they had to interview some 38 witnesses on the subject, amassing a huge amount of information and documentation, some of which was easily dismissed as fanciful, while other aspects would greatly influence their eventual findings.

The first individual to be seen by the committee was Isaac Shoenberg on 8 June 1934. Shoenberg put forward the argument for EMI pointing out that their system was based on just three primary components: the DC component, which had been overcome by Willans; the sending of the synchronization information with the picture, which had been patented by Percival, Browne and White; and the interlaced picture framing, which had been based on the invention by Ballard (Ballard invented the idea of interlacing and one mechanical method, but the exact synchronizing waveform used in the EMI all-electronic system is an EMI invention, and different from that used by RCA). The Baird Television system it seemed was based around four patents, this was the case put forward by Captain West. Baird accused EMI once again of being nothing more than a subsidiary of the American company RCA, to which Shoenberg told the committee every piece of the EMI system had been manufactured in England, including the cathode ray tubes.

This was actually true, but only from a certain perspective. EMI did by now manufacture all their own CRTs, but there was no escaping the fact that seemingly every RCA patent had been made available to them during the process. One of the primary components of the EMI system was Ballard's basic, but much modified invention of interlaced picture framing. Shoenberg claimed, somewhat surprisingly, that EMI had no prior knowledge of Zworykin's iconoscope and that the process for making mosaics had been their own. This was almost certainly untrue, as it would have been inconceivable that RCA had not imparted something of the development work at Camden to EMI. Nevertheless, that is what the committee were led to believe.

This then brought up the question of the developments that had been taking place abroad, especially in the United States by RCA, Farnsworth/Philco and Bell Labs, and in Germany by Fernseh A.G., and Telefunken among others. It was decided that the committee should visit the Telefunken research and development laboratories in both countries and therefore trips to both the United States and Germany were arranged. In the meantime all the interested parties sped up the development of their respective systems hoping to gain an advantage over the others before the committee came to a decision.

August 1934 would prove to be a very productive month for EMI and Alan Blumlein in particular. Two patents were applied for by him that month, No. 445,968 with Edward Cork and Eric White, and No. 446,661 with James McGee. Blumlein was by now obviously considering television on a much wider plane than his counterparts who were working, in the most part, on specific areas of the problem. Blumlein took a different view as we have already seen, and became involved with many aspects of the development problems and Patent No. 445,968 applied for on 17 August 1934, demonstrates this.

For some time there had been discussion between the various members of the EMI team, those who were dealing with transmission and reception of the television signals, regarding the problems associated with aerial arrays. Blumlein had already been granted an earlier aerial array patent (No. 432,978 applied for on 7 February 1934), with Joseph Pawsey, in which they had outlined the potential of an aerial array which could be used for the transmission or reception of radiated signals with the maximum portion of that radiation directed towards the receiving station. This had been done by comparing the gain efficiency of a non-directional radiator with an aerial array where the radiation was kept to a maximum by incorporating good vertical distribution. In other words, the maximum radiation in the direction of the receiver was achieved by ensuring that the power was radiated horizontally rather than just up and down as had been the case previously. This invention was to become known as the 'slot aerial', and some years later Blumlein would later find great use for it when developing airborne radio detection finding systems.

In order to achieve this, the elements of the aerial array (which could be assembled vertically or horizontally), were spaced as to have their electrical centres separated by less than a quarter of a wavelength, with at least two successive elements being electrically phased almost in opposition (so that their radiation or reception characteristics were substantially neutralized in one direction – the unwanted direction), and would work in order to develop the desired directional radiation pattern (in the wanted direction), thus allowing a much improved radiation or reception of the signal and a great saving in area taken up by the array.

In Patent No. 445,968, a different aspect of aerial arrays was considered. In the matter of short-wave reception, an aerial was often connected to a receiver by means of a feeder which might comprise one insulated conductor and one conductor earthed at one or more points along its length. This feeder would then often be arranged centrally within an earthed sheath and insulated from it. Blumlein, Cork and White considered the problem of how to receive both short-wave and long-wave signals simultaneously, such as those which make up a television signal (the picture being transmitted on short wave, while the sound was on long wave). In the past this had been achieved with two aerials which would obviously prove far too cumbersome for a consumer television receiver, so they proposed an aerial system in which the feeder would have two conductors, for example a conductor surrounded by a sheath. The feeder would be at one end with a short-wave aerial at the other end. If the sheath was also a conductor, this could be connected to earth or at an earth point of the receiver, and if this was done through a series of tuneable circuits the sheath conductor would be effectively earthed at a short wavelength and insulated from earth at a longer wavelength.

Blumlein, Cork and White also provided for a short-wave receiving aerial coupled to a feeder consisting of at least two conductors, arranged to transmit the short-wave signals to the receiver in phase opposition on the two conductors (these being insulated from earth), and adapted to operate in parallel and in like phase as aerial and lead in for medium or long-waves by connection to a suitable receiver. This work on the aerial arrays would prove to be invaluable later on, and it was Alan Blumlein again, who would give EMI an even greater edge than their television competitors when, on 3 August 1934, he applied for his patent, with James McGee, for the invention which would become known as 'cathode potential stabilization'. British Patent No. 446,661, is one of Blumlein's most important simply because it solved the problems that McGee and Tedham (and later Broadway), had been having with cathode ray tube secondary emissions.

Although Alan Blumlein was not a vacuum physicist, he had frequent discussions with McGee on the subject, beginning soon after the announcement that McGee and Tedham had succeeded in developing their own tubes. Blumlein and McGee however, had been working with the early tubes that Zworykin had developed, and were now at the disposal of EMI. These were operated with a scanning beam of comparatively high velocity, which in turn gave rise to a large amount of secondary emissions from the mosaic elements as the beam fell upon them. This caused the element to rise in potential if it lost more secondary emissions than primary ones, while alternatively, it could fall in potential if it lost fewer secondary electrons than it gained primary ones.

It was already understood from McGee's earlier work with Tedham and Broadway, that the secondary emission factor was entirely dependent upon the energy of the scanning beam, which would behave differently dependent upon the voltage put on the tube. The secondary emissions were being caused then, by a 1500-volt bombardment of the scanning beam, and that these emissions increased rapidly as the velocity of the bombarding electrons increased.

Blumlein and McGee proposed with cathode potential stabilization that, if a low velocity beam was substituted, then secondary emissions would fall off, with the mosaic element receiving a net positive charge which would be lowered in potential relative to that of the cathode gun. Under these conditions, the photoelectric emissions from the mosaic are efficiently collected so that there is no spreading over to the adjacent elements, and therefore none of the unwanted shading effect.

Though there was great promise in cathode potential stabilization, that was not the end of the story. Because EMI needed a practical, working and efficient camera tube as soon as possible, Browne had been continuing his work on the mechanical film scanner, which by now he had developed to a high degree of refinement. When, however, he had been presented with the first of the electronic camera tubes to use (without cathode potential stabilization), the images that had been produced from it at the receiver were so poor that a rather shocked Browne exclaimed, 'What do you expect me to do with signals like these?'

It could easily have been the end of EMI's interest in the development of the electronic camera tube, and they may well have continued the work perfecting the mechanical film scanner, were it not for the persistence of McGee, Blumlein and Shoenberg, all of whom were fully aware of the potential that such a camera could hold if only they could get it to work properly. Bearing in mind just how much was at stake, with the television committee already in session and about to leave for their respective trips to the United States and Germany, it took a great deal of faith on the part of the staff at EMI to press on with their electronic camera tube development. Had they not, of course, the history of television may have been very different indeed.

Having dealt with the question of 'blacker than black', Blumlein now turned his attention to the question of the synchronization signals that were 'whiter-than-white', in his Patent No. 446,663, of 4 September 1934. A picture signal corresponding to 'full white' was defined in terms of the brightness of the image being scanned, an arbitrarily determined amount, which could in the case of film be considered to be a completely transparent section of the film or in the case of reflected light, it was when the signal was produced from a perfectly reflected part of the object. Whiter-than-white pulses occurred where the component signal was fed to the receiver at a brightness amplitude beyond that of the range of the device, causing flashes on the screen. These pulses could be caused by intense interference and Blumlein's invention was aimed at preventing or eliminating them.

If a whiter than white signal of this kind was applied to the control electrodes of a receiving cathode ray tube, periods of very bright illumination would sometimes occur when the synchronizing pulses were present, over amplifying the signal, and the flashes at the receiver would occur. This situation should be avoided wherever possible, so Blumlein came up with another of his impedance networks, which provided a means of limiting the amplitude of the signal beyond a predetermined amount. That amount was usually set by calculating the 'picture white', that is the most desirable amount of white that makes up the transmitted picture, and this was always be maintained at a low value. Therefore, if a secondary channel containing a limiting device was arranged alongside the standard picture signal channel, in which the limiting action of the second channel was suspended until the amplitude in the primary, picture channel exceeded the given limit, a form of signal attenuation would occur. Blumlein developed a circuit that inverted the tips of the whiter-than-white pulses, attenuating them regardless of whether the peaks had been caused by interference, and re-amplified them to produce black on the screen rather than white.

On the same day, 4 September, that Blumlein had applied for this patent, another was also lodged (which became No. 448,421), which also rates among his most important, the 'cathode follower'. Deceptively simply titled 'Improvements in and relating to Thermionic Valve Circuits', the cathode follower circuit provided a much greater input impedance to the thermionic valve. As M.G. Scroggie points out in his September 1960, article, 'The Genius of A.D. Blumlein', in *Wireless World,*

A form of negative feedback that especially appealed to him was the cathode follower. He did not actually originate this configuration, but was the first to appreciate its great value particularly, in television. His specification 448,421, of 1934, sets forth with characteristic clarity how, by virtue of its very low input capacitance, it can be used to advantage in connecting a high-frequency, high-impedance source, such as a photocell, to an amplifier. Furthermore, it anticipated by 15 years the discovery that it can be used to eliminate almost entirely the shunt capacitance of the source and connecting lead. Note too the use of a screened coupling; long after this, people were still saying that pentodes were unsuitable for cathode followers!

On 18 September, Blumlein, Blythen and Browne applied for Patent No. 449,242, which would later become known as 'DC restoration clamping'. This is another of Blumlein's classic patents, which provides a simple yet absolutely necessary component in the high-definition television story. Essentially, the patent describes the process of using an AC amplifier to amplify electrical variations of frequencies down to zero, or DC, where the DC component represents the average brightness over the television picture. What this patent provided was a novel way of introducing an AC amplifier to do the job, that is to say an amplifier which in itself was incapable of amplifying DC.

This could be done by introducing the DC component as separately transmitted channels of low frequencies down to zero, which, with the aid of a channel incapable in itself of transmitting these components caused the input and output of the channel at suitable intervals to assume predetermined absolute values. This is 'clamping', which restored the level of some part of the signal intermediate between black and white to some suitable constant voltage. In particular, the short bit of black immediately following the sync pulse, is fixed in this way. For some purposes this has advantages over simple DC restoration on tips of syncs, but it necessarily requires the use of an electronic switch which can conduct in either direction on receipt of a pulse, but in the absence of a pulse is non-conducting in both directions (this is best demonstrated in a later Patent Application No. 512,109 of 24 December 1937, by Blumlein and Eric White).

Clamping could be used to accurately determine the DC restoration, which in turn determined the average brightness of the television picture. This then improved the effectiveness of the transmitter because the signals no longer drifted about as the picture changed and the transmitter could be more fully modulated. Reception was greatly improved because constant readjustment of the black level by the viewer was no longer needed and the overall picture itself became far more stable.

October was another productive month for Alan Blumlein with three more applications filed for patent specifications. The first two, Patent Nos. 447,754 and 447,824, were both applied for on the same day, 26 October 1934, and are yet more examples of how diverse Blumlein's thinking could be in such a short space of time. Seemingly unrelated subjects on which he might be thinking about at any given point, could lead to a momentary inspiration and yet another invention specification.

In the first of these, co-written with Herbert Holman, Patent No. 447,754, describes an invention to insulate electrical conductor wires by placing them within a slender glass tube, and goes on to explain how the wire (which should have a diameter of about 0.05 mm), is placed within the glass tube (the diameter of which was about 0.15 mm), which in turn is then sealed at both ends and evacuated of almost all its air. The glass tube is then heated and drawn out with the wire inside it; the heat causes the glass tube to soften and collapse on the wire, thus insulating it.

In Patent No. 447,824, also co-written with Herbert Holman, the insulation idea is again used, this time to overcome the problems associated with the electrically conductive mosaic screen of the type used in a television camera (CRT) plate. Difficulties had arisen when constructing these electrically conductive mosaic screens arising from the fact that the distances between the centres of adjacent holes in the grid needed to be very small. This in turn meant that it was very difficult to coat the grid with an insulating material without filling up the holes of the screen. This was further hindered by the fact that the material had subsequently to withstand being baked in a vacuo, and of course needed to be a good insulator itself. Blumlein and Holman patented a method for coating the weaved wires that made up the mesh of the grid with the drawn glass tube insulation material. These covered wires, now insulated sufficiently, made up a grid, which not only solved the problem of providing a suitable insulator, but also retained the necessary space between the wires to allow the flow of electrons to pass through.

The third patent applied for that October was given the number 449,533, and its application dates listed as 24 October 1934 and 18 April 1935. The reason for the two application dates is unusual but simple to explain. During the application process Patent No. 449,533, which was written by Alan Blumlein and Michael Bowman-Manifold, was altered and improved before the full specification had been left with the Patent Office. Under normal circumstances, if a patent is improved upon, then a totally new patent number results. In this case however, Blumlein and Manifold were working so fast that they improved upon their earlier effort before the original could be granted a specification of its own. Therefore the improved version became a part of the original and hence the two application dates.

Patent No. 449,533 deals with the need for a deflection coil yoke for a cathode ray tube, to produce a magnetic field of uniform flux density. The deflecting method in a CRT often comprised two pairs of deflecting coils which would cause the ray to trace out a desired path (in television this would normally be a number of successive parallel lines, each slightly below the preceding one), on the fluorescent screen associated with the tube. These coils would be arranged to produce deflection of the ray in mutually perpendicular directions, one direction taking place at a higher frequency than the other, by feeding into the coil a current of a suitably chosen frequency (usually of a sawtooth waveform).

Deflection due to one pair of coils commences as the beam enters the coil system,

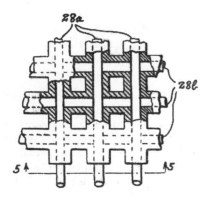

Figure 5.11
Detail from Patent No. 447,824 (1934) co-written with Herbert Holman showing the insulation of electrical conductor wires by placing them within a slender glass tube.

gradually increasing, in a parabolic path, until it leaves. Thus, there is significant deflection before leaving the coils, and unless the magnetic field produced by the other pair of coils is uniform, their deflection sensitivity will vary with the instant-aneous deflection due to the first pair, and vice versa. Typically, what is notionally a rectangular scanned area on the tube face becomes either barrel-shaped or pin-cushion-shaped. The actual shape of the scanning waveform is immaterial, provided its amplitude is constant.

Producing deflecting coils of this kind without introducing distortion had been very difficult up to that point. Blumlein and Manifold's rather clever solution to the problem of producing uniform magnetic fields was to wind coils so that they filled a space between two imaginary cylinders with their axes displaced slightly. Efficiency was improved by encasing the coils in a cylinder of magnetic material, e.g. winding soft iron wire round them. This produced a uniform magnetic sheath around the body of the CRT in which the scanning beam was directed and deflected. When the current was applied to these two cylindrical coils, it flowed in opposite directions through the windings, and because of their slight displacement and shape, at any given point along the cylinders the flow of the beam on the surface of the tube traced out a closed curve. Each turn of the coil was arranged so that it had two active portions that lay on opposite sides of the longitudinal axis of the tube. In this manner, the direction of the current flowing through any one coil was always being produced in an opposite direction to the current flow of the other. This method set up a uniform magnetic field in which distortion was almost totally absent.

Meanwhile, the British Television Committee that visited the United States arrived on 29 October 1934. It was headed by its chairman, Lord Selsdon, with F.W. Phillips, assistant secretary, General Post Office, Noel Ashbridge, chief engineer of the BBC, and Colonel A.S. Angwin, assistant engineer-in-chief, General Post Office. After holding a meeting with the members of the Federal Communications Commission in Washington, explaining the purpose of the visit, they left for Camden and the lab-oratories of Vladimir Zworykin at RCA Victor.

The fact that the Americans seemed very willing to aid, in any way they could, the members of the British Television Commission, is not as strange as at first it might seem. First there was the well-known arrangement between RCA Victor and EMI, and any help that RCA might be able to give the members in relation to EMI would be to their advantage. In Camden, Zworykin showed the committee members a working version of a 343-line system, which could work both indoors and outdoors at 30 frames per second with very reasonable images.

Second, they visited the Television Laboratories Limited of Philco Radio and Television in Philadelphia which, despite Baird Television's constant protestations that their system was 'all-British', had been co-operating with Baird for some time. It seems that the association had developed when it had occurred to Captain West, soon after he had joined Baird Television, that Fernseh had been working on a CRT which might prove suitable for their new cathode ray-based system. Fernseh however, were not very advanced in this field at the time, and so West turned his attention to the only other source of cathode ray tube cameras (other than Zworykin of course), Philo Farnsworth's image dissector at Philco.

With undeniable irony, bearing in mind how vehemently Baird had attacked EMI for dealing with an American company, Baird Television now approached Television

Laboratories Limited, in Philadelphia, to provide them with the cathode ray tube camera that they needed to compete with EMI/RCA. The members of the British delegation were shown the Philo Farnsworth image dissector-based system which also, it would seem, gave a good, clear picture.

The British delegation to Germany left for Berlin on 5 November 1934, it also had four members headed by O.F. Brown, from the Department of Scientific and Industrial Research. They were welcomed in a similar manner to the party which had travelled to the United States, with the German broadcasting authorities, also controlled by the Post Office, the Reich's Rundfunk Gesellschaft, all very keen to impress the British with the achievements of the engineers at Fernseh and Telefunken. By this time the Germans had perfected a 180-line television system which was transmitting every day, though, as Archibald Gill (later Sir), one of the members of the delegation later reported, 'It was plain that very few had receivers at that time'. Although the Nazis had only been in power for just under two years, television was already controlled by the Propaganda Ministry. Goebbels had realized its potential and taken control.

The year 1934 had been an amazing year for television development with an extra-ordinary amount of work produced by scientists all over the world. In England, at EMI, Alan Blumlein's last patent of 1934, was applied for with Edward Cork on 6 December, and became Patent No. 452,713. It relates to the need for suitable cables for high-frequency electric current, and would prove to be the first step and a determining factor in the process to produce video and VHF cables for television, a subject that Cork and Blumlein would return to several times in the years to come.

The cable consisted of a central wire core (preferably of a springy material such as cadmium bronze), which was bent so that it zigzagged its way through the middle of a tubular rubber sheath. The rubber served to protect and insulate the inner wire core from an outer conducting sheath or metal braiding such as copper which wrapped around the entire cable. Any cable of this kind had the advantage that it would stretch both the outer conducting material and the rubber insulation, without breaking the inner central conductor. The bending of this central wire at an angle (suggested to be 45° degrees, producing a slope of 90° degrees), meant that the capacity of the cable could be increased as well as the inductance. In essence therefore, the inner wire core is mostly self-supporting, and substantially surrounded by air, which acts as a dielectric. It sounds so very simple, and indeed, it is; yet, nobody had thought of it before, and of course it was immediately manu-factured and used in many applications thereafter.

On Monday, 14 January 1935, the report of the Television Committee was published in the form of a paper presented by the Postmaster General to Parliament as

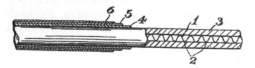

Figure 5.12
Detail from Patent No. 452,713 (1934) co-written with Edward Cork showing suitable cable for high-frequency elec-tric current. This would prove to be the first step in the process to produce video and VHF cables for television.

Command Paper No. 4793. It was available to the general public, as are all such papers presented to Government, at a price (then) of 6d.

The document opened by explaining that the purpose for the commission had been: 'To consider the development of television and to advise the Postmaster General on the relative merits of the several systems and on the conditions under which any public service of television should be provided'. It went on to outline the fact that the press had been notified well in advance of the commencement of the proceedings, that 38 witnesses had been interviewed, that consultations had been entered into with various departments of the British Government, as well as interested parties, and that the members of the committee had travelled to the United Stated and to Germany (articles 1–8).

Following a brief outline of the basic principles of television (articles 9–19), the report explained that the Post Office had always been willing to give support in the form of licences to those able to conduct experiments in television transmission, but that this support had been restricted to a purely research and experimental basis. It was pointed out that any system which had showed sufficient promise to consider it for public transmission, should then come under the jurisdiction of the BBC, who would be responsible for providing suitable facilities on a trial basis (articles 20–21). The report then explained the differences between low (articles 22–26) and high (articles 27–34) definition television, with reference to the fact that Baird Television had been transmitting regular low definition images since 1929, on a 30-line basis, but, that these had ceased on 31 March 1934. Low definition television was described as 'the path along which the infant steps of the art have naturally tended and, while this form of television doubtless still affords scientific interest to wireless experimenters, and may even possess some entertainment value for a limited number of others, we are satisfied that a service of this type would fail to secure the sustained interest of the public generally'.

High-definition television, it was considered, would be determined by a system that produced a definition of no less than 240 lines, at a minimum picture-scanning rate of 25 frames per second. Though the standard attained during much of the experimental work that had been carried out had only been 180 lines, the members felt it was apparent from the trips to the United States and Germany that far higher rates were possible with interlaced picture framing, and therefore they decided that no less than 240 lines should be the rate; though they did temper their objective by saying that this should not preclude the possibility of a higher rate being achieved, and a frequency of 50 frames per second being used (article 28).

Such high degrees of definition could only be obtained through the use of very high modulation frequencies which, it was recognized, could only be used in practice by radio transmitters working on the ultra-short wave band where the effective range was much more restricted than that of the medium wave (article 29). A high definition television system was also determined as one where the receiving device contained no moving parts, as in a cathode ray tube (article 30). The size of the picture produced was considered, with mention given that most cathode ray tubes of the time were producing images around 6 inches by 8 inches (article 31). It was pointed out that various prices had been quoted by the interested parties as to the cost to the public of a receiver (article 32), and that these ranged between £50 and £80. A lot of money in 1935. Though it was also foreseen that when mass production started, once popularity rose, the price should fall rapidly.

The committee members had also considered the issue of television with or without sound transmission, concluding that there would be little point in any service without sound (articles 35–38), the matter of who should control television broadcasts was also outlined (articles 39–40), concluding that any transmissions in the public interest should be the responsibility of the BBC. This gave the BBC quite a monopoly over all aspects of broadcasting in Britain, and therefore the committee tempered their findings with the recommendation that an advisory committee, appointed by the Postmaster General, and working in conjunction with the Department of Scientific and Industrial Research and the BBC, should be set up to exercise control over the actual operation of a television service (articles 41–44).

As a high-definition television service would only be practicable with transmissions on ultra-short waves, the report outlines the problems associated with this, commenting upon the availability of suitable wavelengths between 3 and 10 meters (articles 45–47). The members of the committee felt however, that the eventual goal of a general television service in Britain should be the ultimate aim of all parties concerned, tempered with the addition of the fact that calculations of any costs that would surely be incurred, would, at that stage be purely speculative (articles 48–49). The report then came to the choice of system and the patent difficulties associated with such a service (articles 50–54). These were the key issues as far as EMI and Baird Television were concerned and have been reproduced here in full:

50. We have been furnished with a great deal of information – much of it of a confidential character – concerning various systems of television. Continuous progress is being made in the art; and even during the few months of our investigations, research has brought a number of new and important discoveries. We do not think it would be right at this early stage of development, when practical experience is small and the patent position obscure, that we should attempt to pass final judgement on the several systems of television. A technical description of each system which we have examined in this country, indicating its distinctive features and commenting upon its performance, is however, submitted for your information in Appendix IV. Comments are also made in Appendix III on the systems examined in the United States and Germany.

51. The task of choosing a television system for a public service in this country is one of great difficulty. The system of transmission governs in a varying degree the type of set required for reception; and it is obviously desirable to guard against any monopolistic control of the manufacture of receiving sets. Further, whatever system or systems are adopted at the outset, it is imperative that nothing should be done to stifle progress or to prevent the adoption of future improvements from whatever source they may come. Moreover, the present patent position is difficult: the number of patents relating to television is very large, and in regard to many of them there are conflicting views as to their importance and validity.

52. At the same time it is clear from the evidence put before us that those inventors and concerns, who have in the past devoted so much time and money to research and experiment in the development of television, are looking – quite fairly – to recoup themselves and to gather the fruits of their labours by deriving revenue from the sale of receiving apparatus to the public, whether in sets or in parts, and whether by way of royalties paid by the manufacturers or by manufacturing themselves. It is right that this should be so, and that the

growth of a new and important branch of industry capable of providing employment for a large number of workers, should in every way be fostered and encouraged to develop freely and fully.

53. The ideal situation, if it were possible, would be that, as a preliminary to the establishment of a public service, a patent pool should be formed into which all television patents should be placed, the operating authority being free to select from this pool whatever patents it desired to use for transmission, and manufacturing being free to use any of the patents required for receiving sets on payment of a reasonable royalty to the pool. We have seriously considered whether we should advise you to refuse to authorize the establishment of a public service of high definition television until a comprehensive patent pool of this type had been formed, on terms considered satisfactory by the advisory committee. From the evidence we have received, however, we are convinced that, under present conditions, when the relative value of the numerous television patents is so largely a matter of conjecture, the early formation of such a pool would present extreme difficulty. The Government would have no power to compel an owner of television patents to put them into the pool against his will; and, with the best will in the world, patent holders might find it exceedingly difficult to agree among themselves on the fair basis for charging royalties and sharing the revenue so obtained. An attempt hastily to negotiate a pool under these conditions would in all probability end in failure.

54. While, however, we have been compelled to abandon the idea that the formation of a comprehensive patent pool should be a condition precedent to the establishment of a public service, we are strongly of the opinion that it is in the public interest, and in the interest of the trade itself, that such a pool should be formed. In framing our recommendations we have kept this objective in mind; and we trust that events will shape themselves in such a way as to lead to the formation of a satisfactory patent pool at no distant date.

In determining when the start of such a service should take place (articles 55–56), the committee came to the conclusion that logically, transmissions should begin in London, as it seemed that most of the greater part of the capital city could be covered by one transmitting station, and that two systems of television could be operated from that transmitter. Those two systems were the ones proposed by Baird Television Limited and Marconi-EMI Television Company Limited. The service should be begun with a series of extensive trial broadcasts side by side, transmitted alternately and not simultaneously, and that both companies should be given an equal opportunity to prove the validity of their equipment. Certain terms and conditions were laid out at this point including that the price demanded should not be unreasonable, that the BBC should be indemnified against any claim for infringement of patents and that additional equipment should be allowed to be introduced if the advisory committee recommended it.

The development of the service was also presented (articles 57–59). The report outlined the need for the service, once begun, to grow gradually with the number of stations being set up determined by the growth of the popularity of the service. As each new service came into being, it should be incorporated with all the latest developments in the art, provided these could easily be incorporated within the existing stations as best as possible. If an entirely new system of television was to be invented in the future, the members recognized that it might be necessary to adopt that method if it was proven to be superior to the existing service, though at some point a standardized system of national broadcasting for television transmissions should be attained.

The final articles of the report dealt with the programmes that any television service should carry (articles 60–61), at which point the members felt they should not make any detailed recommendations other than expecting any transmissions to attain a wide public appeal. The financial considerations of the service were covered also (articles 62–71), with an expected set-up cost of providing the London station by the date of 31 December 1936 (when the BBC's current charter expired, and also when they expected the first regular broadcasts to begin), would be £180,000.

No cost inclusion was made for the production of programmes or the duration of the transmission period each day, that was to be left to the participating companies to decide. The committee did however, raise the issue of payment for the service by the public, after the initial outlay for the receiver, by means of a television licence. It was recognized that at first, only a small minority of people would be able to receive television transmissions, and therefore could not be expected to bear the brunt of the cost of such a service through a licence fee. It was suggested therefore, that the existing licence fee of 10s should suffice, at least during the first experimental period of broadcasting.

The question of relaying television broadcasts via radio was considered (article 72), with no reason seen why this should not take place, and, lastly, the fact that the members hoped that private experimentation would continue to be supported with adequate facilities given for such enterprise (article 73). At the end of the report a synopsis of the findings of the committee members (article 74), proceeded a list of the 38 individuals who had been interviewed (Appendix 1). In conclusion (article 75), the members thanked the secretary, J. Varley Roberts for his assistance, the document was dated 14 January 1935, and all signed accordingly.

The Television Commission's findings were made public and the relevant parties made ready for the coming year. Surely 1935 would determine which of the two main contenders would triumph. EMI and Baird Television both felt certain they would succeed. The High-Definition Television period was about to begin.

# Chapter 6
# The high-definition television period

When, in January 1935, Lord Selsdon's Television Commission made their report to Government, it was concluded that the only way a fair determination of which television system should succeed in providing a regular service for the British public, was if both were tested over the coming year. Both Baird Television and Marconi-EMI now geared themselves up for the coming battle which would determine who should finally achieve the goal of a regular high-definition television service, due to commence no later than 31 December 1936.

## The question of a standard

Following the publication of the report of the television commission, the respective companies were given a short period of time to assess the committee members' findings, followed by an invitation to submit their views regarding the standard of definition which they would prefer to adopt. As the committee had recommended that any high-definition system for the proposed public broadcasting of television should have a resolution of no less than 240 lines and 25 frames per second, Baird Television, who were the first of the two companies to respond, expressed their preference for the minimum standard.

This left Marconi-EMI in a somewhat tricky position. Isaac Shoenberg was aware of the fact that Baird Television had opted for the minimum stipulation of the television committee, and felt strongly that in order to further EMI's position he should propose a higher definition system. The question was, how far could he go? Just how much higher in definition could EMI offer the committee members without surpassing the possibilities of their scientific and engineering team?

Shoenberg knew that in order to surpass Baird by any given amount the EMI system would have to push the limits of the current high-definition experiments that they had been working on to the absolute limit. It should be remembered that EMI had only been successfully working with a 243-line, 50 frames per second system since October and that while interlaced picture framing held great promise of much higher definition, there was also the time factor to consider. Shoenberg had no idea if the team could achieve a higher-definition system in the available time in order to beat Baird.

In order to understand how Shoenberg and the team at EMI came up with the eventual figure for the scanning definition of the Marconi-EMI system, it is essential to understand how the number of scanned lines is arrived at, and the consequences of changing, raising this figure.

The number of lines in a scanned television picture obviously relates to the number of components within that picture that make up a single frame. When recalling the method first demonstrated by Ballard at RCA for interlaced picture framing, it should be considered that the number of lines would be chosen as being a multiple of small odd integers, which facilitates the electrical interlocking of the line and frame frequencies required for interlacing. The electronic divider chain, which determines the number of lines in a frame consisted of five stages, with each stage initially divided by three, giving $3 \times 3 \times 3 \times 3 \times 3 = 243$. Various other odd number combinations could be chosen which is how, by the end of 1934, RCA had raised their system from 243 lines to 343 lines using the formula $7 \times 7 \times 7 = 343$.

This had all been discussed one Sunday afternoon in November 1934, when Blumlein had invited several of his colleagues to lunch at his house at 32, Audley Road, Ealing, a house he and Doreen had only recently moved to. At the luncheon that afternoon Eric White, Edward Cork, Michael Bowman-Manifold and William Percival, discussed with Blumlein his thoughts about the possible design of an experimental pulse generating system capable of being rapidly switched from 243 lines to a number integer higher than that. Shoenberg now seriously proposed that a much higher number of scanned lines would be achievable and he wanted to surpass not only Baird, but RCA as well. The next obvious odd number integer series meant substituting the last of the series of threes in the 243-line formula with a five; this then gave, $3 \times 3 \times 3 \times 3 \times 5 = 405$. That was it then. The Marconi-EMI system would have 405-lines. In the following two months the system was constructed (mostly in the laboratory, though some parts in the workshop). A basic demonstration of the 243-line system versus the 405-line system was given to Shoenberg just before he made the announcement to the television committee.

When Shoenberg returned to the television committee members in mid-January 1935, just under two weeks after Baird Television had made known their offer, and proposed to them Marconi-EMI's 405-line, 50 frames per second (alternate frames being interlaced to give 25 picture frames per second), fully-electric system, many, including most of the people at EMI, had grave doubts about whether it was attainable. Not only had the stakes been raised some 67% higher than any scanning rate yet achieved up to that point, but the frequency bandwidth had also been raised 2.8 times higher than that of the 243-line system. Another consideration that Shoenberg had to bear in mind was that none of the receivers then available or experimented with were capable of dealing with the bandwidth that a 405-line system would require.

In one of scientific history's most far sighted decisions, Shoenberg felt justified in offering a system which he was sure would be achievable, would offer significantly higher picture definition in order that later developments in receiver design could be employed without changing the standard and would in the process guarantee success for the Marconi-EMI system over that of Baird, RCA or anyone else for that matter might come up with. Many years later he would recall, 'The choice as to the number of lines was no longer limited by mechanical considerations, but by the bandwidths which could be dealt with at the time in the transmitter and the receivers. Great differences of opinion existed in the laboratory, but finally, early in 1935, I took my courage in both hands and chose 405 lines.'

When the members of the television committee were made aware of the Marconi-EMI offer of 405 lines they too, like many in the scientific world, felt that Shoenberg had taken a most unnecessary risk, and so in accepting the offer they proposed that EMI produce a 240-line system to work along side the 405-line system, just in case, pointing out that they were happy with a level of definition that their inquiries had led them to believe would be sufficient. Why therefore face all the difficulties that the much higher standard of 405 lines would present? Regardless, Shoenberg stood by his offer, and in so doing stood by the men who made up the scientific and engineering team at EMI who were about to embark on a process taking them into the unknown.

## An unnecessary risk?

Now that Shoenberg had set the target of a 405-line, 25 pictures per second, 50 frames per second fully interlaced system, it had to be designed and built. With the figure of 405 lines set, it now became necessary to generate a frequency that was the result of the number of lines times the number of picture frames per second: $405 \times 50 = 20{,}250$ cycles per second, which was then divided by 2 to give 25 pictures per second. This frequency had to be generated by an oscillator and then divided into steps again using odd integers as the basis for the calculation: 20,250 divided by 2, then divided by 9 = 1125, divided by 9 again = 125 and finally divided by 5 = 25. In an interlaced system the word 'frame' is used to mean the whole set of lines, 405 in this case, and 'field' for a half set. Therefore in the 405-line system, each field consisted of 202.5 lines, which was 405 divided by 2. The complexity of the task was not lost on those who had chosen to undertake it, though several of them had serious misgivings at the outset as to whether the number was beyond achievable limits. As late as 4 April 1935, an internal confidential report drawn up at EMI and marked 'Not circulated to Committee' (Television Committee), discussed the merits of a 240-line system as well as a 400-line system at 25 pictures per second.

Sadly, the report has no author, though it was very probably written by Condliffe after lengthy discussion with Blumlein, Browne and others. The comparison of a 240-line picture and 400-line picture transmitted on a 1250 kilocycle modulation band resolved that, provided a frequency band that high could be handled, it would pay to transmit on 400 lines rather than 240 lines as regards the immediate quality of the picture. It was also pointed out that another advantage would be that the transmission of a 400-line picture would allow 'the system to remain unaltered, through continuous improvement, for a large number of years. Increase in the picture size is inevitable, with consequent demand for better and better definition.'

EMI had a grasp of the future of television in their hands even before the first high-definition television system had been built.

A day later, 5 April 1935, Shoenberg received the report he had asked for from C.S. Agate in the Television Development Section, Works Designs Department, who would be responsible for the co-ordination of the production of EMI television receivers. Agate voiced a somewhat concerned, though none the less practical opinion about the relative advantages of these two line frequencies:

'A fortnight ago I was doubtful as to whether satisfactory reception could be carried out on 400 lines under conditions approximately similar to practical use. As a result of tests made during the fortnight, I now believe that a 400-line receiver will work without too much skilled attention under ordinary circumstances of reception when tuned up exactly right.

As I read the report of the Commission, the future of a television service is likely to be determined by the results obtained in the 18 months trial period between the summer and the end of next year. Although I believe that technical developments are likely eventually to make 400-line television as reliable a job as 240-line is today, I cannot help feeling somewhat worried lest an attempt to establish 400 lines now may lead to the production of rather unreliable receivers by ourselves, which may involve us in heavy expense and that undesirable heavy service troubles which may get high fidelity television in such ill repute that the decision of the Committee at the end of the trial period may be adverse to continuing experiments.

The prospects of television might not be badly affected by them if most makes of receivers on the market were bad, and ours were good, but there seems to me a real danger that if other makes are terrible, and ours bad, the opinion may grow that high fidelity television is fundamentally too unreliable for practical use. There is a final point, that if Baird is allowed to transmit on 240 lines during the trial period, while we transmit on 400, our 400 transmissions are unlikely to be much better than Baird's 240, while in the opinion of the public, trade etc., the higher number of lines should give very much better results. It seems therefore, that a quite unjust judgement will be formed to the effect that our system is inferior to Baird's because it needs a considerable increase in number of lines to give the same result as Baird's.'

It had been established that 405 lines required a 67% increase in scanning lines and a three-fold increase in bandwidth, it also required a five-fold decrease in signal-to-noise ratio in the signal amplifiers which would present problems all of their own. The various departments at EMI and Marconi were now set specific tasks of the job to complete. Blumlein would be responsible for all aspects of the circuitry design and integration; Browne dealt with the vision input devices, the general design layout of the studio equipment and control surfaces; McGee, the vacuum physicist, dealt with all aspects of the cathode ray tubes for transmission, Broadway dealt with the receiver tubes; and from Marconi came N.E. Davis and E. Green who would be responsible for the radio transmitter specifications. Practically every department of the research laboratories at Hayes was working in some form or another on the development of the 405-line television system.

Shoenberg and Blumlein were undoubtedly the essential players at this time. It had been Blumlein who had called everyone together in November 1934, to consider raising the number of scanned lines in the first place, and it was the belief in Blumlein's electronic genius upon which Shoenberg had ventured so much on the project. There is no doubt that Shoenberg placed great faith in Blumlein; he had, after all employed him to produce the Columbia recording system, and now, these two men, who were close friends as well as colleagues, had weighed up all the possibilities, considered the options open to EMI, and concluded that the system they now proposed could indeed be achieved.

Speaking years later Shoenberg would recall:

'There was a complex of three branches: there was picking-up, the link between picking-up, and the receiving end. As one did very much affect the other, it was necessary that whilst each of them wanted a specialist, it was also necessary that they should work together with the other people. In the beginning, Mr G.E. Condliffe was my right hand man on many problems in many ways but, the supervision and the day-to-day direction of the work on the three elements which I have mentioned before was something which I myself maintained in every way.

The man chosen for the pick-up was Dr McGee, the man for the link and circuitry generally was Mr Blumlein, and the man for the tube was Dr Broadway. These were very brilliant people, who had under them also people who were in their circles brilliant. I would like to say that it is impossible for me to mention all the names who were there, but I mention only the leaders because the great point about them was the close co-operation between them. Owing to the fact that the team worked so closely, of which I was a part on a daily basis, the possibility of getting decisions quickly as to which piece of research should be gone into, which should be dropped, these problems from day-to-day resulted in an atmosphere which is very rare because of its complete absence of any frustration. A very rare thing. We had complete confidence in each other, the complete desire to give all at one head, and that, I believe, is the reason why we had been able to do that very complex job in such a short time, with a very small number of people and comparatively little expense. I was saying to my boys, "We are lighting a candle that will not be put out." '

McGee, reflecting upon this period some 35 years later would recount:

'Shoenberg made what was probably the biggest – and I consider the most courageous decision – in the whole of his career. The cynic may say that this was a piece of gamesmanship planned to overwhelm our competitors. But no one who knew Shoenberg or who was aware of the real state of technical development at that time would give this idea a moment's credence. No! – it was the decision of a man who, having taken the best advice he could find, and thinking not merely in terms of immediate success, but rather of lasting, long-term service, decides to take a calculated risk to provide a service that would last. I confess that even now, 35 years later, the television receiver in my own home provides only a picture of that original, 405-line standard!'

Throughout 1935, EMI had been demonstrating their progress with television to a number of individuals both connected and unconnected to the Television Commission. At first the 243-line system had been shown to Sir Frank Smith (11 January 1935), Lord Inverforth (22 January 1935), Messrs Lusk, Sporborg, Herod and Macdonald (12 February 1935), and Colonel Angwin, Messrs Gardner, Sambrook and Robinson (13 February 1935), all at Abbey Road. Lord Selsdon seems to have been the first independent witness of a 400-line demonstration (of what must have been a very rudimentary manner), at Hayes on 18 February 1935, with Sir Frank Smith, Noel Ashbridge and Mr O.F. Brown, also given a demonstration of a 400-line system at Hayes on 25 February.

By March, a regular series of demonstrations was taking place both at Abbey Road, and mostly at Hayes. Among the guests were Sir Stafford Cripps (7 March 1935), the

R.M.A. Television Committee (8 March 1935), a Mr Escolano and two gentlemen from the Spanish Navy (27 March 1935), with the major demonstration being to the Television Committee at Abbey Road on Friday, 12 April 1935. It was for this demonstration that much of the equipment which was being used by Blumlein for his binaural film experiments was borrowed for several days.

On 24 May 1935, Blumlein sent a memorandum to the television team in which he outlined the decision that the Marconi-EMI system would use standard picture definition of 405 lines:

> The Baird Company propose for their system the adoption of a standard picture definition of 240 lines sequential scanning, 25 picture traversals per second, 25 complete frames per second; whilst the Marconi-EMI Company propose for their system a standard of 405 lines, 25 pictures per second interlaced to give 50 frames per second each of 202½ lines. Subject to satisfactory tenders being received, the Advisory Committee recommend the adoption of the standards for a public service during the trial period.

## A frantic period of work

There now followed a frantic period of work where patent after patent was applied for as the elements of the new system began to fall into place. It had been realized very early on that interlacing was the key to the system working, as 25 frames per second gave an intolerable flicker. In America RCA had also arrived at the same conclusion with their 343-line system, though they had only raised the frame rate to 30 frames at first, by now they decided that 60 frames per second, which coincided with the same number of cycles per second as the standard power source, would produce a much better, flicker-free picture.

Back at Hayes the sheer number of patents applied for during the period January 1935 to July 1936 is so staggering, that it is now hard to believe so much could be achieved in so short a period of time when one considers that every aspect of the work being done was totally new. Blumlein (working alone and in conjunction with others), applied for, and was granted, 27 patents during this period, which works out at roughly one every three weeks. These included work with Edward Cork, John Hardwick, Eric Nind, Joseph Pawsey, Eric White, William Connell and Cecil Browne and some of the more significant elements of the high-definition television system were worked out at this time.

It should also be remembered perhaps that while many of Blumlein's patents are groundbreaking in many ways, others are quite simplistic and even obvious. This was not lost on Blumlein himself who, constantly bemused that nobody had thought of an invention previously, often only applied for a patent after some period had passed, during which it was discovered that he had an entirely new specification on his hands.

During the fourteen-month period between the beginning of 1935 and the opening of the 1936 Berlin Olympics, Blumlein's personal patent count rose from 37 to 62. Patent No. 452,772, applied for on 25 February 1935, sees Blumlein and Cork returning to the subject of multicore cables, whereby improvements were made to the internal core wire by changing the shape and configuration of the bend that

Figure 6.1
Detail from Patent No. 452,772
(1935) co-written with Edward Cork
demonstrating how, by
providing kinks in two axes at right
angles within the central wire
core of the cable the loss of high frequency signals could be minimized
or even prevented.

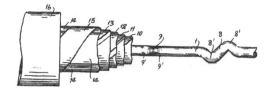

had been proposed. By this time it had been discovered that this type of cable construction could be applicable to feeders for short waves of 40 MHz or more, or for
medium waves where carrier frequencies were required to travel long distances with
a minimum of attenuation.

In this specific instance the central core wire, which was about 0.06 inch diameter,
was bent (again into a zigzag), so that the apices were rounded to form two kinks of
continuous sharp bending at 45° angles (which in effect produced a sinusoidal
shape), with a substantially longer piece of straight wire between one set of kinks
and the next set. The length of the straight wire between kinks could be varied in
order to produce different effects from the cable. Blumlein and Cork had found
that in a flexible cable of this nature, if the cable was flexed the high frequency
losses were increased due to the inner conductor lying over to one side of the cable.
By providing the kinks in two axes at right angles within the central wire core of the
cable this loss of high frequency could be minimized or even prevented. The degree
to which the new cable improved transmission of the signal was astounding, even
over the earlier patent (No. 452,713). The invention was destined to become the
basis of the balanced pair of cables carrying video signals, used to transmit the television pictures of the coronation of King George VI in May 1937.

Having produced this cable, Blumlein then turned his attention back to directional
aerials again in Patent No. 452,791 applied for three days later, on 28 February 1935.
The aerial was constructed from a series of tubes of copper arranged in half wavelength (of the wave to be transmitted) sections. Each tube had two or more conductors shielded from the others using a similar arrangement to that of the VHF
cable described in Patent Nos 452,713 and 452,772. If five such tubes were used (the
example used by Blumlein in the specification), this would give a total length of
2½ wavelengths with short insulating gaps between the tubes.

At each of these gaps, the inner conductor of the lower element is joined to the
outer tube of the element above and, similarly, the inner conductors of the upper
element are connected to the outer tube of the section below. The combined

Figure 6.2
The London Television Service
television cable as designed by
Blumlein and Cork and
manufactured by Siemens Brothers.
(*Courtesy of Felix Trott*)

arrangement forms an aerial 2½ wavelengths of the feeder in which the inner and outer conductors interchange each half wavelength. This interchange was effected by inserting an insulator with a number of slots in it, through which the inner conductor could be coupled to the appropriate outer tube with a series of copper strips. The entire array was fed by connecting a concentric balanced feeder to the inner and outer conductors of the bottom radiating element.

The net result would produce a vertical array which would radiate most strongly in the horizontal direction provided all the half wave radiators were operating in the same phase, that is the upper ends of all the half waves should be positive together when their lower ends are negative together. Previous aerials, which had used half wavelength sections, had been arranged in a line one above the other with the top half of each element joined to the bottom half wave element above by means of a half wave auxiliary element. This was folded to neutralize their radiation or to ensure that their radiation added to that of the straight half wave aerial elements. The problem was that when this type of aerial was fed at one end, for example the lower end, the radiation from the lower end seriously attenuated the currents flowing to the upper elements so that they were not fully effective. Blumlein's arrangement overcame this because the outer of each element acted as a radiator, with the inner and outer elements each feeding those beyond.

On 7 March 1935, Blumlein and John Hardwick applied for a patent (No. 455,492), for an electric signal transmission line in which signals of a wide frequency band were to be handled. This was achieved with a loaded transmission line for transmitting electrical currents over the bandwidth where inductances and resistances in series with one another were connected in shunt with the line at intervals to increase the attenuation over the desired frequency range.

This was followed on 20 March, by Patent No. 458,585 (which ranks as Blumlein's third longest patent, running to some 22 pages with 3 diagrams), 'Improvements in and relating to the Transmission of Electrical Signals having a Direct Current Component'. The circuitry within the specification was for the automatic gain control of the television signal (compensating for varying attenuation due to the direct current changing from the effect of the picture signal), by inserting black into the waveform.

In some respects Blumlein was returning to the problems Browne, Blythen and he had tackled in Patent No. 449,247, in September 1934. Television technology had progressed a long way since then, and Blumlein evidently felt that a better technique for DC restoration could be found and that is what he outlines in No. 458,585. This was done by transmitting, at spaced intervals along the same transmission channel as the direct current signal component, a check signal which had a portion of fixed predetermined amplitude and which could, at the desired point in the channel, cause an influencing correction upon the amplitude of the DC component signal to compensate for the variations.

Blumlein drew out the picture signal waveform showing the peak position of the white or bright element of the picture signal, with an indicator showing the amplitude level of the black signal and the synchronizing impulses shown in the blacker than black direction. As each frame traversed the inserted check signal, spaced at regular intervals, it controlled the generation of sawtooth oscillations at the line and frame frequencies in a given, predetermined way.

Figure 6.3
Bernard Randolph Greenhead with
the prototype Emitron camera in
1935. (*Courtesy of EMI*)

On the same day, 20 March 1935, Leonard Klatzow applied for a patent (No. 458,586), that improved the efficiency of the iconoscope mosaic because of the thin layer of silver that was deposited on top of the silver caesium photosensitive layer. Klatzow found that this layer had two distinct characteristics, firstly it increased the sensitivity of the mosaic from about 7.5 microamps per lumen to about 25 microamps per lumen and, secondly, it provided the tube with very good response to visible light but not beyond a wavelength of 700 nanometers. This would eventually be incorporated into the prototype Emitron camera tube the first of which were already being made. These were hand blown by specialist glass-blowers that EMI employed for the purpose.

Two days later the German Post Office opened the country's first regular public television service which transmitted from Berlin on 180-line, 25 frames per second stan-

Figure 6.4
Harry Neal, one of the specialist glass
blowers at EMI, here putting the
finishing touches to an Emitron tube.
(*Courtesy of EMI*)

dard, and while the service was intended to reach private individuals, no actual receivers were to go on sale until August, so the public had to make do with watching the transmission in theatres and cinemas. Then, in August 1935, a mysterious fire destroyed the transmitter postponing again the sale of receivers, though the transmitter was rebuilt and re-commenced transmissions on 15 January 1936. By then the Nazis had decided that television was too powerful a propaganda weapon to allow it into the hands of the people.

There was great excitement among television developers in Britain at this time as it had been announced that the long-awaited site for the London Television Station would be made public on 3 April 1935. On that same day Blumlein and Eric Nind applied for Patent No. 456,135 which was a circuit for modulating the carrier signal to reduce it to zero at the peaks of the synchronizing impulses. This was a difficult thing to achieve because of the curvature of the bottom bend of the characteristic of most valves used as modulators which necessitated an undesirably large amplitude of the synchronizing pulse to reduce the anode current to zero.

What Blumlein and Nind did was to modulate the picture signal (with or without the synchronizing signals superimposed), at a point in a radio frequency amplifier system, in which an additional modulation consisting of only the synchronizing signals, would have already been introduced. In this way, the additional modulation would serve to ensure that during the synchronizing signals, the carrier is reduced to substantially zero (ensuring cut-off in the blacker-than-black region).

Blumlein had obviously grasped the importance of the full understanding of a television waveform (which, incidentally, with the new CRTs that EMI were now working with, could finally be displayed and studied on a screen, rather than only imagined at or worked out on paper), his next patent (No. 455,858, applied for on 24 April 1935), demonstrates this as it deals with synchronization of television signals which can be combined on one channel such as line and frame pulses. The specification provided means for transmitting these two kinds of spaced electrical impulses along a common channel using a switching arrangement, in which either kind of impulse (line frequency or frame frequency impulse), could be fed into the channel such that any changeover from one kind to the other was automatically prevented from taking place.

In June 1935, the British Broadcasting Corporation drew up the 'Specification of

Figure 6.5
The prototype Marconi-EMI transmitter assembled at Hayes in 1935. (*Courtesy of EMI*)

Ultra-Shortwave Television Transmitter for London' (Specification No. TV/2), a 21-page document which contained the extent of the work, conditions of the contract, time of completion, as well as all the technical specifications for the Marconi-EMI studios at Alexandra Palace. A similar document was sent to Baird Television. At the back of the Marconi-EMI document is a schedule of prices and deliveries; it lists:

Price of complete radio transmitter tested at works and delivered to site (excluding power plant, water system, aerial feeders, valves and erection, £8,035; Price of two complete sets of valves, £1,416; Price of complete water system, £940; Price of power converting plant, £3,706; Price of aerial and feeder, £436; Price of erection of all apparatus, £1,560; Total price, £16,093. Time in which all work to this contract can be completed: Assembly of transmitter at works, 16 weeks; Testing transmitter at works, 4 weeks; Delivery of material to site, Time required for erection at site, Time required for tests at site, 6 weeks; Total time from date of placing order, 26 weeks.

It was signed and dated 4 July 1935.

Meanwhile, back at EMI headquarters in Hayes, the final stages were taking place in the negotiations for the erection of a prototype of the mast designed to go on top of the south east tower at Alexandra Palace. The 200 foot-high aerial mast was also to be used for television transmission tests which were then taking place. The aerial was needed to give the low power transmitter which had been used for these tests, a service area equivalent to an 18 kW peak transmitter which should allow for good demonstration conditions at Abbey Road where the EMI system had been provisionally set up for the committee and other interested parties. Unfortunately, these negotiations had taken longer than anticipated; considerations such as contacting the Air Ministry to check that it would not interfere with low-flying aircraft had to be taken care of first, and the Air Ministry only gave their verbal agreement for the construction of such a mast on 5 June. Shoenberg, having been informed of this by J.G. Robb, development manager for the project, contacted the Hayes and Harlington Urban District Council on 2 July 1935, to inform them of the proposed aerial mast and construction on the ground at the back of the Hayes research laboratory building.

Blumlein's next patents, Nos 461,004 and 461,324 (applied for on 4 July and 12 August 1935), are both variations of the closely coupled inductor ratio arm bridge. In the first example (which is a power-engineering version of the bridge) Blumlein takes the two or more loads fed from the same source of an AC current and de-couples them from the common impedance, the circuit being designed to remove some or all of the internal impedance of the source as regards to its effect in causing interaction of one load on another. It was of considerable use in the 500 cycles-per-second specially dedicated power supply to the modulator for the Alexandra Palace transmitter.

In the second example, Blumlein looked at the problems associated with transmission lines in which a number of loads were distributed along the length of the line. It had been known that the connection of one load across the line should not disturb adjacent loads, but at power frequencies such as that of the main grid, i.e. 50 cycles per second, this could only be achieved by making the impedance of the generator low, compared to the impedance of the load, and by employing a line of low inductance and resistance.

The invention in Patent No. 461,324, comprised of a transmission line with a feeder adapted to be fed at one end from a generator, having an impedance of the same order of magnitude as the surge impedance of the feeder, which is provided with a tapping point. At this point, a tapped load is connected to the feeder by means of two tightly-coupled inductive arms and a loading impedance, the arrangement being such that a balance is established between the incoming part of the feeder and the loading impedance. This establishes that the power supplied to the succeeding feeder is unaffected by the impedance of the tapped load.

Blumlein's next patent is curious in that it was applied for on three separate dates, 8 July, 9 September and 5 December 1935. Patent No. 462,530, 'Improvements in or relating to Electric Circuits for Reducing the Effective Shunt Capacity introduced by Circuit Elements such as, for example, Electric Batteries' became known as the 'constant resistance capacity stand-off circuit', and it is effectively a network designed to feed power into points that cannot normally tolerate high capacitance to earth. In many resistance devices, such as amplifiers used in television, trouble can be experienced owing to the stray capacity introduced by certain circuit elements such as DC coupling between two valves and the effect of the output of one valve on the input of the next. Similarly, the capacity of a cathode heating battery to ground may be excessive when the load is put in the cathode circuit of a valve.

Blumlein's specification provided a means by which the effects of these stray capacities could be eliminated by arranging that such devices were connected to the points at which they are required through high impedances (or at least through high frequencies), to a closely coupled inductive circuit of pure resistance throughout the entire frequency range, which effectively eliminated or at least reduced the greater part of the unwanted stray capacities to ground. Two versions were illustrated namely two-terminal arrangements of an inductor $L$, a capacitor $C$, and two equal resistors $R$, having the property that the impedance measured between the two terminals is purely resistive, of value R, at all frequencies, provided $L/C=R^2$. This property was known, but Blumlein adapted the circuit as a means of removing from critical points in a circuit e.g. a wide-band amplifier, the stray capacity to earth of, for example, floating power supplies. In another example, and a much used circuit, the application of the idea was put to the filament supply for the cathode follower output valve of the vision modulator for the original Alexandra Palace transmitter. This hardware is now preserved in the Science Museum, London.

This specification directly led to two further patents, Nos 462,583 and 462,584, both of which were applied for on 8 July 1935, in which one resistor required to dissipate high power and to have low capacitance to earth was replaced with two resistors, one of which dissipated the high power while the other one dealt with low capacitance to earth (462,583); and a battery with a high capacitance to earth was used to supply a DC bias between points where the capacitance could not be tolerated (462,584).

The Telecommunications Department of the General Post Office informed Marconi-EMI on 24 July 1935 that the frequencies of 45 megacycles and 41.5 megacycles for vision and sound respectively, which were those proposed for use at the projected London Television Station at Alexandra Palace, could be used for the experimental station at Hayes which had, until that time, been working at 44 megacycles and 40 megacycles respectively. It was pointed out that such an arrangement was for experimentation up to the time when public transmissions from Alexandra

Palace commenced, at which time it would be necessary to arrange a time schedule for the different transmissions from Hayes.

In early August, Isaac Shoenberg took a vacation in Austria, staying at the Grand Hotel, St Wolfgang, until 15 August 1935, when he moved to the Hotel Bellevue, Thumersbach. His absence however, came just as a minor crisis occurred with the Television Committee. The national press had been giving the committee members a rough time because of the speed, or rather lack of it, with which they were moving. J. Varley-Roberts, secretary of the committee, wanted to appease the press and so drafted a press statement in which a certain paragraph said: 'Such technical information regarding the characteristics of the television signals radiated by the two systems as will facilitate the designing of television receivers capable of picking up these signals, *will be supplied to manufacturers on application to the respective contracting companies named.*'

Condliffe, who had spoken with Varley-Roberts on the telephone, was horrified at this, and immediately sent a telegram to Shoenberg in Austria, on 8 August, saying that: 'We can only agree to the publication of such a statement if the words underlined were changed to "...*will be published in the technical press at an early date*". Secondly, we can only agree to the publication of such a statement, provided that the Baird Company are also willing to allow specification of their waveform to be published simultaneously.'

At the same time as Condliffe wrote his telegram on 8 August, Blumlein and Browne also sent letters to Shoenberg outlining much the same objection. On 9 August, following another conversation with Varley-Roberts on the telephone earlier in the day, Condliffe again wrote to Shoenberg explaining that Varley-Roberts had agreed to modify the committee's press statement according to EMI's wishes, but that they wanted to distribute the press statement as soon as possible, and could Marconi-EMI give them the go-ahead by the end of the day? Condliffe had pointed out that they could not reply immediately as 'no executive member of the Marconi-EMI Television Company Limited will be available before 26 August at the earliest.' Condliffe went on to add:

I am passing over all the correspondence to Broadway and explaining to him the position, and leaving the Committee to look after itself. I think that Mr Roberts is beginning to understand that when we said that we would be out of action for any decision during August, we meant it. Furthermore, I see no particular reason for wishing to help the Committee in this manner as they are only doing it because they are being upset by newspaper publicity which is decidedly unfavourable to the speed at which the Committee are moving. Again, apologizing for interrupting your holiday, and hoping that it is still going well. Yours sincerely.

While Shoenberg was in Austria, Blumlein applied for another patent specification, No. 462,823, on Monday, 12 August 1935, which is in effect an improvement on his previous specification, No. 421,546, of June 1933. Whereas in the earlier patent Blumlein had devised a circuit which essentially compensated a power supply with reactive elements, making it appear to load as pure, constant resistance, the new variant allowed for the same using resonant circuits arranged to prevent currents of an undesirable frequency from reaching the load. On 17 September, just after

Shoenberg had returned, Blumlein applied for Patent No. 463,111, in which a coil of fine insulated wire, cut of a length so as to act as a resonant rejecter, was wound round a centre conductor to form a frequency-selective coaxial cable which could be considered as forming a tuning circuit. This offered the advantage over an attenuating cable loaded with high permeability magnetic material in that the copper wire loading did not require the annealing needed by magnetic alloys, and the copper wire loading was not affected by direct or low frequency currents in the cable.

Upon his return from vacation, Shoenberg did agree to the publication of the Marconi-EMI technical specifications, but only on his terms. Douglas Birkinshaw, chief engineer for the BBC, in a meeting with Marconi-EMI on Thursday, 19 September 1935, agreed a list of wireless and trade publications that should be informed of the details for publication; these were: *Wireless Trader, Broadcast, Wireless World, Wireless Magazine, Practical & Amateur Wireless, Television & S.W. World, Practical Television, Popular Wireless, Electrician, Electrical Time, Nature, Wireless Engineer, World Radio, Electrical Review* and the *Journal of the Television Society.*

Subsequently, Blumlein sent out a series of letters on 23 September to the various editors of these magazines in which he enclosed a complete specification of the radiated waveform. Blumlein pointed out that it was being arranged that Baird Television Company would forward the corresponding information on their system on the same date, and that the information should not be published until 1 October 1935. *Wireless World* published their article, 'Television Transmissions, details of the Baird and Marconi-EMI systems', in their issue of 4 October, with the three-page article as comprehensive as both companies would allow.

Perhaps to be expected, the published details of their system prompted a series of letters requesting more details, including one from D. Grant Strachan, director of the Radio Manufacturers' Association, on 7 October, who listed no less than 11 further technical specifications he wished to have more details for. Blumlein replied on 14 October, pointing out that most of the questions needed to be answered by the BBC for contractual reasons, though he did give vague answers to a few including the bandwidth of the sound transmitter, given as ±10,000 c.p.s. – probably not the precise answer Mr Strachan was looking for. Meanwhile, Blumlein continued to apply for more patents. On 19 October 1935, Patent No. 464,443, was sent for consideration; written with Joseph Lade Pawsey and Edward Cecil Cork, it comprises an aerial built out to a substantially pure resistance over a considerable frequency range forming a correct termination for a feeder. The aerial fed from this long feeder which then matches the wideband of frequencies to the transmission line.

There is an interesting letter in the files of G.E. Condliffe on 24 October 1935, in which an enquiry had evidently been made about 'Television Transmissions from Arsenal Football Ground'. Condliffe was making arrangements for one of the Marconi-EMI engineers from the research department to visit the Arsenal football ground, where new buildings were being erected, 'in order to examine the possibility of television transmissions being made from there'. He says:

> It would be very useful if you, or one of your engineers could visit Hayes so that he would have some idea of the type of equipment involved, and, in particular, the type of cable which it would be necessary to run from the field to the portable transmitting equipment. I wish to point out that it is very early for con-

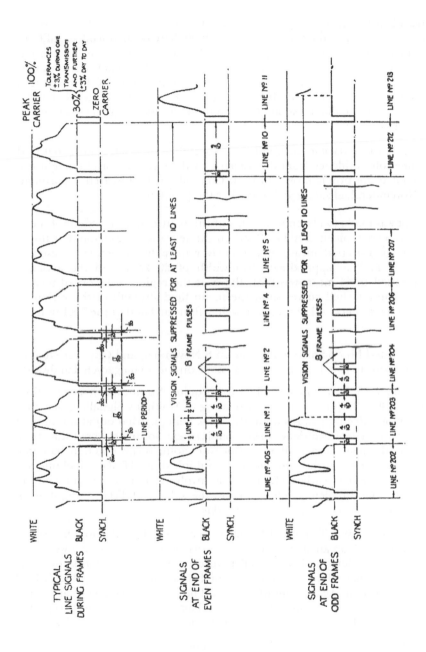

Figure 6.6 The Marconi-EMI television system transmitted waveform.

crete plans to be made with regard to the installation of television transmitting equipment, but collaboration on the lines suggested should be extremely useful.

At the end of October, Turnbull needed to write to Angwin at the Television Committee to complain about a transmitter which had been traced to GEC in Wembley which was working on the previously allocated frequency of 45 megacycles, and so interfering with the test transmissions from EMI. When test transmissions were being sent to Abbey Road, it had been found that the GEC field strength was 'considerably greater than the field strength from our transmitter. I shall be glad if you will examine the question of wave-length allotment with a view to eliminating the interference.'

Colonel Angwin wrote back to Turnbull on 1 November 1935, suggesting that EMI produce some solid evidence from the tracking of the signal before he might act. Turnbull handed the problem over to W.H. Connell, in the research department who, by 7 November, had succeeded in tracing the problem:

An attempt was made to obtain some idea of the nature and the extent of the interference experienced at Abbey Road on the television waveband. For this purpose, a signal strength measurement receiver and a dipole aerial were used. The receiver had an approximate bandwidth of 20 kc/s, and the aerial, although tuned to 44 mc/s, was fairly flat over the region of recorded measurement. The attached curves show the mean field strength of the various sources of interference. All of these are well within the band of the EMI transmitter.

Turnbull presented his evidence to Angwin on 8 November, and GEC were told in no uncertain terms to switch off the transmissions on the 45-megacycle bandwidth.

On 14 November 1935, Blumlein applied for a quite extraordinary specification, Patent No. 470,495, 'Improvements in or relating to Multiplex Signalling Systems'. The specification effectively draws upon the work being carried out at the time in the laboratory in the television waveform; what is so extraordinary about this specification is that Blumlein, despite all the frantic work going on around him, had time to apply this knowledge to his time division multiplexing (TDM), telephone signalling ideas of a few years earlier. This long and involved patent specification would not be the last time Blumlein would return to the subject of TDM, using it again later when developing systems for airborne radar.

Blumlein's last patent specification of 1935, No. 466,418, dated 22 November, shows a multicore cable with an auxiliary line wound outside in order to give lower interference over a great length. The coaxial nature of the cable employs the same kinked core wire that Blumlein had used in his specification No. 452,713, of December 1934, producing a circuit which would be free of interference at low frequencies, as the two outside conductors act as a balanced line effectively rejecting the interference, while the screening gives protection at higher frequencies.

As 1935, drew to a close, Marconi-EMI had much of their equipment already constructed and were preparing to move it to the rapidly developing studios at Alexandra Palace. On Friday, 20 December 1935, Turnbull wrote to Condliffe that:

Electrical design: The electrical design of the system is complete. Mechanical Design and Drawing. The remaining work outstanding for the drawing office comprises some seven items, in addition to the sound power switchgear, which we are handling in collaboration with Mr Sage. I estimate this work (including switchgear) will take some five weeks for the two draughtsmen we have at present. This means that DO (Drawing Office) work on our equipment will finish about the end of January.

Manufacture: So far, with the exception of an incomplete B amplifier which we have obtained from Mr Durdle for preliminary tests, we have obtained nothing from the workshop. I understand that a number of amplifiers are practically complete, but assembly cannot be finished until certain components, mainly radio frequency filters from the factory, have arrived. Unless these items arrive soon, our estimated delivery dates will be seriously prejudiced. Provided that these components arrive within the next few days, we shall be able to obtain from the shop some eighteen amplifier panels, and one monitor panel, about Jan. 8th these will comprise the backbone of the panel equipment. Assuming that some panels can be delivered before the above date, I think the wiring can be completed on this batch by Jan. 18th. This date, however, depends very largely on how much vision wiring has also to be accommodated. Assuming this date can be kept, testing should be complete before the end of January. Providing the power supply, which is being handled by Mr C.O. Browne, is ready, we should be able to begin shipment of this part of the equipment to Alexandra Palace by Feb. 1st.

The RF filters to which Condliffe refers were found to be necessary at all the power supply inputs to the equipment panels, due to the fact that the equipment would be near (both at Hayes and at Alexandra Palace) to a quarter wavelength 200-foot mast at Brookmans Park. This was near enough to these medium wave transmitters to induce considerable RF current in the mast and thus cause interference with the video-frequency band.

The year 1935 had been a busy one for EMI. The year 1936 would prove to be even more so.

## Alexandra Palace and the London Television Station

When the television commission had published its findings to the Government in January 1935, there had been the first mention of a public television service that should initially serve the London region, before expanding, through popular demand, to the rest of Britain in the years that followed. There was, however, no mention of exactly where this London Television Station should be situated, a considerable factor to consider with the stringent demands such a service would undoubtedly place on such a facility in terms of equipment, broadcasting coverage area, and the effective range of the ultra-short wave transmissions that were deemed necessary for a high-definition television system to operate efficiently.

There were in fact three major criteria for consideration before such a site to be constructed. First, it was, as has been mentioned, necessary for ultra-short waves to be used for the transmission of high-definition television pictures. Ultra-short waves however, have a limited effective range, which at one time was thought to be restricted to an area not further than the optical range determined by the curvature of the earth. Though this was subsequently disproved, ultra-short waves none the less should be transmitted from a site as near to the centre of the area of population for whom reception is desired, in order to achieve the maximum disposition of the transmitted signals.

Second, the height of the transmitting aerial above the surrounding territory would dictate to a great extent the distance that those signals would travel (beyond the optical range of the curvature of the earth), as well as overcoming any lowering of the transmission signal power due to geographic features such as hills. This presented quite a problem in London, especially as the city, while appearing to be ostensibly flat, is in fact, a series of undulating hills separated by flat valleys. It was therefore of extreme importance that any television transmitting aerial that was to serve London be situated on as high a plot of ground as possible with no restrictions to the erection of the aerial mast.

Third, the television studios and control rooms would undoubtedly occupy a large area. There were, after all to be offices for both companies concerned, as well as the BBC, coupled with quite specific power requirements, as well as the various studios, viewing rooms, stages, changing rooms, eating facilities, etc. It would obviously be desirable in a city such as London to place a television station right in the centre of the city where it would have the optimum range for transmission. London however, and especially the centre of London, does not offer satisfactory solutions to the either the first, second or third essential requirements for a practical television transmission station, which are as follows:

First, the centre of London was not the most highly populated area of the city during the peak viewing hours when television reception would be at its height. The centre of London was, and still is, most highly populated during the working hours of the weekdays; during the weekend however, the City of London especially, was practically deserted. It would not therefore be practical to broadcast to a densely populated audience during the weekdays when very few people could conceivably be considered able to watch television.

Second, the centre of London does not really offer an ideal location for a high transmitting mast essential for television broadcasts. Apart from the fact that it would need to be of the order of 300–400 feet higher in the centre of London to clear the effect of any surrounding buildings (such as St Paul's Cathedral), the centre of the city also lies at the basin of one of London's many valleys. This factor alone would raise the necessary height of any aerial on such a site yet again, in order to achieve a broadcasting range suitable.

Third, the centre of London, because of its very nature as a business Mecca, has the highest price to ground-space ratio in Britain, a preclusive factor for such a large building requirement as a television station.

Attention was therefore turned to sites on the higher ground to the north and south of the city centre, with careful studies made of all potential areas of interest. In the north, Hampstead and Highgate were both considered, but once again due to factors such as cost and the severe restrictions that would be made on the erection of very high masts, these two areas were ruled out fairly quickly. In south London, places such as Shooters Hill and the high ground near Crystal Palace were considered. In both cases, it was considered that the greater part of the service area of the station would be too much displaced in a southerly direction, and again both were quickly ruled out.

The best option seemed to be the site of the Alexandra Palace on Muswell Hill, which possessed some outstanding advantages over all the other sites considered.

Alexandra Palace had been built in 1873 as a 'Palace for the People', being used for all manner of exhibitions and activities which it was deemed were of suitable interest to the general public. By 1935, for various reasons, the building had fallen into disuse and therefore offered an ideal location for a television station as very little cost would be involved in the construction process (the building itself being very sound and only requiring conversion rather than re-building).

Not only was Alexandra Palace quite vast inside, with rooms easily capable of housing all the needs of the television companies, the building stood on ground that was already 306 feet above sea-level, and there would be Government permission to build a suitable mast on the site if it were chosen. Muswell Hill is also a geographical feature, upon which the palace surmounts, with ground falling away rapidly in all directions to comparatively low-lying ground below. This meant that the transmitter aerial would not be impeded by any structure or geographical feature in the near vicinity. The location was to the north of the centre of London in one of the highest suburban population areas that served the city, though the palace itself was isolated from the immediate surrounding houses by the very fact that it was built on the top of the hill.

On 3 April 1935, it was announced that Alexandra Palace had duly been chosen as the site for the new London Television Station, with Terence C. Macnamara placed in charge of planning the new station and Douglas C. Birkinshaw given the job of chief engineer. Then, on 7 June 1935, it was also made public that the new television service would have dual line standards of 240-lines at 25 frames per second for the Baird Television system, and 405-lines interlaced at 25 frames per second for the Marconi-EMI system. On 12 June 1935, Marconi-EMI received from the BBC Installation Department the provisional drawing (Dg. No. AP/LO/1148), of the space that would be made available to them for their studio and control room layout.

Construction work was to commence immediately the plan specifications had been drawn up, with the building of the 220 foot high transmission mast given the highest priority on the top of the 80 foot high south east tower of the building. This aerial mast, which was constructed by Marconi to the BBC specifications, and erected by J.L. Eve Construction Company, was to be almost identical to the one which was being used at Hayes for the test transmissions between Abbey Road and the EMI headquarters. Adjacent to the aerial site, approximately 30,000 square feet of space had been made available for the station proper, with another 25,000 square feet adjacent to the north east tower for the television theatre. The work on the aerial mast commenced in mid-April with the work of adapting the areas for the needs of television beginning in September 1935, and space being sub-divided to create offices for all the technical, production and administrative staff.

A vast army of workers now descended on the site and began work on the erection of the mast which required the removal of the existing pylon top and the internal floors in the south east tower, to be replaced with steel members, and the main brickwork of the tower being strengthened with steel ties. This provided not only a

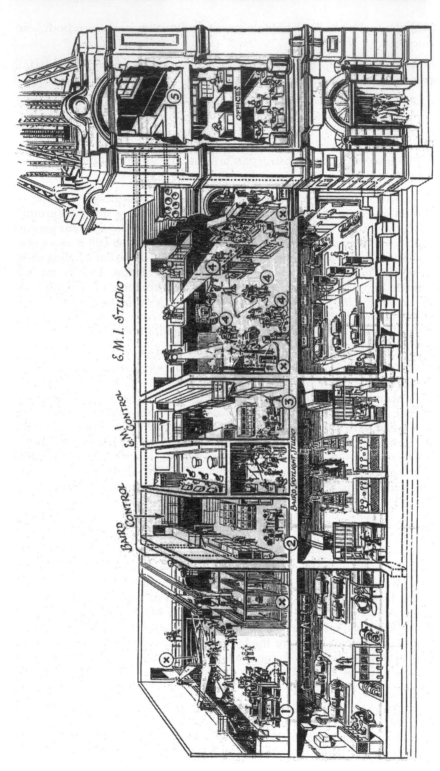

Figure 6.7 Cut-away detail of the television studios of Baird and Marconi-EMI at Alexandra Palace from an article of late 1935. (*Courtesy of Television & Short-Wave World, November 1935, pp. 656–60*)

structure of great solidity, but also an additional 8000 square feet of available floor space for offices. Once completed, Douglas Birkinshaw's office was fitted out just beneath the mast, at the top of the tower, thereby ensuring spectacular views of London for the BBC chief engineer from his vantage point.

Part of the conditions of use of Alexandra Palace had been that the outfacing structural facade should be changed as little as possible, that any major structural work should be, wherever possible, contained within the building itself, and these guidelines were adhered to as best as possible. The palace frontage required little cosmetic changes to be made, and the wings of the building had only slight alterations where structural improvements were necessary to support the heavy equipment being installed.

The respective companies who were to occupy the building were each allocated space for their equipment, studios and transmitters. Baird Television and Marconi-EMI each occupied one of the two large halls on the ground floor for their vision transmitters, in a space measuring 70 feet by 50 feet. These rooms were painted a battleship grey colour and once the transmitters had been started up, emitted a deep continuous hum from the electronics within. In the centre of each vision transmitter room was a large flat panelled control desk into which 16 meters (VU-type) had been mounted for checking the various levels to and from the camera feeds to the transmitter relay racks. There was also a central small monitor for visually checking the picture quality.

At the back of the control panel, on the left-hand side, was the modulator apparatus in the form of a single, large metal cabinet with doors that opened outwards to reveal the electronics. To the right of the control panel was the radio apparatus, which consisted of three large metal cabinets. Between these would sit a consumer receiver to monitor the picture signal as it was actually transmitted. The entire room had been designed to be operated by no more than two engineers who would spend their time checking the vast array of knobs and dials, and inspecting the enormous valves as they gently glowed encased within their tubular water-cooled anodes.

A central third hall on the same floor housed the Marconi sound transmitter, which again consisted of a large flat control panel in front of which were several large metal cabinets. In these were housed the drive unit and low-power, high-frequency stage, a final power amplifier, a modulator unit and a power switchboard. Behind the Marconi sound transmitter room was the cinema which was fully equipped for the transmission of films, in a space 40 feet by 15 feet. This facility was to be used by the production staff of either company for the selection of excerpts from films which they proposed to transmit by television; though, when film cutting was not taking place, it could be used as a miniature cinema, seating about thirty people.

At the west end of the ground floor was a scenery storage area some 52 feet by 22 feet with a wide entrance door capable of handling large pieces of scenery and props. It was considered that this space could also be used as a spare or temporary studio if needs be, for the televising of objects too heavy to transport to the first floor. The remainder of the ground floor was taken up with boiler space for the central heating system and a restaurant for all the personnel using the station.

On the first floor of the south east wing the two studios of the respective companies were located directly above their vision transmitters, Baird's nearest the central

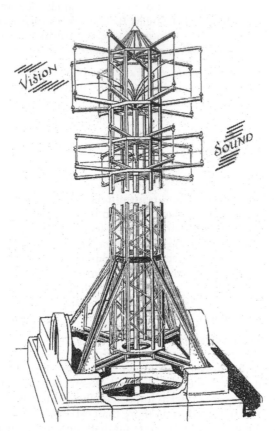

Figure 6.8 Alexandra Palace transmission tower detail showing vision and sound aerial differences. (*Courtesy of* Television & Short-Wave World, *November 1935, pp. 656–60*)

entrance hall and Marconi-EMI's nearer the south east tower. The Marconi-EMI vision transmitter had the radio part constructed by Marconi, while the modulator had been built by EMI. For Baird, the entire vision transmitter had been constructed by Metropolitan-Vickers. Each studio was 70 feet by 30 feet in dimension and 27 feet high. Between the two studios were the respective control rooms, the Baird complex requiring four separate rooms, a control room, a spotlight studio for announcements, another spotlight scanner and an intermediate film scanner room. The Marconi-EMI control room was a single enclosure with the film scanner apparatus built on the first floor balcony colonnade. A series of artists' dressing rooms, both male and female, a make-up room and a separate room for the band were also provided.

The north east wing of the building was dominated by the auditorium and stage which had a foyer (at the far north east end of the building), adjacent to the entrance hall behind the tower. The entire Alexandra Palace complex required a new electricity supply to be taken from the North Metropolitan Electric Supply Company's system, who installed a ring main in the form of two feeders fed from the local Alexandra Palace sub-station (which was based at the side of the south east tower). There was also a small electricity control room occupied by personnel from the NMSC inside, just below the south east tower and next to another small entrance hall.

The second feeder came from the Wood Green traction sub-station via the supply company's own sub-station in Ringslade Road. This in turn was fed by several alternative routes from the company's power station at Brimsdown, and thereby assured a continuous supply of electricity at 11 kV, 3-phase, 50-cycles per second and distributed at 415 volts, 3-phase, 4-wire with earthed neutral.

As the rivalry between the two protagonists, Baird Television and Marconi-EMI, grew, the job that Macnamara and Birkinshaw had been given became ever more difficult. Neither Baird nor EMI wanted the other party to be aware of any of the work that they were doing, and it became the job of Birkinshaw as chief engineer for the site, to keep the two companies apart from one another as the installation process progressed. This led to innumerable difficulties as each engineering team desperately tried to conceal from the other the essence of the work that was being carried out and the nature of the equipment being installed. There were even petty

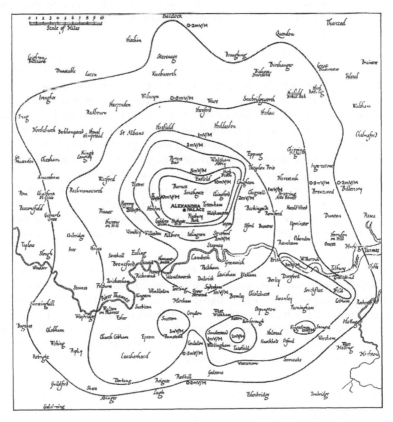

Figure 6.9 Projected field strength map of the London Television Service.

Figure 6.10
The transmission tower at Alexandra
Palace looking straight up. (*Courtesy
of EMI*)

wrangles between the companies about such minor details as the fact that the Marconi-EMI studio, on the first floor, had been designated Studio A, while the Baird studio was designated Studio B.

Despite the increasingly difficult situation the work progressed, with the studios being treated acoustically for the first time. This was quite a novelty in 1935, as most previous recording studios had been treated only lightly to ensure that sound did not escape rather than to prevent sound from the outside penetrating within. Both Studio A and Studio B were to be made into acoustically accurate environments, with properties desirable for television, rather than those which had previously been associated with sound broadcasting.

In each case, it was found that 2 foot square slabs of asbestos felt, about 1 inch thick, would provide the kind of absorbent material needed to counteract the reverberant nature of the rooms, which was quite noticeable, despite the presence of all the scenery which acted as huge baffles. In both cases, one of the long walls, which had been originally covered with a lath-and-plaster material, was found to be unsuitable for the mounting of the asbestos felt slabs (which were suspended from wooden battens nailed to the walls), and so these had to be glued to the plaster instead. The windows had specially designed sound-absorbent shutters, made of two plies of board with an air-gap between. In this gap, a layer of sawdust was used to fill the space, and the boards were then covered, on both sides, with a canvas-covered acoustic quilt. Though the work was carried out separately in each of the studios, by each company's own workmen, it is worth pointing out that the reverberation co-efficiency of both studios, once completed, gave remarkably similar results when plotted.

This similarity ended however, with the lighting systems that were employed. The Marconi-EMI studio used AC studio lighting which did not introduce any flicker into the television pictures, while Baird Television opted for DC studio lighting which was needed for its film process and the electron camera. In both studios, apart from floor mounted movable arc lamps, the main lighting system was mounted on a scaffold at a height of 14 feet above the ground, which covered three sides of the studio floor forming a lighting gallery on a gantry around the working area.

While all the work took place out at Alexandra Palace (all of which, incidentally, was filmed for posterity, and regularly reported upon in the cinema newsreels of the day), back at EMI, the engineers were frantically putting together the rest of the design elements of the 405-line television system. They had always had Lord Selsdon's date of 31 December 1936 to aim at (the day by which the Television Commission had said a regular British television service should have commenced transmissions), but now with just over a year to go there still seemed an enormous amount of work to do.

## The pace quickens – 1936

At the beginning of January 1936, Blumlein had worked on a simplified explanation of the entire HDTV system in the form of a series of explanatory notes for general reference within EMI and also for the technicians at the BBC, working at Alexandra Palace. The Marconi-EMI system was obviously now going to work, of that there was little doubt. The contingency plan to produce a simultaneous 240-line system, should the 405-line system not work, had been scrapped by late 1935, as Shoenberg became more and more confident in his original decision. On 28 January 1936, Blumlein published his internal document which would form the basis of the technical notes for many of the periodicals who were still desperate for any information on what was the most advanced television system in the world.

Condliffe was probably under the most pressure at this point, as it was he who was responsible for co-ordinating each department and its activities. In February, he issued a directive to Blumlein, Browne, Cork and Tringham regarding television test transmissions:

> In order to assist the Works Designs Television Receiver Group with their urgent production programmes, the following transmission times will be arranged each week: – Monday afternoon, sound and sight (bars); Tuesday afternoon, sound and sight (bars); Wednesday afternoon, sound and sight (studio or film). These transmissions are to be considered as an absolute minimum. Every endeavour should be made to provide additional transmissions when work on the transmitting equipment permits.

Blumlein's first patent application of 1936, No. 470,408, is dated 13 February. It is a method for reducing the inductive interference of coaxial transmission lines by driving them from a high-impedance source such as a tetrode or pentode valve. However, Blumlein terminated the receiving end of the cable by an impedance of the same order of magnitude as the characteristic impedance of the cable. In this way, frequencies at which the cable is electrically short, say not more than 10 to 15 miles, would not be affected by the disturbing electromotive force in the circuit which was causing the resistive potential of the conductor to drop. In his next patent, No. 473,276, dated 10 March 1936, Blumlein designed a rectifier for the provision of DC from an AC power supply in order to compensate for the leakage reactance of the transformer in a tuned circuit in the AC supply path. The AC supply to be rectified was fed through a series inductance and capacity to the primary winding of the transformer, while the cathodes of two rectifying valves were also connected, but to the secondary winding of the transformer. The anodes of the rectifying valves were connected together with one to the DC leads, while the other lead was centre tapped on the secondary winding of the transformer. A shunt condenser was then connected across these two DC leads. In such an arrangement the condenser in series with the primary winding of the transformer could be adjusted so that, together with the added inductance, it will tune out the leakage inductance of the transformer as well as the AC supply if so desired.

Later that same month, Blumlein modified one of his earlier patents, No. 461,004, with Patent No. 475,729 dated 19 March. In his original design for a power regulating system, Blumlein took the two or more loads fed from the same source of an AC current and de-coupled them from the common impedance, the circuit being designed to remove some or all of the internal impedance of the source as regards to its effect in causing interaction of one load on another. He now replaced the generator with a synchronous alternator which effectively meant that the two loads were now represented by the mains, which he considered a negative load, and the true load. In this way it was possible using unequal ratios, to achieve a low impedance facing the true load and possible to obtain a substantially exact balance for either impulsive voltage variations, or slow voltage variations.

In April 1936, two further patents were applied for, both on 29 April, No. 474,607 and No. 479,113. Specification No. 474,607, was co-written with Eric White and relates to the need to stabilize the high tension voltage supply for the triode valves being used in the high-definition television system. Two circuits were designed which would make use of triode valves to smooth and adjust the supply across the terminals of a source in order to control the voltage from the most effective point, thereby ensuring that impulsive changes of the potential of the source of supply would not cause corresponding impulsive variations of the voltage at the load terminals. In one arrangement, this was done by replacing the first coupling condenser with a battery to determine the working point on the characteristic of the first valve with reference to the tapping point on the potentiometer. The battery replacing the condenser coupling on the second valve mainly controlled the DC potential of the output terminals because it determined the bias potential on the grids of the second and third valves. In another arrangement, the potentiometer comprised an ohmic resistance connected in series with one or more neon discharge tubes connected across the source to be smoothed. The discharge from these tubes was known to give a constant voltage for varying currents passing through them, and this property was utilized in the circuit.

Patent No. 479,113 was a design for a drive amplifier for cathode ray tube deflection coils. It is an improved circuit employing negative feedback which could be used to correct the increase in voltage from the valve when a saw-tooth waveform is produced. Blumlein designed a circuit in which the negative feedback source was taken from a separate network, rather than from a resistor in series with the scanning coils. In so doing, the valve did not build up the undesirable increase in voltage that the introduction of such a resistance would normally produce.

In May 1936, Alan Blumlein found his time occupied by another milestone in his life. On 11 May 1936, his first son, Simon John Lane Blumlein was born. The Blumleins had recently moved from No. 7 Courtfield Gardens, West Ealing, where they had lived since 1933 just after their marriage, to a new and larger house at No. 32 Audley Road, Ealing. This was just the other side of Hanger Lane from the house in which he had lived at Woodville Road.

Two more patents were applied for in May 1936. Patent No. 476,935, dated 15 May, relates to a design for an automatic gain control for a television waveform when the television signals are to be transmitted to or from a moving object such as an aeroplane, where a rapid gain control effect is required. Blumlein devised a circuit in which both blacker-than-black and whiter-than-white signal levels could be introduced, thereby including all the characteristics of the picture. The blacker-than-black elements represented an amplitude equivalent to zero, while the whiter-than-white element represented an amplitude equivalent to the peak value of the modulating signal. By defining the maximum amplitude in this way, all that remained was to devise a synchronization signal with an impulse that extended between this range, and that could alter within a defined period, in this case 10 microseconds.

Patent No. 477,392 of 22 May 1936, is once again co-written with Eric White. It is a design for an improved diode-triode voltmeter for use with negative feedback to give high input impedance. Blumlein and White designed a circuit in which the diode rectifier which could cover a large range of input voltages, with a direct current thermionic amplifier in which provision had been made for introducing the negative feedback to the input of the amplifying valve. This is fed in the form of a potential derived from the current in the output circuit and is used to increase the input impedance of the amplifying valve. Such an arrangement would have a high impedance at all frequencies, and is therefore able to use a high value of load resistance in the diode rectifier circuit which will produce improved and more sensitive readings from the voltmeter.

Meanwhile, at Alexandra Palace, engineers from Baird and Marconi-EMI were working practically side-by-side by this time, installing the vast amount of equipment into their various allotted studio spaces. Naturally relations, which had never been anything other than fiercely competitive, now became very strained, and Macnamara and Birkinshaw had trouble at times keeping the peace between the two companies. In April, an argument had developed between them regarding the exchange of universal information, which Baird said they required from EMI, in order to determine whether certain adjustments needed be made to the proposed television receivers. The argument as to whether this would infringe upon one or other patents developed to the point that by May, both Isaac Shoenberg and Harry Greer were involved. On 26 June, EMI wrote: 'As we think you will appreciate on further reflection, we should not care to adopt the course of giving opinions to others as to possible infringements of our Patents in their proposed designs and we regret therefore that we are not able to oblige you by expressing an opinion on the particular case in question.'

Harry Greer wrote back:

I fully appreciate your attitude that your Company cannot be expected to express an opinion as to the infringement of its Patents by the designs proposed to be used by others. It is not, however, in relation to a specific design that the questions contained in my letter were addressed. Having regard to the fact that the transmitter to be supplied by your associated Company to the BBC will require a particular type of adjustment to be made to an ordinary cathode ray receiver to enable the latter to receive the transmission, my company desired me to attempt to ascertain whether the use of any and every such receiver to receive your associated company's transmissions would be regarded by the owners of Patent No. 420,391, as an infringement thereof in the absence of a licence under such Patent. Unfortunately your letter does not assist me in ascertaining your attitude in what will clearly be an important matter both to my Company and to our licensees and other members of the trade.

Clearly, both Baird and Marconi-EMI were doing anything that they could to be as obstructive to the other as they could. It must have been an extraordinary set of circumstances with engineers from both companies working in the close confines of Alexandra Palace, arriving every day, some parking in the same car park, eating in the same restaurant and of course working so close to each other.

There are many stories which originate at about this time demonstrating Blumlein's character and humour. EMI had taken delivery of a particularly large oscilloscope from the National Physical Laboratory on which all manner of things were to be tested. Unfortunately, because of the sheer size of this piece of apparatus it became evident as it was installed that it was obviously quite immobile. On seeing the arrival of this huge screen Blumlein was heard to remark 'Ah, Mohammed has at last gone to the mountain', and from then on every oscilloscope in that series was named after one of the Caliphs who followed Mohammed.

Blumlein was also well known for quoting Thévenin's Theorem (which states that an equivalent circuit contains an equivalent voltage source in series with an equivalent resistor). This allows us to reduce a circuit to a single voltage source with a series resistor. As Philip Vanderlyn explained,

'It has been said of him that he could fly into a rage over unimportant details: if this was so I never saw evidence of it. He certainly had a sense of humour. One of his favourite devices was to bring a discussion to its close by quoting Thévenin's Theorem, and he would often be half way down the corridor, puffing his pipe and shouting "Thévenin" over his shoulder before we were able to gather our wits together. On one occasion, he missed an opportunity to carry out this manoeuvre, and the following day was presented with the Thévenin Medal, fabricated from the lid of a cocoa tin. After a moment of hesitation it was gracefully accepted. Such are the touches that turn geniuses into great men. They also speak more eloquently for their relations with their staffs than any mere statement of fact.'

On occasions, as the television work neared completion, Blumlein would telephone his wife who was at home with their newly arrived baby son Simon (Doreen had a receiver in the living room for just such tests). As he made various adjustments, he would ask her what she could see on the screen:

'He was working very hard on television and we had a set where you looked in the mirror, you see, the looking glass of the lid; and the first thing I ever saw, before there was any programmes, was when we were in Audley Road in Ealing. He would ring me up and say, "Hold the telephone darling, and look in the set. Now tell me what you can see". And I would say, "Well, I can see the smoke from your pipe", of course he was never departed from his pipe, except in bed; and I said "A stripe in your suit." '

On yet another occasion, Blumlein had climbed the aerial mast on the south east tower of Alexandra Palace to make some adjustments of some kind to the aerial array, and he dropped the pipe out of his mouth. So he promptly came down, drove

Figure 6.11
The south east tower at Alexandra
Palace showing the completed sound
and vision transmission aerials.
(*Courtesy of EMI*)

off to the nearest tobacconist, bought a new pipe and ascended the aerial again to continue his work. All of those who recall Blumlein tell of how he was never without his pipe: 'Oh God. Couldn't, couldn't move without his pipe', Doreen would recall, and J.B. Kaye too:

'Yes, he did smoke a pipe, he smoked only corn cob pipes for a while. I don't think you can get them now, but he was quite a sight with his straw hat on and a corn cob pipe walking down Kingsway. He always swore that smoking a pipe related to the cortex of the brain and that cigarettes intended to dope you. So he never smoked cigarettes – I can't recall him smoking cigarettes. I don't think he smoked cigars. He stuck to his pipe smoking and he smoked a pipe a lot, and he never seemed any the worse for it.'

Maurice Harker recalled how 'he was a compulsive pipe smoker and the metal "No Smoking" notice, which hung in Ham Clark's laboratory was unceremoniously removed and bent up to form an ashtray for him to empty the contents of his pipe into!' Yet another compulsive pipe-smoker was Edward Cork. Eric White recalled that one day Cork, who used to walk from Hayes railway station in the morning to the research laboratory 'tried to light his pipe while facing a westerly wind. Stopping to light the pipe, Cork found that he could not do so, and turned around with his back to the wind. Having successfully lit his pipe, Cork then proceeded to walk back to Hayes station, caught a train, and so to home!'

## The long-tailed pair

In July 1936, Blumlein applied for another of his many famous patents, No. 482,740, dated 4 July. It has become known as 'the long-tailed pair'. This now familiar and much used circuit was first needed in the valve amplifiers for the original video cable between points in central London and Alexandra Palace. The cable used was not the now common co-axial type, but a shielded pair, and the problem which Blumlein's circuit overcame was how to obtain the 'push-pull' signal, uncontaminated by 'push-pull' interference pick-up over the distance to be covered. Blumlein knew from his early days that in telephone practice a transformer serves this purpose, but transformers that were capable of handling the video frequency range were not then available, and a different solution would have to be found.

The long-tailed pair circuit, which was incidentally given its name by Blumlein himself, and which subsequently stuck, was originally devised from the need for a long video frequency cable which could carry television pictures over great distances. When a very long cable or 'tail' is wanted, the so called 'push-pull' video signals can be much attenuated, especially at the high frequency end. So, it is necessary to amplify these frequencies considerably while, at the same time, attenuating any 'push-pull' interference picked up on the cable. The resistance in the cable can be replaced by a third valve in the circuit with a cathode feedback resistor to make the impedance seen at its anode very high. It is then natural to connect a fourth valve to make a 'long-tailed pair' with the third valve, and so on throughout the length of the cable dependant upon the distance the video signals are to be carried. The resulting cable (devised by Blumlein and Cork), with the long-tailed pair circuit would see use in May 1937, during the coronation of King George VI.

Four days later on 8 July, Blumlein and William Horace Connell applied for Patent No. 479,599, a means of providing an automatic or manually controlled selection circuit for modulated carrier wave receivers. Blumlein and Connell obtained the variable selectivity in the carrier wave by providing means for the production of negative feedback. This negative feedback was, in turn, frequency selective for the purpose of reducing the remote side bands relative to the carrier frequency. The net effect of this would mean that feedback could now be adapted to alter the required variation in selectivity. In order to obtain automatic variation of the effect of the feedback, the slope of the valve to which the negative feedback is applied is also varied. If, for example, the invention was applied to a superheterodyne receiver,

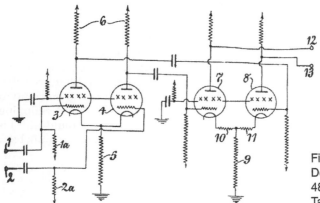

Figure 6.12
Detail from Patent No. 482,740 (1936), the 'Long-Tailed Pair'.

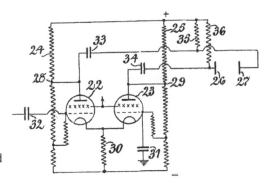

Figure 6.13
Detail from Patent No. 482,740
(1936), variation on the 'Long-Tailed
Pair'.

the negative feedback could be employed in conjunction with the variable amplification to produce a delayed automatic volume control. On 18 July 1936, Blumlein and Rolf Edmund Spencer applied for Patent No. 480,355. They had invented an extra high tension supply for a two-anode cathode ray tube which had a resistance network within the circuit in order to ensure that a constant ratio was maintained between the voltages on both anodes.

Much has been written and speculated of the atmosphere at EMI during this quite extraordinary period. Many years later it was recalled by J.D. McGee saying that:

'It has sometimes been suggested that our team was a "one-man band", with Blumlein as the genius surrounded by a group of competent but quite ordinary men. This I would deny and, I am sure, so would Blumlein. I am sure that we who survive him would willingly admit that he was *primus inter pares*; he was so brilliant and at the same time of such a generous nature and integrity that it has always been easy to acknowledge his primacy in the group. It was in this sense that Shoenberg regarded him; I am sure that Blumlein's views on policy carried more weight with Shoenberg than those of any one of the rest of us. There was ... a good understanding between Shoenberg and Sir Louis Sterling, the managing director. A similar bond extended down the chain through Blumlein to the rest of us. Thus, we had an excellent communication channel and firm decision-making machinery: very necessary in any organization – and quite indispensable in a research organization. It would be invidious to omit the part of G.E. Condliffe in this hierarchy; he had the unenviable task of managing this large group, co-ordinating our activities, calming our quarrels, providing the facilities we required and maintaining a critical eye on the whole programme of work.'

## From Radiolympia to the official opening

The building and installation process at Alexandra Palace took the best part of a year to complete, and so, by the early spring of 1936, both Baird Television and Marconi-EMI, despite all their wrangles and bickering, were busily engaged in the testing of their respective systems. The achievement is all the more remarkable when one considers that Marconi-EMI, for example, were only working with about 30 senior staff with about 120 assistants in their laboratories. There was no time for mistakes and the correct decision about all aspects of the work had to be made at each point of the development to allow the practical results to be obtained with the limited manpower available.

The summer of 1936 was to prove a very important one for television as the eleventh Olympic Games, held in Berlin, Germany, from July to August, were to become the first major event in history to be covered 'live'. German broadcasting had made enormous technological steps forward in the three years since Hitler had come to power, driven by the belief that all forms of propaganda could be used by the Nazis to further their ends, and by the fact that most of its television companies had shared patents in the early days with British and American firms.

Coverage of the Olympic Games was given to the German Post Office, Deutsch Reich's Post (DRP), using three iconoscope cameras furnished by Telefunken. The cameras were used to broadcast the events that were considered the most interesting to the public, and, it should probably be pointed out, where the Nazis expected German athletes to achieve most, though Jesse Owens, among others, rather spoiled that for them. One camera was situated at the swimming pool and the other two at the main events in the vast stadium, specially constructed for the event. Fernseh also had equipment at the games, based on the Farnsworth Electron camera.

The television pictures were seen in 25 'viewing parlours' and regular cinema theatres in and around Berlin, as well as at the Olympic Village itself. This was done primarily because private reception was almost unheard of at this time due to the lack of consumer receivers. The recorded viewing results from the televised events varied from 'very good', with 'sharp, clear signals', to 'rather poor', with complaints of 'shaky pictures with considerable strain being placed on the eye'. Nevertheless, with the 1936 Olympic Games being televised throughout the events, as Hitler had expected, the exercise was a major propaganda coup for the Nazis.

It had been announced as long ago as January 1935, in the report of the Television Commission that the start date for the public television service was to be no later than 31 December 1936. It did not specify however, that an opportunity, should it be found, to broadcast on an experimental basis earlier than that date could not be taken. Moreover, it so happened that the organizers from the British Radio Manufacturers Association were having trouble selling space at the forthcoming Radiolympia show, due to open on Wednesday, 26 August 1936. They had appealed for experimental transmissions, and in the interests of promoting television to the general public, invited the two companies to take space at the exhibition and to both transmit from Alexandra Palace, assuming they were ready to do so of course. The Television Commission saw no reason why either company should not take part, though they did insist on one proviso, that it be done fairly. It was decided therefore, that the flip of a coin would determine which company would broadcast on the first day, followed by the other on the following day, and so on. Both companies agreed, and accepted the offer.

When Radiolympia opened in London on 26 August, Baird Television won the toss of the coin and so had the honour of starting the first day's transmissions from Alexandra Palace. At the show itself, there were three brands of receiver on demonstration, one from Bush Television (manufactured by Baird Television), one from the Marconiphone Company, and one from HMV, with all three tuned to work on both the Baird and Marconi-EMI standards. They varied in price from £50 to £100, which represented quite a considerable sum to the average man in the street who probably only earned between £4 and £5 a week at the time. The large proportion of the first day's transmissions were of course 'live' feeds from the show itself as well as newsreel excerpts and motion pictures. The broadcasts, which were scheduled to run each day from noon to 1.30 p.m., and from 4.30 p.m. to 6.00 p.m. were, overall, received very well by the public attending the show.

There were numerous breakdowns of the Baird equipment on the first day, and again on the third, some of which they even tried to attribute to sabotage, though it was by now very obvious that the flying-spot scanner only worked well when transmitting film. Under studio conditions however, it required almost complete darkness except for the intense strobe-like scanning beams on the main subject, conditions which were totally impractical for outdoors or, as in the case of Radiolympia, from the show floor. When it was the turn of the Marconi-EMI system on the Thursday, the second day, the broadcast began at noon as it had the day before, on a glorious sunny summer's day.

The following is a transcription of the opening speech by the BBC's announcer Leslie Mitchell, which was panoramic view across London being transmitted to Olympia from Alexandra Palace:

'Ladies and gentlemen, you are looking at a view from the balcony of our studios at Alexandra Palace and this is the signature picture of the BBC demonstration to Radiolympia by the Marconi-EMI system. We are looking out over the grounds of the palace. Beyond is the main railway to the north, with Wood Green in the foreground. It is a perfect afternoon. A superb panorama lies below us. On the horizon, the Lea Valley, and now we are passing over Hornsey and Finsbury Park, and I wonder if you can see the smoke that is just by the Harringay greyhound stadium?

Now we are coming down over the grounds themselves. On the road below you will notice someone passing (Mitchell pointed this out in case anyone disbelieved that the scene was not being televised). There goes a car. Another one is just coming in. This is the road that leads up to the studios and main offices. On the horizon beyond the smoke and the white roof, are the Hackney Marshes, and in the foreground Hornsey and beyond the old City of London with St Paul's hidden in the mist. The camera is now getting over to the west,

Figure 6.14
Leslie Mitchell.

but I'm afraid it is almost hidden in the heat of the day. I wonder if you can see on the extreme right of the picture, quite a way over, just coming in now, the spire of Hornsey church?

That ridge before the camera is looking over the west part of the grounds. Just coming to the top of the picture now are the actual grandstands of the Alexandra Palace racecourse. There you see children playing in the sunshine. This is the view from the studios and offices. There goes the van which travels between Broadcasting House and here each day. This view we see every day, and I think you will agree there is not a finer view to be seen anywhere in London.

Hello Radiolympia. This is the British Broadcasting Corporation experimental demonstration transmitted by the Marconi-EMI system. It is as I said an experimental demonstration of television especially arranged for Radiolympia. We want to stress that it is the first of its kind, and we have not had the opportunity of showing you anything like it before. Now, before they fade me out, I just want to emphasize that this is a specially arranged television programme. The programme consists of a news reel, and an excerpt from a film, and half an hour of variety followed by some excerpts from the outstanding films of the year, directed from the studios at Alexandra Palace – about 10 miles away from Radiolympia – Here's looking at you.'

Leslie Mitchell then handed over to co-announcer Jasmine Bligh (Elizabeth Cowell would have made the announcements probably, but she had been taken ill on the first day of the show), and she in turn introduced Helen McKay, who had often been heard on radio broadcasts, but who was appearing at Radiolympia for her first address on television. Following Helen McKay there was entertainment from The Three Admirals, Carol Chilton and Maceo Thomas presenting 'Tapping in Rhythm', and Elizabeth Bergner reading 'Rosalind' from Shakespeare's *As You Like It*. An excerpt from a Jessie Mathews film followed, as well as Miss Ellissa Landi showing Douglas Fairbanks Junior in his first Independent British Production of Jeffrey Farnol's book *The Amateur Gentlemen*.

Jack Buchanan was seen in an except from 'When Knights were Bold'; following this the British and Dominion Corporation production of a new comedy at Elstree was

Figure 6.15
Elizabeth Cowell.
(*Courtesy of APTS*)

Figure 6.16
Jasmine Bligh.
(*Courtesy of APTS*)

seen with Ned Sparks and Gordon Harker. After this Charles Laughton played Rembrandt in a London Film Production, produced by Alexanda Korda, none of which had been seen in public at that point. Finally Paul Robeson sang 'Old Man River' from 'Showboat 1936'. Leslie Mitchell closed the proceedings: 'Well, Ladies and gentlemen, that ends the demonstration to you at Radiolympia. We hope you have enjoyed your first experience of television and that you will become regular viewers and so good-bye for the present.'

Despite the problems, which there undoubtedly were, and incidentally which were by no means limited to the Baird system, in general, things went well throughout the two weeks of the event, and succeeded in interesting the public in television, which had been the purpose of having both companies attend in the first place. On the very day that the first transmissions took place at Radiolympia, 26 August, Condliffe sent a delightful memorandum to Blumlein, Connell, White, Tringham, Broadway, McGee, Percival, Browne, Cork, Turnbull and others, which clearly demonstrates the pace at which Marconi-EMI were working. It would seem even the internal engineering staff were fascinated by the work that they were carrying out, as Condliffe points out:

> 'Engineers in the Works Designs Department have suffered severe inconvenience as a result of a large influx of visitors who have crowded into the laboratories to see the reception of the Alexandra Palace transmissions. I shall be pleased if you will insist that all engineers and assistants should obtain permission before going into the Works Designs building. A receiving set will be placed in the Research Department in the course of a day or so, and the staff will be given an opportunity of seeing reception there.'

The Radiolympia Show proved to be a very useful experiment for both Baird and Marconi-EMI now readying themselves for the opening of the London Television Service on Monday, 2 November 1936. On Thursday, 27 August 1936, the day after Radiolympia opened, Eric White and Alan Blumlein applied for Patent No. 482,370 in which a circuit was employed to effect economy in power consumption in an oscillatory circuit and to prevent waste of power. This was done by making use of the pulses produced when current through a cathode ray tube deflector coil was sharply cut off. This became known as the 'economy diode circuit'. Blumlein applied for

Patent No. 482,195 on 21 September 1936, in which he outlined a CRT deflection coil arranged as a single, and continuous winding so that tapping across one diameter would give line deflection, while tapping across another diameter would give frame deflection. Blumlein's next patent, No. 483,650, a month later and dated 20 October 1936, again returns to CRT deflecting coils. Blumlein now devised a method of providing a cheap and efficient electromagnetic coil deflection system for a CRT by arranging it so that the entire circumference of the neck of the tube was occupied by windings lying substantially longitudinally with respect to the tube axis. Thus the separate coils providing the line and frame deflecting forces were arranged to lie side by side at the surface of the CRT instead of one set of coils lying on top of the other set as had previously been devised.

By this time it had been confirmed and announced in public that the trials of the two systems were set to last for three months, until February 1937. At that time, the BBC would make a further announcement as to which company had been successful, based upon the decision of the chief engineer reporting to the members of the Television Commission and the senior management at the BBC. Once again, the honour of transmitting first was to be decided by the toss of a coin, and as before, Baird Television won. When the great day finally came, it has been estimated that less than 1000 receiving sets had been sold in the whole of Britain (sales did not pass 1500 until the summer of 1937), and of course the vast majority of those were in the central London area.

Nevertheless, despite the somewhat small 'audience', everything was readied for the momentous day, when the service would begin. While there are no records that list exactly who attended from Marconi-EMI (other than the official guests of course), it is reasonably safe to suppose that most of the major development engineers were present. There are, in the EMI archives, invitations from the BBC to some of the department heads, including Broadway, Lubszynski, Browne, Condliffe and Shoenberg. We also know for certain that Blumlein was present at the opening ceremony as his reply to the BBC on 24 October 1936 reads: 'I have your invitation to be present at the opening of the Television Service at Alexandra Palace on the 2nd November, and shall be very pleased to attend.' By late September the BBC were conducting the last of their 'acceptance tests' on the Baird and EMI equipment which were necessary before they would allow any public transmissions to take place. These tests began on Monday, 28 September 1936, when the BBC engineers arrived at the palace to witness test transmissions from both the studios, while additional BBC engineers were to watch the reception of the pictures, in the case of Marconi-EMI at Abbey Road.

At Alexandra Palace, all the last minute preparations were taking place; explanatory notes were typed up by D.H. Munro, BBC Television Productions Manager on 30 October 1936, and the entire event was rehearsed on Saturday morning, 31 October (without the dignitaries in attendance of course, and the artists were not made up). The rehearsals went well, and everything was set for the following Monday. Three camera positions were used for both the Baird Television system and the Marconi-EMI system. For the Baird transmission Camera No. 1, which was the Electron camera, was placed directly in front of the table at which the honoured guests would sit and speak; this was operated by Mr Truck and Mr Bliss. Camera No. 2, operated by Mr Wright, was for announcements. Camera No. 3 was a reserve, fitted with a telephoto lens, and was operated by Mr Tong. At precisely

Figure 6.17 The Baird system at the opening of the London Television Service, 3.00 p.m. Monday, 2 November 1936. Left to right: Lord Selsdon, chairman of the Television Committee; Major G.C. Tryon, Postmaster-General (standing); R.C. Norman, chairman of BBC, Sir Harry Greer, chairman of Baird Television.

3.00 p.m. on that Monday afternoon, the service commenced with an introductory speech from Leslie Mitchell (who had Jasmine Bligh in reserve, just in case anything went wrong), followed by 15 minutes worth of speeches from R.C. Norman, Chairman of the BBC, Major Tryon, the Postmaster General, Lord Selsdon, Chairman of the Television Commission and Sir Harry Greer, Chairman of Baird Television.

Though John Logie Baird was present, he had not been invited to take part in the opening ceremony itself, and had been relegated to a seat in the audience, a fact which he later reported to his wife, Margaret, saddened and annoyed him greatly. Another dignitary who was also conspicuous by his absence on the podium was David Sarnoff of RCA. Sarnoff had been invited, but decided to bide his time and watch the British television 'experiment' from afar, despite his evident financial interests in the success of the Marconi-EMI venture. Sarnoff would eventually proclaim 'the birth of television', at the New York, World's fair on 20 April 1939, in front of an NBC Iconoscope camera, but by then of course, the 'birth' he referred to only meant 'in America'.

The transmission at Alexandra Palace on 2 November, to the great credit of the Baird engineers, had passed off without a hitch. The entire broadcast took just thirty-four minutes to complete, at the end of the Baird transmission, Leslie Mitchell explained to the viewers: 'You have been watching the opening programme of the London Television Service, by the Baird system. Will you now please switch your sets to the Marconi-EMI system, by which a vision signal will be radiated at a quarter to four. From now until ten minutes to four, there will be a musical interlude in sound only, by the Television Orchestra. At four o'clock the opening programme will be repeated by the Marconi-EMI system.'

Indeed at 4.00 p.m. precisely, the entire programme was repeated; everything was planned and timed to the minute. Here is the entire inaugural address, as it was broadcast on the Marconi-EMI system at 4.00 p.m. on Monday, 2 November 1936. Leslie Mitchell once again appeared on screen (this time with Elizabeth Cowell in reserve):

**Leslie Mitchell:** Good afternoon once more everybody. This is the BBC Television Station at Alexandra Palace. At three o'clock this afternoon, the television service was opened by the Postmaster General using the Baird system. The opening programme will be now be repeated on the Marconi-EMI system. First, the Postmaster General will be introduced by Mr R.C. Norman, Chairman of the BBC, and will be followed by Lord Selsdon, Chairman of the Television Advisory Committee, and Mr Alfred Clark, a director of the Marconi-EMI tele-

Figure 6.18 The Marconi-EMI system at the opening of the London Television Service, one hour later at 4.00 p.m. Monday, 2 November 1936. Left to right: Lord Selsdon, chairman of the Television Committee; Major G.C. Tryon, Postmaster-General; R.C. Norman, chairman of BBC, Alfred Clark (standing) chairman of EMI. An HB1B microphone can clearly been seen at the end of the microphone boom arm. (*Courtesy of EMI*)

vision company. The speeches are followed by the latest edition of the Movietone News, British Movietone News that is, then Adele Dixon and 'Buck and Bubbles', both accompanied by the Television Orchestra.

**Mr R.C. Norman, Chairman of the BBC:** Mr Postmaster General, Lord Selsdon, and viewers. We arc met, some in this studio at the Alexandra Palace, and others at viewing points miles away, to inaugurate the British Television Service. My first duty is to welcome you Major Tryon in the name of the British Broadcasting Corporation, and to say how happy we are that you should have done us the honour of performing the inaugural ceremony. We at the BBC are proud that the Government should have decided to entrust us with the conduct of the new service. We are very conscious of the responsibilities which that decision imposes upon us.

At this moment, at the starting of television, our first tribute must be to those whose brilliant and devoted research, whose gifts of design and craftsmanship have made television possible. We are honoured by the presence of some of them here today. We wish also to record, Lord Selsdon, the guidance and encouragement that we have received from the two television committees over which you have presided. As for the future, we know already that television is much more complicated than sound broadcasting. We are encouraged in facing its intricate and fascinating new problems, by the patience with which the public and the press have waited for this day, and by the interest, and let me add, at times the indulgence, which they have shown during the recent test and trial transmissions. We hope that their interest and tolerance will continue for we shall certainly need both. We are however, confident that television in its special combination of science and the arts, holds the promise of unique, if still largely uncharted opportunities, of benefit and delight to the community. We are happy to think that some of its earliest opportunities will have as their setting, the historic pageantry of next summer.

The foresight which secured for this country a national system of broadcasting, promises to secure for it also a flying start in the practice of television. At this moment, the British Television Service is undoubtedly ahead of the rest of the world. Long may that lead be held. You may be assured that the BBC will be resolute to maintain it. Today's ceremony is a very simple programme. In every respect, it will doubtless seem primitive a few years hence, to those who are able to recall it. But we believe that these proceedings, for all their simplicity, will be remembered in the future as an historic occasion, not less momentous, and not less rich in promise than that day almost exactly fourteen years ago, when the British Broadcasting Company, as it then was, transmitted its first programme from Marconi House.

In that belief, Mr Postmaster General, we asked you to take the leading part in this ceremony, and I now invite you to inaugurate the new service.

**Major Tryon, The Postmaster General:** Lord Selsdon, Mr Norman and all who are watching this ceremony from afar. It is a great privilege to be invited to inaugurate the British Television Service. For we are launching today a venture that has a great future before it. For me, it is also a new and extremely interesting experience. Though I have had experience of speaking into the microphone, many times, this is the first occasion on which I have faced the television camera.

Few people would have dared fourteen, or even ten years ago, to prophesy that there would be nearly eight million holders of (radio) broadcasting receiving licenses in the British Isles today. The popularity and success of our sound broadcasting service are due to the wisdom, foresight and courage of the governors and staff of the British Broadcasting Corporation, to which the Government entrusted its conduct ten years ago. The Government of today, is confident that the corporation will devote themselves with equal energy, wisdom and zeal, to developing television broadcasting in the best interests of the nation, and that the future of the new service is safe in their hands.

I was very glad, Mr Norman, to hear your reference to the guidance and encouragement which you have received from Lord Selsdon, and the members of the Television Advisory Committee. We in the Post Office know well how unsparingly Lord Selsdon has devoted his great ability and high personal qualities to the public interest, both as Postmaster General, and on the television committee. I am very pleased that under his guidance, the Post Office had been able to co-operate, through the Television Advisory Committee, in the development of this new service.

I also should like to pay a tribute to all those who have devoted their talents and their time to solving the very difficult problem of television. We owe it to their skill, and their perseverance in research, that television has passed from the region of theory to the realm of practice. As you said Mr Norman, television broadcasting has great potentialities. Sound broadcasting has widened our outlook and increased our pleasure by bringing knowledge, music and entertainment within reach of all. The complimentary art of television, contains within it, vast possibilities of the enhancement and widening of the benefits we already enjoy from sound broadcasting.

On behalf of my colleagues in the Government, I welcome the assurance that Great Britain is leading the world in the matter of television broadcasting, and in inaugurating this new service, I confidently predict a great and a successful future for it. Now, I have pleasure introducing Lord Selsdon, Chairman of the Television Advisory Committee.

**Lord Selsdon, Chairman of the Television Advisory Committee:** Mr Postmaster General, Mr Norman and viewers. I stand before you as representing both the television committee, which originally investigated the possibilities of this new field, and also the Television Advisory Committee, which continues to advise regarding its development. My colleagues and I much appreciate what has been said about our work, and I only wish that time and space permitted them to appear before this instrument today. In their name, I thank you.

It has rightly been said that the potentialities of this new art are vast, and it is possible for instance to conceive of its being applied, not only to entertainment, but also to education, commerce, the tracing of wanted or missing persons, and to navigation by sea or by air. All these and more, will no doubt in due time, be tested, and some of them will arise. The patient industry of inventors has helped us so far, now we hope that the kindly interest of the public will help us further.

From the technical point of view, I wish to say that my committee hopes to be able, after some experience of the working of the public service, definitely to recommend certain standards as to number of lines, frame frequency and ratio of synchronizing impulse to picture. Once these have been fixed, the construction of receivers will be considerably simplified. But meanwhile, do not let any potential viewer delay ordering a receiving set, for fear that a change in these standards may put it out of commission almost at once. It is an essential feature of the development plans that for two years after the opening of any service area, no such change shall be made therein. For at least two years therefore, today's receivers, without any radical alteration, will continue to receive Alexandra Palace transmissions.

Just how wide this London service area will prove to be, is difficult to say with absolute certainty. Roughly speaking, it will cover Greater London, with a population of about ten millions, or again, roughly speaking, a radius of more than twenty miles, with local variations. There may be some surprising extensions. For instance, I should be unwilling to lay heavy odds against a resident in Hindhead, viewing the coronation procession. In the light of experience here, we shall proceed with the location of second and subsequent transmitting stations, according as public interest justifies this course. Technically, Britain leads today.

Following the speeches a film of British Movietone News ran for approximately 9 minutes, then the short entertainment from Adele Dixon, who sang the specially commissioned signature tune written for the start of television, 'Here's looking at you'; Buck and Bubbles, the two coloured American comedians, and the Television Orchestra which ran for approximately 7¼ minutes; this was followed by Leslie Mitchell again with the closing announcement: 'So ends the opening programme of the BBC Television Service. The next programme will be televised at nine o'clock this evening, by the Baird system. Good afternoon, everybody.'

The transmission was faultless with most independent observers noting afterwards that the EMI pictures were somewhat better than the Baird ones. While we know that Alan Blumlein attended the ceremony, we only have the recollections of Doreen, his wife of the events of that day:

'Then of course they started broadcasting, we went up to Alexandra (Palace), C.O. Browne and his wife, and Alan and I. We had a look at Baird's, and of course, as you know, Baird's didn't work, but of course it's all double Dutch to us women! I mean, it was just a lot of machinery – and we had to be thrilled. I said to Isaac Shoenberg, "I don't know anything about it you know", and he used to say, "That is why you are so good for him. That is why you are so good for him". I only just learnt to say "yes" and "no" in the right places you see, because I couldn't possibly understand what Alan was doing.'

The evening broadcast, at nine o'clock, contained the first edition of the programme 'Picture Page', devised by Cecil Madden, in which Joan Miller, operating from a (dummy) telephone switchboard, appeared to be receiving calls from viewers requesting to see certain celebrities. Naturally, these were produced, and among the first were Jim Mollison, the aviator, Kay Stammers, the tennis star, 'Bossy' Phelps, the King's Bargemaster, George Whitelaw, the Lord Mayor's coachman, Alexander Shaw, the pearly King and Queen of Blackfriars and Algernon Blackwood. On this programme too, the two lady announcers, Jasmine Bligh and Elizabeth Cowell, were first introduced to the viewers.

It was an historic day, 2 November 1936; the day that had finally arrived when the two systems would be put to the test. It was now up to Marconi-EMI or Baird

Television to prove to an audience, the BBC, the Television Commission members and of course to each other, just whose system was the better. Regular service therefore, continued the following day, Tuesday, 3 November 1936.

# Chapter 7
# From television to radar

## There has to be a winner

Almost at once, following the first few days' broadcasting, there were some serious problems associated with the Baird Television system. During the initial test period it had been observed that Studio B, the spotlight studio, was hopelessly out of date for the form of programme that was required. The studio had to be operated in near darkness apart from the lights on the subject being televised. These dazzling lights and the practically immobile wet-film camera that was being used became so unwieldy that they were only used on occasional evenings, with the Farnsworth Image Dissector camera being used on most others. The best option for Baird was still the film scanning process, which relied heavily on the fact that it did produce very good images, but only of scanned film, not 'live' footage.

The film scanning method had been around for some time, and both companies had experimented with it at various points, though only Baird persisted with the method. EMI had two film scanning units at Alexandra Palace, which were used for existing film transmission, unlike the intermediate film process that the Baird system used. Essentially a photographic process much like that of cine-film development, only much faster, the image to be televised was captured with a camera and the film immediately transferred within the apparatus to a series of tanks for developing and printing. The film, which was still wet and dripping when it came out to be immediately televised, took 64 seconds to process, which was very quick, but not exactly 'live' television. Though mechanically very ingenious, the film scanning system was no match for an all-electric system and besides, the development tanks had a nasty habit of leaking corrosive chemicals all over the nice new floor of Studio B. One curious side effect of the process involved the artists themselves who, upon completing their act, would make a dash for the film screening room just in time to see themselves televised. Later, the orchestra got to hear about this and so you had the extraordinary scene of large numbers of people, instruments and all, dashing for this small screening room to see their performance, or at least a few seconds of it, as it was televised.

On 10 November 1936, just a week after transmission had begun, Blumlein and Cecil Browne applied for Patent No. 490,205 in which the elimination of the unwanted frequency which caused the effect known as 'tilt' in the picture signal was devised. 'Tilt' and the 'tilting effect' manifested itself as secondary images, superimposed on the picture image on the receiver. It was caused by the spread of secondary electrons released by the scanning beam from the mosaic, which also compromised performance as they tended to neutralize the charges stored on the mosaic elements. Blumlein and Browne eliminated tilt by means of an equal and opposite sawtooth waveform to counteract the sawtooth wave upon which the picture signals were imposed. This was done by constructing a grid placed between the lens and the mosaic of a television camera tube, which generated a series of black lines at right angles to the direction of the scanning, which interspersed the picture signals on the camera screen.

These black lines were then fed to an amplifier which had such a low frequency cut-off, that the frequencies relating to the 'tilt' element were not transmitted with the picture signal. By working out the number of black lines produced at the grid and comparing this to the line frequency in cycles per second, the frequency of the black lines and their association to the tilt effect could be determined. All that then remained was the removal of the tilt effect frequencies at a suitable point in the picture signal transmission process by means of a DC reinsertion device.

When Baird returned to the flying-spot scanner for their 'true' live broadcasts, the pictures were poor by comparison. One engineer who attended the trials described the Baird system as 'clockwork', compared to the Marconi-EMI system next door. Baird had also been working to perfect the Electron camera (which had been developed from the license agreement between Baird and Philco-Farnsworth), so that by the time the trials began, the new apparatus had started to show real promise. There was however, yet one final twist of fate for Baird Television when, on the night of 30 November 1936, the great fire which totally destroyed the Crystal Palace, also took with it much of Baird's equipment, including the Electron camera system, which had been kept there. For John Logie Baird it was the final blow.

Blumlein's last patent of 1936 is dated 2 December. The year had not only been one of the busiest of his career, it had also been the second most productive year of his life with a total of fifteen patents (only surpassed by 1935 in which 16 patents were applied for). Patent No. 491,490 improved the coupling between a cable and its termination which allowed a wide band of signals to pass but prevented any undesired direct current from passing and attenuated low frequency signals. The coupling circuit included separate windings which served to prevent the passage of any undesired low frequencies. The ends of these windings were at like potentials at high frequencies being connected together through a bridging circuit including a condenser and a resistance connected in series.

## Trials and improvements

The trials continued with the transmissions being broadcast for two hours a day, on a basis of the Baird system televising on one week, followed by Marconi-EMI system the following week. This swapping over of the system went on for the entire three months. It was plain for all to see however, that the EMI system was far the superior. Long before the BBC made their announcement that they had chosen the Marconi-EMI system over that of Baird Television Limited, anybody could have guessed the conclusion. It was as evident to those who viewed in their homes as much as those who attended the studios daily. In the end, on 4 February 1937, the BBC gave three reasons for their ultimate choice. First, the 405-line system was superior in definition to the 240-line system. Second, there was a very noticeable lack of flicker due to the interlaced framing system; and third, there was a greater scope for future

Figure 7.1
HMV Model 905 television receiver
available in 1936. This was a
combined television 'sight and sound'
model with a six-valve 'all-world' radio
receiver. The cathode ray tube screen
was just 6¼ inches by 5 inches. The
Model 905 cost 35 guineas or was
available on hire
purchase.

development with the Emitron (and later Super Emitron) camera. The BBC announced that they did not expect the standards that had been reached and so successfully demonstrated by Marconi-EMI to be abandoned at least until the end of 1938. This was as Lord Selsdon had promised in the opening ceremony at Alexandra Palace the previous November. What nobody could have known then was that as the television service resumed on Monday, 8 February 1937, with the Marconi-EMI system alone, those standards would not be altered for another 49 years.

Blumlein's next patent was applied for during the week that the BBC made its announcement that the EMI system had won the right to continue the television transmissions from Alexandra Palace. Patent No. 489,950, dated 5 February 1937, is a circuit in which a tapped coil is used to couple between valves for use in power amplifiers or generators which feed power into relatively large capacities (such as the input capacity of a further thermionic valve). In the usual form of coupling used at the time, the value of the coupling resistance was limited by the input capacity of the following valve or other load. Where the amplifier was required to handle high frequencies the coupling resistance became unduly small if attenuation of the signal was to be avoided. In effect, this meant that the optimum power output of the valve would not have been achieved.

What Blumlein did was to provide a potentiometer or auto-transformer constructed to produce a constant voltage ratio at all frequencies within a predetermined range, with the output frequencies being lower at higher frequencies than at the lower ones. In this way, a reduction in the voltage due to the decreasing impedance of the load at high frequencies could be prevented. The coupling resistance could then be increased to a value closely approximating the desirable resistance for which the valve is designed, whilst at the same time the output impedance of the higher frequencies is not so large that the input capacity of the following stage produces undesirable attenuation.

## The coronation of King George VI

In the early hours of 20 January 1936, King George V passed away at his home of Sandringham, and his eldest son, Edward, Prince of Wales, assumed the title of King Edward VIII. The coronation of this enormously popular member of the royal family was planned for May 1937, and it was hoped that a long and peaceful reign might follow. The King, however, had decided to marry a divorcee, Mrs Wallis Simpson, which was against the constitutional rights of the Monarch, and so, on the evening of 11 December 1936, in one of the most dramatic acts of devotion of the century, Edward VIII abdicated his throne in favour of his younger brother George, Duke of York. The coronation therefore, which had been planned meticulously for over a year, took place on Wednesday, 12 May 1937, with George VI crowned King of England.

The BBC had pushed hard within Government circles for the right to televise the coronation despite some considerable attempts to block what many considered a sacred ritual. When, after some months of debate the Government decided that the BBC would be able to have discreetly placed cameras along the processional route, the BBC accepted the invitation immediately, and plans were drawn up to find suitable sites for the positioning of three cameras along the route which would feed pictures to an outside broadcast vehicle that EMI were busily modifying for the purpose. At Marconi-EMI, everyone was still enjoying the success of beating Baird to become the only television system for the London Television Service. Now more frantic preparations were needed to ready the outside broadcast units for the coronation to be held in just over three months' time.

The first reference in the EMI archives to the coronation is a small memorandum from Cecil Browne to Condliffe regarding the mobile scanning van; it reads: 'The mobile scanning equipment should be complete and tested by May 1 1937. This date should allow us some time (1–2 weeks) for any experimental work which we might like to do with the van.' The main van would contain the mobile control room and so needed to be quite large in size. In fact the finished vehicle was 27 feet long, 7 feet wide and over 10 feet high, it weighed over 9 tons and was fitted with equipment to control up to three Emitron cameras at a time, each of which could

Figure 7.2
The Marconi-EMI outside broadcast unit showing three Emitron cameras and their operators in early May 1937. It was finished just days prior to delivery to the BBC.

Figure 7.3
Detail of the interior of the Marconi-EMI outside broadcast unit first used at the coronation of King George VI, 12 May 1937. (*Courtesy of EMI*)

have up to 1000 feet of multicore cable attached to it. There were two other vehicles on site, both much smaller than the control van, one had an ultra-shortwave vision transmitter built into it working in the region of 64 megacycles and delivering a power output of 1000 watts peak white picture modulation to a special antenna which had been devised by Blumlein (see Patent No. 452,791). The third vehicle supplied electric power to the other two. These were used as standby transmission apparatus should the video cable fail in any way, though they were all used a month after the coronation at the Wimbledon Lawn Tennis Championships.

On 21 January 1937, Marconi-EMI sent a quotation to the BBC for this mobile apparatus amounting to a cost of £987-0s-0d. The full specification for this had been drawn up by Blumlein and sent to Birkinshaw on 23 January, and comprised of an eight-stage, super heterodyne receiving link operating on 64 megacycles, an automatic volume control panel, amplifier with picture-synchronizing signal ratio adjustment and included a DC re-establishment circuit from black level. There was also a picture synchronizing signal level indicator, meter panel, distribution amplifier, line jack panel and one signal monitor with an adjustable time base. All of this was to be supplied and mounted in 7-foot-high racks which would fit inside the van itself and completed with all the inter-panel wiring in lead-covered CMA-grade cables. On 19 February, Blumlein again wrote to Birkinshaw informing him that an additional £137-0s-0d would be required to fit a balanced output stage to the mobile scanning unit, something the BBC had requested.

On 4 March 1937, Blumlein applied for Patent No. 490,150 (strangely, EMI are also credited as having co-written this patent), which refers to the earlier Patent No. 455,375 in which the specification disclosed a circuit arrangement which depended upon the components having shorter time constants between pulses than during pulses. Actually, the time constant between the pulses was much shorter than the gaps between them and this later patent aimed to provide a pulse separating circuit which would give correct separation for interlacing frame pulses, while employing time constants which are longer during the intervals between pulses than during pulses. What Blumlein was doing was effectively separating the relatively long duration frame pulses from the line pulses of relatively short duration despite the fact that they both were of the same amplitude.

In the earlier patent, a circuit arrangement had been provided for effecting selection from a mixture of line and frame synchronizing pulses which are of differing duration. Of the frame pulses which exceed the duration of the line pulses, this circuit rendered the intervals between the pulses effectively short by means of a unilaterally conducting device such as a diode valve. In the new circuit, Blumlein described how on application of a synchronising pulse to the grid of the valve, that valve was rendered conductive and the condenser tended to charge negatively. In the intervals between the synchronizing pulses the potential of the anode of the valve tended to return towards high tension voltage. The time constant in this instance was such that the discharge of the condenser occupied a time somewhat greater than that of the line period. On application of the frame pulses to the grid, the anode potential tended to fall further than in the case of the line pulses, passing a limiting value at which the frame scanning apparatus was operated. The time taken for the potential to pass this point depended upon the amount of charge in the condenser left over from the preceding line synchronizing pulse.

By the first week of March Marconi-EMI were coming under increasing pressure from the Television Advisory Committee to ensure that promotion of television receiver sales were not harmed by the possibility of an alteration to the system then in operation. The Television Advisory Committee were preparing a public statement to that effect and required assurance from Marconi-EMI that they would not be altering their system until the end of 1938, at the earliest. J. Varley Roberts wrote to Shoenberg on 5 March 1937, pointing out that:

Figure 7.4
The morning of the coronation of King George VI, 12 May 1937. On the plinth at Apsley House Sir Noel Ashbridge, chief engineer, BBC is showing camera position No. 1 manned by camera operator Bliss to the gathered VIPs below. These include Lord Selsdon, chairman of the Television Advisory Committee, F.W. Philips, GPO, Sir Charles Carpendale, controller, BBC, C.G. Graves, director of programmes, BBC and Isaac Shoenberg, EMI. (*Courtesy of EMI*)

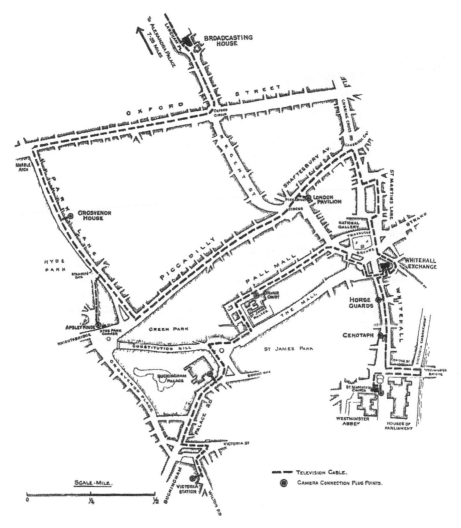

Figure 7.5 The 8-mile route of the balanced television video cable.

... while television sales at the moment show most excellent promise, there still seems to be some doubt in the minds of both trade and public concerning the stability of the present transmission system, and until this is finally removed I feel that it will have a detrimental effect upon future business. I am therefore going to ask you if it is possible for the Postmaster General to issue to the press a statement emphasizing the fact that the present transmission standard will remain unaltered until the end of 1938. The publication of this would I am sure re-act most favourably upon the buying public, and would enable the industry as a whole to embark upon its production programme with the feeling of perfect security.

The outside broadcast units were now being readied and Turnbull informed Birkinshaw on 18 March, of the specifications for the sound equipment that would

Figure 7.6
Siemens Brothers installing the balanced television video cable at Alexandra Palace prior to the coronation of King George VI.
(*Courtesy of Felix Trott*)

be required in the OB truck. This comprised of one fading and mixing console suitable for use with four HB1C moving coil microphones, two microphone amplifiers (one of which was a standby), one main control panel, two line amplifiers capable of raising the mean power level to +10 dB before sending the signal over the line, one switch panel and a suitable power supply for everything. Four 600-foot lengths of microphone cable were required to feed from the microphones themselves which were to be placed right by the television cameras on the procession. A complete set of valves was also to be supplied and the entire equipment installed in the mobile unit for a cost of £100-0s-0d with delivery 10 weeks after acceptance (in other words, just in time for the coronation). Turnbull also pointed out to Birkinshaw: 'With regard to your suggestion that the sound controls should be mounted in a position remote from the sound racks, we feel that since it will be necessary to mix and fade at very low levels, any unnecessary risk of pick-up in these leads should be avoided. We therefore prefer to locate the controls on the sound rack, but if necessary raise the operator so that he will have a good view of the picture monitor above the heads of the vision control engineers.'

Figure 7.7
The installation by Siemens Brothers of the balanced television video cable at Broadcasting House.
(*Courtesy of Felix Trott*)

Figure 7.8
Camera position No. 2 (Apsley Gate)
on the cold and wet morning of 12
May 1937, waiting for the coronation
of King George VI. (*Courtesy of EMI*)

The mobile television units, while late (it had originally been intended to hand over to the BBC for testing by 10 March), were ready by 24 March. What was needed now was a way of relaying the pictures from the remote unit (which would have the pictures and sound fed to it from the camera positions), the eight miles back to Alexandra Palace. This was solved by using the ingenious cable which Blumlein and Cork had come up with the previous year. Manufactured under licence by Siemens Brothers of Woolwich, the television cable was twin core with two self-locating conductors 0.080 inch diameter, paper-insulated and sheathed to a diameter of 0.91 inches with a ternary lead alloy. It was flexible enough to be wound on huge wooden drums which were then driven by lorry and positioned at convenient Post Office man-holes on the proposed route. These already contained several telephone and telegraph cables. Installations of this type took place at the Houses of Parliament, Westminster Abbey, Buckingham Palace, St James's Palace, Victoria Station, Broadcasting House and of course Alexandra Palace. The television cable was fed through the Post Office cable conduits connecting at joining chambers, several of which had television camera connection plug points. This would allow up to nine television cameras to be positioned along the route, at Victoria Station, Horse Guards' Parade, St Margaret's Church, Westminster, the Cenotaph, Buckingham Palace, Apsley House on Hyde Park Corner, Colour Court on Pall Mall, Grosvenor House on Park Lane and at the London Pavilion on Shaftesbury Avenue. Each of the possible nine camera positions could be connected in circuit if required though, in the end, it was decided just three would be needed.

The additional problem of where to put the cameras so that they would have an uninterrupted view of the new Royal Family, also presented itself to the BBC, who were having to scour the proposed course of the procession to find suitable positions. Of course, the ideal place would be inside Westminster Abbey itself; the job fell to Gerald Cock, who became the first director of television and who was in charge of running the outside broadcast aspect of the BBC Television Service (having already run outside broadcast for BBC Radio for a number of years). Gerald Cock now approached the Archbishop of Canterbury, Cosmo Gordon-Lange, as well as the Duke of Norfolk, who had gathered a meeting of the press, radio and Movietone media at the abbey to discuss the matter. Cock asked Cosmo Gordon-Lange if he would allow the television cameras inside the abbey itself, but despite the fact that Cosmo Gordon-Lange knew Gerald Cock well, and apparently quite liked him, he felt that the solemnity of the occasion precluded intrusion by televi-

Figure 7.9 King George VI passing by and waving to camera position No. 3 (right foreground) 12 May 1937.

sion and refused saying: 'Well, look you've only got a few viewers, it really isn't justified.'

So Cock had to find somewhere else to site the cameras to make the event exciting: 'I found the one place in London,' he later explained, 'where you could get a close-up of the King and the Queen and the Princesses without any interruption, and that was the central plinth of the Hyde Park Corner archway, where 'Bridgy' (Bridgewater) later on took the camera and did the camera action. Not content with that, I went to 'Buck' House and asked the King if he would be good enough to smile into the TV camera. And damned if he didn't! Actually, I was told afterwards, his secretary told me, that he'd had a slip of paper inside the coach, and he'd got on this 'look right outside the window at Hyde Park Corner and smile', and they both did, and we got a beautiful close-up.'

Wednesday 12 May 1937 was a damp and rather dull day which threatened at one point to ruin the procession not only for the massed crowds, but for the television cameras, which required reasonable amounts of light in order to transmit decent pictures. In the morning, Isaac Shoenberg arrived with dignitaries of the Selsdon Commission to see the camera positions at Apsley House on Hyde Park Corner. The three cameramen were, Mr Bliss (Camera No. 1: North face of the main arch at pavement level), Mr Tonge (Camera No. 2: North face of the main arch on a timber plinth extension) and Mr Price (Camera No. 3: South face of the main arch on a timber plinth extension), the sound engineer was Mr Ward, and in the control room in the OB truck was Bernard Greenhead from EMI who was in charge of the vision rack relays and the vision mixer was Mr Mair. For the BBC, Gerald Cock was the director of television, either Douglas Birkinshaw or Terence Macnamara were to be the engineer-in-charge (depending upon who was at Alexandra Palace or with the outside broadcast team), D.H. Munro was the television production manager and Tony Bridgewater the supervisory engineer.

The Marconi-EMI equipment had all just been delivered and set-up the day before, with the television vans only just completed in time for the event. On the day of the coronation it nearly all worked perfectly first time with no problems reported from any of the three camera positions or the engineers being relayed the pictures back at Alexandra Palace. The problem that did arise nearly cost the entire transmission however. Just as the royal procession was nearing the first camera position a dry joint in the vision relay circuitry in the control truck shorted out and the picture transmission stopped. Bernard Greenhead, who had been responsible for much of the equipment (including its installation and operation on the great day), took a guess at which panel would be the cause of the problem. Giving the offending relay rack a sharp kick with his shoe, the circuit was re-established and with barely three minutes to spare, picture transmission commenced. Blumlein of course had installed a television set at home long before the coronation in order to carry out some of the tests from Alexandra Palace the year before and so the Blumlein household were among the privileged few who were able to watch the procession as it happened.

Figure 7.10
Gerald Cock, the first BBC director of television. (*Courtesy of APTS*)

During the television development period Blumlein would naturally attend IEE meetings, usually with his friends L.H. Bedford and J.B. Kaye. Recalling one such evening J.B. Kaye explained:

'Of course we used to go to the IEE meetings. The one I remember possibly most, I think it was on the phototubes, and Blumlein, L.H. Bedford and I went along, and we, as on other occasions, we'd possibly dine, the three of us, at the Stone's Chop House just off Leicester Square. It was knocked down in the Blitz as far as I recall, and then rebuilt, and is rather a flashy sort of place now. One got lovely steaks at that restaurant. That particular evening, it was the "odd route", we called in at the funfair at the Haymarket, and they had little racing cars on tracks. And I should think they must have thought they'd got three prize lunatics in.

It would only run two cars at a time, because one raced against one another. Blumlein and Bedford, being just about on a par of explanation in the mechanics, proceeded to explain the way to get these little cars round in the shortest possible time. It wasn't the maximum speed of the wheels, because if you overdid it, the wheels spun on the track. You had to keep it on the track, going at the maximum speed of the little car. Anyway that was quite a merry evening, although I think the other people in the place had thought three lunatics had arrived, we survived that one and headed for Stone's.'

It seems that L. H. Bedford had equally fond memories of such evenings:

'After IEE meetings we retired to Stone's Chop House to partake of moderate food and excellent philosophy. We used literally to sit at the feet of the master, and listen to what he had to say; it was always interesting. I remember he was always cracking up Thévenin's Theorem and how he would interview job candidates using trick circuits to test them. I also remember him telling us that he was trying for a formal proof of an intuitive "Blumlein's Theorem", to the effect that one cannot improve signal noise by feedback! Once, we had met up at EMI, when he had been there only a little time, and he said, "I don't count for anything in this place. I'm much looked down on here." I said, "Well, how's that?" He said, "I have a drawing board in my office!"'

Did Blumlein seriously doubt his worth as an engineer and a scientist? Despite his legendary modesty, it seems that from time to time, as with all great men, Blumlein did question his abilities. Philip Vanderlyn recalled one such occasion:

'One of his colleagues once remarked to me that Blumlein had one fear – that his inventive streak would fail him. I myself find it impossible to imagine him without his characteristic flair for turning the multifarious properties of valves and circuit components to good account, and for doing so in an elegant way, so that his circuits operated reliably and repeatedly. It seems to me that if you have this gift, you always have it.'

Blumlein had not applied for a patent since February until No. 496,139 on 24 May 1937. The invention was for a self-starting synchronous induction motor which would be started by a squirrel cage winding and could be operated as a synchronous motor with a series-wound DC excitation circuit. The purpose of the invention was to produce and adjust a value of negative resistance in sufficiently small steps so that the effective resistance of the circuit could be brought to a positive finite value which was substantially less than the resistance of the windings alone. In this way the induction motor could improve the synchronization characteristics when running up to speed. June 1937 was a much busier month for Blumlein and he applied for four further patents. The first of these, dated 5 June, is another which is widely considered to be among his most important. Patent No. 496,883 is for an invention which later became known as the 'ultra linear amplifier'.

As with so many of Blumlein's circuits, the design is deceptively simple. It shows a pentode output stage of a single-ended audio frequency amplifier with a tap on the primary winding of the output transformer. This tap provided feedback to the second grid to improve the linearity of the amplifier. Blumlein realized that if the tap was placed at the anode end of the primary winding, the valve would then be connected as a triode, and if the tap was at the supply line end, it would be a pure pentode. It was well known that when a pentode was connected as a triode it would be far less efficient and therefore provide a lower power output. If however, the tap were placed at a distance say 15–20% down from the supply end of the transformer, the valve would combine positive features of both the triode and the pentode.

The ultra linear amplifier circuit in its push-pull form became widely used in high-fidelity valve amplifiers during the Second World War, though it was a good many years before general practice caught up with Blumlein's thinking. While Blumlein did not consider the problems associated with distortion resulting from feedback to the screen, he regarded it mainly as a convenient alternative to the control-grid feedback for reducing the undesirably high impedance of pentodes while retaining their efficiency. He did suggest lowering the output impedance of the valve to approximately equal the optimum load. This would have been done in order to prevent damage to the valve or the output transformer if, for example, the load became disconnected.

On 7 June 1937, Patent Nos 495,724 and 501,966 were applied for. The first of these is a simple method for producing a non-circular spot from a cathode ray tube in order to eliminate line structure and other unpleasant visual effects. The second patent, co-written with Rolf Spencer, is a rather clever circuit design to improve the method of 'overlaying' signals from two television cameras in order that the image viewed by one camera could be monitored by the other. Two days later on 9 June 1937, Patent No. 497,060 was applied for. Here Blumlein describes a gamma-correction amplifier for television signals which used a non-linear element (diodes or triodes in this case) in the negative feedback path.

During the first twelve months of the London Television Service (and for some years after that until war broke out in September 1939) Marconi-EMI charged the BBC quite considerable sums for the services of their engineers during training and outside broadcast transmissions. The BBC were engaged in a massive training programme of their own by this time, but engineers did not grow on trees. EMI, realizing that this might prove to be a lucrative period for them, engaged the BBC in a contractual agreement whereby the very engineers who had designed and built the high-definition television system, were now being used to train the BBC engineers in its use.

One such occasion is highlighted in a letter from Condliffe to the BBC dated 29 July 1937, 'We beg to inform you that our charge for the services of engineers during the television of OBs of the Davis Cup covering July 15th, 16th, 17th, 19th, 20th, 23rd, 24th, 26th, 27th and 28th will be £72-0-0 (Seventy-Two Pounds).'

October and November 1937 were to be equally active months for Blumlein as six more patents were applied for. On 4 October Blumlein produced his only patent which has no diagrams whatsoever, No. 505,079. The invention however, is designed to reduce co-channel interference from adjacent television stations by polarizing them vertically and horizontally. It might have seemed odd that Blumlein was thinking in terms of adjacent television stations causing such interference less than a year after the initial service had begun, but this is a typical example of the foresight he could have.

Four days later on 8 October, Patent No. 503,765 describes a high-impedance twin feeder aerial using extremely thin wires which were constructed to be kinked and then supported in the tubes of the aerial frame. In this way, should the aerial

Figure 7.11
Marconi-EMI transmission aerial at
Wimbledon 1937. (*Courtesy of EMI*)

become stretched through adverse weather conditions for example, the wires in the tubes would 'concertina' rather than break.

Just under a week later on 14 October, Blumlein applied for Patent No. 503,555. He defines a system whereby the picture detail in a high-definition television system (of the type which Marconi-EMI had designed and invented and the BBC were now using) would be increased thereby allowing the detail in the reproduced image at the receiver to be correspondingly increased. This was done by using a much finer scanning spot with carefully synchronized spot-wobble allowing the detail to be increased on special receivers to an effective definition of 800-lines or more in a single frame.

Two more patents were applied for on 6 November 1937. The first of these, No. 505,480, describes an invention for a pentode grid with a cathode follower while Patent No. 507,239, which was co-written with Eric White, details a black-level clamp for television. This clever circuit used slightly delayed, synchronized pulses to open a valve that would then clamp the black level. Finally on 18 November 1937, Blumlein co-writing with Cabot Seaton Bull (with whom he only wrote this one specification) applied for Patent No. 507,665. Here, a valve is used as a potential divider for high voltages by having the grid and the input to the anode earthed, so producing a reduced output at the cathode.

## Building confidence in television

During the remainder of 1937, BBC television cameras had started appearing at

Figure 7.12
An Emitron camera and camera operator at Wimbledon 1937. (*Courtesy of EMI*)

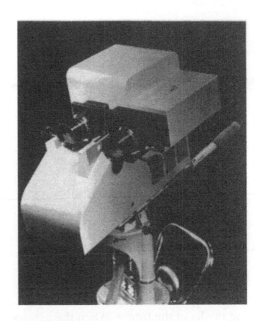

Figure 7.13
The Emitron camera front detail.
(*Courtesy of EMI*)

many outside broadcast locations, starting with Wimbledon and the Lawn Tennis Association Championships which took place in June. Marconi-EMI built and delivered to the BBC a mobile outside broadcast unit complete with radio-link transmitter vehicle and portable directional aerial. The championships were transmitted on ultra-shortwave signals the twelve miles from Wimbledon to Alexandra Palace using this system. It was the first time that any outside event had been televised without a cable link (even though there was provision on the vehicle to carry 600 feet of cable on drums), another first for the London Television Service.

The 1937 Derby was also televised for the first time, with one camera positioned right by the winning post. As the horse that won flashed past at the last minute, coming from nowhere in the pack, there was a fabulous picture, in perfect clarity, of a long banner which said 'Booths Gin' on it. This put the whole of the commercial world in a spin – television, what a great way to advertise for free, the advertis-

Figure 7.14
The Emitron camera side elevation
with the casing and lenses removed.
(*Courtesy of EMI*)

Figure 7.15
Armistice Day, 11 November 1937.
The laying of wreaths at the
Cenotaph. (*Courtesy of EMI*)

ing media of the future. When the new soccer season began in August, so the television cameras were there, and again when rugby football started a month later, that too was televised.

Being able to televise the procession of the coronation, various sporting events and the like, was naturally a major coup for television, but there was to be yet another milestone event just a few months later in November; one which would open peoples' eyes to more far-reaching possibilities. The occasion was Armistice Day, the laying of the wreaths at the Cenotaph to mark the anniversary of the ending of the

Figure 7.16
Emitron camera on roof looking
towards Big Ben, 11 November 1937.

Great War on 11 November. The television cameras were there again and Gerald Cock had refused to have a BBC commentator during the programme, as he wanted the service itself to tell the story to the viewers. During the event there were several intervals where the cameras would change views from Whitehall to a scene looking at Big Ben as eleven o'clock approached, and then back to Whitehall again. For this, the BBC had placed television cameras on the rooftops near Whitehall, which could pan across the mist of the November day in London, towards the famous tower. The whole event conveyed exactly the sense that Cock had wanted, it was all very moving and sombre, and it worked, with many TV viewers reportedly in tears because of the whole event.

At one point however, a man who it was later discovered had been a shell-shock case during the First World War, hurled himself behind King George VI, near to the base of the Cenotaph, right in front of the television cameras. Of course, the police all charged at this poor fellow and wrestled him to the ground to get him away from the King; but what the event had shown – for the very first time – was that television would be able to show things as they happened, regardless of what it was, because nobody knew what was going to happen. In this case the viewers were privy to an event which could have been calamitous. Television had come of an age, and in so doing that one event began a rush on the sale of television receivers.

## From Emitron to Super Emitron

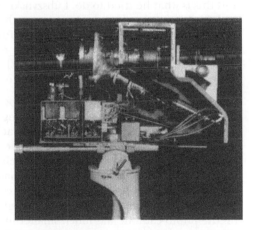

Figure 7.17
The Super Emitron camera side elevation showing interior detail. (*Courtesy of EMI*)

The King laying a wreath at the Cenotaph in November was also most significant for EMI as it was the first occasion when the new Super Emitron cameras were used for an outside broadcast. Developed by Lubszynski, the Super Emitron replaced the original Emitron (Patent No. 455,123 applied for by Lubszynski and McGee on

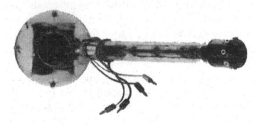

Figure 7.18
An Emitron tube. (*Courtesy of EMI*)

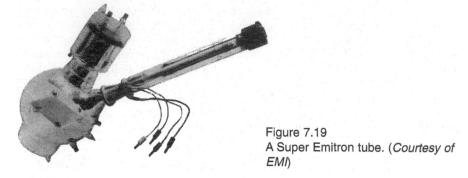

Figure 7.19
A Super Emitron tube. (*Courtesy of EMI*)

9 February 1935) tubes, which, while they worked, only had an efficiency of about 5% of the theoretical maximum. The reason for this low efficiency was the lack of saturation of the photo-emission from the mosaic during most of the frame period. The spread of secondary electrons released by the scanning beam from the mosaic also compromised performance as they tended to neutralize the charges stored on the mosaic elements, and would generate spurious signals such as 'tilt', a low frequency component super-imposed on picture signals.

Lubszynski considered the possibility that if the two functions, photo-emission and charge storage, could be separated, the Emitron tube could be made more efficient, and this is what he tried to do. Lubszynski devised a new tube with the electron gun and mosaic predominantly the same as the original Emitron, the chief difference being the addition of a projecting neck for the electron optical conversion of the light picture into an electron picture. At the end of this neck was an optically flat window in front of which was placed the photo-cathode, and this was surrounded by a short cylinder for concentrating the photo-electron beam.

The silver coating on the walls of the tube forming the second anode of the scanning gun was extended into this projecting neck, the potential on it serving also to accelerate the electrons from the photo-cathode, which were focused on the mosaic by means of a coil forming a magnetic lens. The photo-electrons arriving on the mosaic each released several secondary electrons, thus forming an amolified electron image. The magnetic focusing also had the effect of producing an angular displacement of the whole image about its centre, so that the gun had to be set at a corresponding angle by adjusting the angle of the whole tube. The optical picture on the photo-cathode was much smaller in the new tube than that on the mosaic of the original Emitron and therefore smaller aperture lenses could be used for the same illumination degree resulting in a greatly increased focal depth on the transmitted images.

Lubszynski called the new tube the Super Emitron and it was almost immediately put into operation by the BBC who were especially pleased with its response in low light situations which would have previously been considered marginal for television picture transmission. Later the sensitivity of the electron beam forming the pictures was increased even further with the addition of an amplification stage of the secondary emissions before they reached the mosaic. In the normal Emitron the photosensitivity was limited to about 12 µA/lumen. However, with the main feature of the new tube being that the tube area itself was a secondary amplification stage at the mosaic surface the amount of charge stored was multiplied according to the sec-

Figure 7.20
The prototype of the Super Emitron
camera being daylight tested outside
the Works Designs Research and
Development building at Hayes in
1937. (*Courtesy of EMI*)

ondary emission coefficient. This measurement gave the Super Emitron for equal sensitivities about 10 to 15 times the signal amplitude of the original Emitron tube and, in a few cases, this ratio was considerably greater. As the size of the optical picture on the photo-cathode was much smaller in the Super Emitron than the Emitron, lenses with a shorter focal length could be used. Furthermore, owing to the increased sensitivity of the new tube, smaller lens apertures could be used for the same illumination resulting in greatly increased focal depth of the transmitted pictures.

During this period, EMI still enjoyed a relatively good working relationship with RCA in America who were keen to benefit from the British television developments. A constant stream of visitors from the United States had come to Alexandra Palace to marvel at the uniform high quality of the pictures, the regularly scheduled programmes and the coverage of outside events. Some of the comments at the time were 'an operative system giving good stable pictures of acceptable detail, brilliance and interest', 'stability, freedom from fault sync and a wide contrast range' and 'of remarkable contrast and detail; exactly like a movie'. These and others were a genuine tribute to Isaac Shoenberg, Alan Blumlein and the engineers of Marconi-EMI.

It was decided that Blumlein should travel to America to visit RCA in New York to negotiate a free exchange of knowledge with Marconi-EMI. Naturally, Blumlein was happy to agree to this and he travelled in the style of the day on a transatlantic liner. During the voyage Blumlein evidently entertained himself in the manner which he had become accustomed to with his friends in London, winning first prize at the onboard shooting gallery. The certificate he won on his transatlantic excursion in 1937 is still retained by the Blumlein family to this day.

In December 1937, Blumlein applied for three further patents. No. 507,417 is a rather clever anti-ghosting device in which Blumlein returned to his old favourite – networking systems. Having explained in the Patent how television signals can arrive at a receiving antenna in multiples, all within a milli-second or so of each other, Blumlein goes on to demonstrate how this annoying effect known as 'ghosting' could be eliminated. The main television signal was fed through a network and delayed as appropriate before being fed back as required so effectively removing the ghosting effect. Patent No. 508,377 is a long and complex work comprising some 21 pages and 12 diagrams. It describes a method of combining the output from a television camera (giving the AC level of the signal) with the output from a single photocell (giving the DC level) in order to remove spurious signals during the re-insertion of the DC or low-frequency component of the signal. The final patent of 1937, No. 512,109, was co-written with Eric White. The invention described is a bi-directional switch adapted to be controlled by pulses utilizing three diodes with strapped anodes. The patent was re-worked and applied for again with a few amend-ments on 25 April 1938.

Towards the end of 1937, the first plans had been made for experimental airborne television trials to begin. These were continued for a while during 1938, but although some progress was made, the technical difficulties were considered to be uneconomical at the time, and so were shelved. Baird Television, having lost the race to perfect the system used by the London Television Service had turned their attention away from mechanical televising processes by the end of 1937, and were investigating the principles of all-electronic television at last. Early in the New Year 1938, Baird announced that it planned to provide its own programming service based at Crystal Palace. By 4 February, John Logie Baird was able to give a demon-stration of 120-line colour television at the Dominion Theatre, London with wire-less transmission from Crystal Palace. Though this system still required 20 revolving mirrors, rotating at 6000 rpm, the image was effectively interlaced through a lens with 12 concentric slots at different distances from its periphery. By this means, the field given by the 20-line drum was interlaced six times to give a 120-line picture repeated twice for each revolution of the disc.

On 21 February 1938, Alan and Doreen Blumlein's second son, David Antony Paul Blumlein, was born. Though the family were still living at 32 Audley Road, Ealing, it was becoming obvious that the growing Blumlein household would soon require more space. By the end of 1938, Alan and Doreen had found the perfect home for their family quite literally just around the corner. The new house was large with a garden, and was within the same fairly new housing development that Audley Road was part of. No. 37, The Ridings, was on a road which was bisected by Audley Road to the south. It was just far enough away from the busy interchange at Hanger Lane to be secluded from the noise of the traffic, yet it was perfectly situated for easy access to a main road for Alan to travel to Hayes. Most probably he would have driven down Hanger Lane to the junction of the A4020 Uxbridge Road, and from there it was a straight road through to Hayes. Though Alan Blumlein had been born in Hampstead, he must have had quite an affection for this part of London as he spent most of his adult life living there. In all, the four houses in Ealing in which he lived are separated by a matter of just a few streets, with Courtfield Gardens the fur-thest away from The Ridings at just over one mile.

On 7 March 1938, Blumlein applied for Patent No. 515,684 which though he was not to know it at the time would prove in the years ahead to be one of his most useful inventions. The design has become known as the 'slot aerial' and comprises a longitudinal tube with a slot running its entire length. This forms a condenser and thus provides, with the inductance of the conductor, a tuned circuit for controlling the propagation properties of the conductor. During the war this slot aerial design would be used over and over again for short-wave radar design, and of course they are commonplace today. When Blumlein put forward his patent for the slot aerial however, even bearing in mind his known freedom from convention, it must have seemed quite revolutionary.

Patent No. 515,044 was applied for on 23 March 1938, and was co-written by Blumlein, Cecil Browne and John Hardwick. It describes a circuit for measuring the instantaneous value of pulses on a cathode ray tube, in conjunction with a variable bias potential or current by moving the trace past a datum line. The system was used to visually identify television signal pulses (though it can equally be used to measure any other type of pulse for that matter) and then to give a graphic representation of the potential of a pulsating waveform. It was of particular use in a sweep circuit where any changes at a given instant could be identified and displayed during the cycle of change through which the potential passed.

One week later, on 30 March 1938, Blumlein applied for Patent No. 515,348 – a variable gain amplifier in which the automatic gain control kept the level of a pilot signal constant at the output, while the amplification varied inversely as the amplitude of the pilot signal altered at the input. In April two patents were applied for. The first of these, dated 13 April, was No. 514,825 in which Blumlein co-writing with Eric White invented a protection system to ensure against surges caused by the final anode of a cathode ray tube should it short-circuit to earth. The second patent, No. 514,065, dated 25 April 1938, describes an application of the long-tailed pair in which electrical pulses from different trains can be mixed. Many of Blumlein's 1938 ideas are based on aspects of the high-definition television system which by this time was in full service and justifying everyone's hopes in it. Though his inventive fervour never once diminished, by April 1938, Blumlein had prepared what would prove to be only the second document of his professional career to be published. It was a masterpiece in its own right, and with similar works from his colleagues, would be presented before the Institute of Electrical Engineers, in order to explain the sum of their achievements.

## The process of publication

Having completed the development of the Marconi-EMI high-definition television system at Alexandra Palace, the men responsible spent part of 1937, and much of 1938, producing a series of definitive publications read before the various sections of the Institute of Electrical Engineers, in order to explain the sum of their achievements. The BBC's Macnamara and Birkinshaw were first with a paper entitled 'The London Television Service', which was originally received by the IEE on 1 December 1937, and then read before the north-eastern centre on 11 April 1938. It was, however, ten days later before the Institution members gathered at their Savoy House headquarters with Macnamara and Birkinshaw in conjunction with the Marconi-EMI development team that the main readings took place.

The evening of 21 April 1938 commenced with an introduction to the new television service, which took into account such things as a brief history of the development of television, the decision for the site at Alexandra Palace, the design and building of the studios, and an outline of both the Baird and Marconi-EMI equipment as well as the aerial mast on the south east tower. This was followed by a three-part, in-depth explanation of the Marconi-EMI system to be given by Alan Blumlein, Cecil Browne and Norman Davis with Eric Green.

The first part of the system to be explained, 'The transmitted wave-form', was given by Alan Blumlein, in which he spoke about unidirectional constant-velocity scanning, various forms of modulation, 'infra-black' pulse synchronizing, DC transmission, 405-line scanning, intervals between lines and frames as well as an explanation of the Marconi-EMI transmitted waveform. During his speech before the Institution members, an unknown cameraman, possibly a journalist associated with the IEE, sitting to Blumlein's right, and at the back of a quite crowded room, stood up and took his photograph. Inadvertently, this photographer provided us with the only known picture of Alan Blumlein giving a presentation of his work and one of just four known photographs in total of the man. The photograph appeared first in the journal of the IEE in 1938, and later in Stanley Preston's article 'The Birth Of A High Definition Television System', in the *Journal of The Television Society*, in 1953.

Following Blumlein, Cecil Browne gave the second part of the lecture on the Marconi-EMI system, talking about the 'Vision input equipment'. Browne explained about the studio layout, the cameras, cables and fading mixer, as well as the amplifiers, and power generators. Finally, Green and Davis, gave a joint lecture on the specifications of the transmitter, valves, the fundamental circuit and modulated amplifier construction details. The entire published document runs to 34 pages in total, with Blumlein's contribution totalling nine pages. The contribution to this paper represents only Blumlein's second published account of his work (other than patent specifications), and the first since he and Professor Mallett had co-written a paper for the IEE in 1924.

Following the completion of the reading of the papers, the IEE opened up the floor to the traditional questions session. Sir Noel Ashbridge first paid tribute to the

Figure 7.21 Alan Blumlein giving a lecture to the members of the IEE on the Marconi-EMI high-definition television system on the evening of 21 April 1938. (*Courtesy of EMI*)

Marconi-EMI development team for the work that they had achieved, and then went on to point out differences in economics between sound broadcasting and vision broadcasting, with specific reference to studio equipment and staffing levels. Then A.J. Gill, one of the members of the Selsdon Committee delegation who had travelled to Germany in 1934, remarked on the achievement of the production and installation of the 8-mile length of cable by EMI, drawing particular attention to the fact that reception was no different from the direct pick-up or through the cable itself.

G.E. Condliffe, who had of course, played such an important role himself as the man responsible for managing the EMI development team, commented on the decision to adopt the 405-line system, pointing out that is was unfortunate that (at that time) Germany and America were going to adopt a 441-line system (the next suitable odd integer), and that to achieve any noticeable difference, the line rate would have to be raised to around 600 lines (which, of course, it eventually was, and still is). Among the other members in the audience who raised points was Blumlein's old friend, L.H. Bedford, who praised Blumlein for his three outstanding sections on the waveform, the compensation of spurious signals by means of tilt-and-bend signals, and the choice of the 405-line standard.

The papers on 'The London Television System' and 'The Marconi-EMI Television System' went on to be read at the various IEE centres throughout Britain, Mersey and North Wales (Liverpool) at Liverpool, on 28 November 1938; South Midland Centre, at Birmingham on 5 December 1938; North Midland Centre, at Leeds on 17 January 1939; and the Western Centre, at Newport on 13 March 1939. At each reading, praise was given to the members of the team not only for the achievement of placing Britain at the forefront of world television development, but also for the manner and speed with which it had been completed.

Other members of the EMI development team also read papers giving details of their work before the members of the IEE, before having them published in 1938, and 1939. Edward Cork and Joseph Pawsey wrote a paper titled 'Long Feeders For Transmitting Wide Side-Bands, With Reference To The Alexandra Palace Aerial-Feeder System', which along with a paper titled 'EMI Cathode-Ray Television Transmission Tubes', written by James McGee and Hans Lubszynski, were both read to the Wireless Section IEE members on 7 December 1938. This was followed by Henry Clark and Ivan Turnbull, who wrote a paper titled 'The Marconi-EMI Audio-Frequency Equipment At The London Television Station', which was read before the wireless section of the IEE on 3 May 1939.

## The transversal filter and others

Blumlein had several notable patent achievements in the year of peace that remained. In May 1938, a series of patents were applied for with consecutive numbers by several of the EMI television team, four of which bear Blumlein's name. Each dated 30 May, Patent No. 515,360 is a method of establishing a black-level datum in signals from a television camera. In Patent No. 515,361 (Blumlein alone) and No. 515,362 (co-written with John Hardwick and Frank Blythen) various improvements are made to Patent No. 458,585 of April 1935. Last in this series is Patent No. 515,364 in which Blumlein makes another improvement to his original idea for DC restoration circuits for television. These specifications prove how, as the months since the high-definition television system had first been put into service, the development team at EMI had continued to improve many aspects of its potential and application.

In America on 3 June 1938, the Television Committee of the Radio Manufacturers Association (RMA) approved a new set of American television standards. These included a 441-line frequency so that the synchronizing generator could be simplified ($3 \times 3 \times 7 \times 7$ as integers) at 60 fields interlaced. Other specifications included a 4:3 aspect ratio, negative picture modulation, horizontal polarization, a constant black (DC) level and the adoption of the serrated type of vertical sync pulses with additional equalizing pulses. With the exception of the 441-line frequency, most of these standards had been 'given' to the Americans by the British. Many of the final forms of the circuit designs had been brought by Alan Blumlein to America during his transatlantic voyage the previous year. This was all part of the free exchange of ideas that Marconi-EMI shared with RCA. It has to be said though that there was a natural disappointment felt by the British television development team when the Americans chose to 'raise the stakes' by adopting the next available odd integer above 405 lines.

Within weeks of this announcement, the Germans adopted a 441-line frequency television standard also. By the time the Berlin Radio Exhibition opened on 5 August 1938, Fernseh were ready to demonstrate 441-line systems on a large screen projector, 10 feet by 12 feet. Though by October they had also previewed home receivers with a picture some 8 inches by 9½ inches costing the equivalent of £35-0-0, there was little chance of the average German being able to purchase a television receiver. The Nazis had seen to that.

Following the series of modifications to existing specifications, Blumlein's next patent was entirely new. Originally dated 28 June 1938 (also 15 February 1939, when certain amendments were made), No. 517,516 embodies the invention which would eventually become known as the 'transversal filter'. It is another of Blumlein's classic circuit designs. Co-written with Heinz Kallman and William Percival, this specification represents an important development in linear network theory. It approaches the problem of shaping waveforms from a radically different point of view from that used in the usual wave filter theory of the time. It also happens to be Blumlein's largest patent by far with 26 pages and 28 diagrams to support.

Described by Eric White many years after the patent had first been applied for, the transversal filter was explained thus:

> The basic idea of the circuit is to put the input into a delay line with a series of taps along it, and to add weighted signals from all the taps in order to form the output signal. Some weights will usually need to be negative, and to achieve this the resistors are taken to a separate bus-bar whose output is reversed by a phase-reversing device before adding to the remainder. To show that the transversal filter can do anything a conventional filter will do, consider the effect of the conventional filter on a unit impulse. If the amplitude/frequency and phase/frequency characteristics of the filter are known, then the effect on a unit impulse can be calculated. It will produce some arbitrary waveform of finite duration but, proper choice of the weighting of successive taps will enable the transversal filter to reproduce the same arbitrary waveform from a

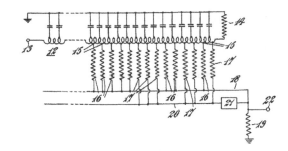

Figure 7.22
Detail from Patent No. 517,516
(1938), the 'Transversal Filter'.

unit pulse. It will then have the same amplitude/frequency and phase/frequency characteristics as the original filter.

A particular virtue of the transversal filter is that it is easy to make the equivalent of a linear-phase filter, simply by ensuring that the weights are symmetrical about the centre tap. In this case it is only necessary to provide half the delay line, and the first half of the set of taps and weights; if the line is left open-circuit then reflection will do the rest. Rudimentary forms of this are often to be found used for aperture-correction circuits.

The final patent of 1938, No. 520,646, dated 27 October, describes a television camera tube in which the mosaic is scanned by a light spot instead of using an electron beam. Blumlein had spent some time investigating the problems of 'tilt' in television transmitting tubes and had proposed several designs to alleviate the problem where possible. By using a light spot instead of an electron beam, the effect of tilt would almost be completely cancelled because it was the effect of the electrons themselves on the mosaic, which caused the phenomenon in the first place.

Television had indeed come of age. Sales of receivers were doing well and by the summer of 1939, following very successful Radiolympias in 1937, 1938 and 1939, total figures for all television receiver sales numbered approximately 20,000. Broadcasting had increased from the original one hour a day, to several hours, with repeated programmes in the evenings. There were newsreels, films, entertainment, sport of all kinds including tennis from Wimbledon, the Boat Race, soccer, rugby, cricket, horse racing, even athletics. Wherever there was an event, it seemed the television cameras were there. When, on 30 September 1938, Prime Minister Neville Chamberlain arrived back at Hendon aerodrome from his historic meeting with Adolf Hitler in Munich, Marconi-EMI television cameras were there. Chamberlain made his now famous 'Peace in our time' speech, into an Emitron camera. EMI were doing well from licensing the technology they had developed to other manufacturers keen to sell television receivers which, by mid-1939, numbered more than two dozen varieties.

As Lord Selsdon had hoped and predicted in the Television Committee report of 1935, the price of television receivers was beginning to come down. Marconi-EMI wisely announced that the (then) current television system, which it had been promised would not change at least until the end of 1938, would remain the same for the foreseeable future.

By then however Alan Blumlein and many of the team who had been directly involved with television development were mostly engaged on other projects. EMI were confident that television would have a bright future – provided of course the increasingly precarious peace held in Europe.

## The link towards radar

Blumlein had found his life far too busy after 1935 to return to the subject of binaural sound and audio engineering in general. He did however, make one final use of the lessons which he had learned during those first few years at Columbia and EMI. In June 1938, George Condliffe had suggested that the outputs of over-shuffled microphones could be attached to the input plates of a cathode ray tube in order to indicate the direction of a sound source. The idea was simple enough, in Blumlein's binaural sound scheme two directional microphones were required with 'figure-of-eight' (cosine) polar diagrams, ideally co-located, with maxima of their lobes 90° apart. Such microphones, though developed eventually, were not initially available, so Blumlein used two pairs of omni-directional microphones.

Considering just one pair of these microphones, the sound reaching them from a given source, away from the plane bisecting the axis joining the two instruments, gives signals which are similar, but with a time difference. For sound wavelengths greater than say, six times the spacing of the microphones, the difference between the two signals is substantially the mathematical time differential of either of them, with an amplitude approximately proportional to the cosine of the bearing relative to the axis. If then, this difference is integrated, the resulting waveform is similar to the original signal, modulated by the required cosine-type polar diagram. Blumlein later applied this principle to aircraft sound locators. Using the sum signal from the two receptors as a reference, to remove variations of the source signal, $cos\ \theta$ and hence $\theta$ is obtained as a ratio, e.g. by applying the sum and the integrated difference signals respectively to the two pairs of deflecting plates of a cathode ray oscilloscope.

Henry Clark was set to work designing a rudimentary circuit to test the theory which, following a series of outdoor tests in July, proved to be a viable proposition. Clark now handed over the basis of his calculations and the rudimentary circuit design to Blumlein who proceeded to correct the circuit's deficiencies. These tests were relatively simple in concept. A listener wore headphones connected to the microphones modified with the new circuit. The headphones detected the sound source with the system to an accuracy of 4°; this was achieved with two microphones, each facing at 90° offset to a loudspeaker, which in turn was placed 100 feet away from the sound source. If the difference signal was boosted this could be raised to an accuracy of 1.5°, and when a CRT screen was connected this could be improved once again to an accuracy of 0.5°. The trials were demonstrated (under the supervision of Alan Blumlein), to the Air Defence Experimental Establishment (ADEE), Sir Frank Smith and a Mr Forse of the Air Ministry on 1 October 1938. Further demonstrations to AMRE, the radar establishment, were conducted on 15 October. Not surprisingly, the authorities were impressed by what they saw. They contracted EMI, under direction from Blumlein and Clark, to manufacture a series of units consisting of a pair of microphones in a single parabolic reflector to be an add-on unit for the existing sound detector locators being used.

It so happened that the War Office had been engaged in an attempt since mid-1937, to find a way of designing an anti-aircraft gun-laying radar system in which height could be accurately determined. However, no satisfactory way of achieving this had by that time been found. So the authorities now turned to the EMI sound locator system that Blumlein had demonstrated as a suitable remedy to the situation. The drawback of course was the proximity of the guns to the sound locators and the possibility of the operators suffering from shattered eardrums. Shut-off valves were provided, though this did not eliminate the risk entirely. Blumlein gave the problem some thought and came up with the idea of a visual indication system known as VIE, in which the elevation information, as well as the bearing, were displayed on the cathode ray tube. It was this work that opened up the way for EMI to get involved with the many radar projects that were then being proposed.

## When war came – September 1939

When the scheduled television service was due to be broadcast as usual on Friday, 1 September 1939, there were approximately 23,000 television receivers in the country, though the vast majority of these were still in the London area. War with Germany was however, inevitable despite all the belated efforts of the politicians. Douglas Birkinshaw recalled how he received some ominous news that Friday evening,

> 'I can never forget my feelings when, at 10 o'clock on 1 September 1939, Broadcasting House telephoned me to close down Alexandra Palace at noon, and send the staff to their war duties. So we went away. Many to distinguish themselves in the sister sciences of radar and radio-navigational aids, and many to stay in the BBC to contribute to the service which it rendered to the nation in maintaining morale, in maintaining contact, for all sorts of purposes with overseas countries, and in defence.'

The following Sunday morning, a sombre Prime Minister Chamberlain spoke to the nation on the radio explaining that as a result of the refusal of the German armed forces to withdraw from Poland, Britain was at war with Germany.

The Government had already prepared for this eventuality and closed down all transmissions from the London Television Service at Alexandra Palace for the duration. It is even possible that the Government were somewhat relieved at this. In a strange sort of a way; the enormous cost of running the service (which was still not generating anything like the revenue it needed from licence fees), would be deferred until the war was over. Blumlein and many of the team at EMI who had designed the high-definition television system, were already engaged in work on a whole host of projects, many of which were of a military nature. EMI had contracts for spares with the BBC for Alexandra Palace, which ran through well into 1940, though it made little sense for these to now be delivered to a television station that was off the air. The contracts were cancelled and the spares put to other uses. Quite how much money EMI lost is hard to estimate. Looking through their archives at the long lists of valves, transmitter modulators, monitors, amplifiers, microphones, leads and cables that were all on order with the BBC from contracts signed in the summer of 1939, it is obvious that quite a considerable amount of revenue was instantly lost to the company.

During those first few months of what became known as the 'Phoney War', Blumlein was still working on television Patents which needed to be applied for. On 13 October 1939, Specification No. 577,817 was one of these. Co-written with Eric Nind, they had devised a generator which used voltage doublers to produce a DC output for valve amplifiers, and an AC output to produce extra high tension for cathode ray tubes. In Patent No. 554,715 dated 16 November 1939, Blumlein describes a tubular FM modulator where a coaxial line with a diaphragm housed in a cylindrical shield was driven to follow the frequency modulation waveform thereby varying its length. The apparatus was designed specifically to deal with very short wavelengths and, as the diaphragm vibrated with the effect of the frequency modulated waveforms acting upon it, a resonant circuit of very low damping was effectively tuned and adjusted. This made the apparatus particularly stable in operation.

Though regular television services did not resume for another six and a half years, Alexandra Palace, and perhaps more importantly the transmitter station based there did not however, remain entirely dormant throughout the hostilities. It so happened that during 1941, the German Luftwaffe were employing a radar guidance system for their aircraft called Y-Gërat, which British Intelligence had calculated operated on a frequency of between 42.5 and 46.5 megacycles per second. If it could be proved that this was the case, the German transmissions guiding the bombers to British cities could be re-radiated, with the effect that the ground station would think that the aircraft was at a false distance from the target.

Though the system was known to work (several confused exchanges between German aircraft and ground controllers proved this), British Intelligence could not use the BBC transmitter to immediately jam German transmissions or the fact that their frequency had been discovered would be given away. For several nights in succession therefore, during prolonged raids over Britain, the power of the transmitter at Alexandra Palace was gradually turned up in an operation code-named 'Domino', and the German bombers became more and more confused. They finally gave up on the Y-Gërat system. Purely through luck the British had chosen to produce a transmitter for peaceful purposes which, due to a poor choice of frequency on the part of the German designers, made Alexandra Palace and its transmitter an invaluable step towards winning the air war.

# Chapter 8
# The story of radar development

EMI and Alan Blumlein were not directly responsible for the invention of radio direction finding (RDF) or radar (RAdio Detection And Ranging), as it would eventually be known. They did however, play an increasingly important role in the development and application of certain elements that were used in the original form of the invention, as well as the improvement and refinement of H2S, one of the most widely used forms of the device. Blumlein's association with the assembly and testing of what became the most effective of all the radar systems used during the Second World War, would ultimately cost him his life.

The work that EMI and Alan Blumlein carried out with radar arose from an almost natural progression of the development of the high-definition television system. It is a fact that Blumlein's services were brought to bear on the problems of radio direction finding almost by accident, and were it not for the fact that EMI were the producers of many of the early receivers being used by the radar developers, he might never have become involved at all. Ultimately, any understanding of his involvement and activity within the sphere of radar development requires an understanding of the circumstances that brought about the invention and application of radar in the first place. As with so many scientific breakthroughs, the history of radar was a result of a long, complex and often troublesome process.

## Origins

Following the air raids on London during the First World War, it became obvious to the Air Ministry that any future conflict would undoubtedly include the use of enemy bombers flying over British cities, attempting to demoralize the civilian population below. In November 1932, a speech was given in the House of Commons by Stanley Baldwin, then part of the Coalition Government headed by Prime Minister Ramsay MacDonald (though in most people's eyes Baldwin was a 'virtual prime minister', following MacDonald's fall from grace as leader of the Socialist Party). Part of this speech centred on the problems facing the armed services with regard to the air defence of the nation; Baldwin concluded that in any future war, expenditure on deterrents against attacking aircraft was a waste of time as 'the bomber will always get through'.

This was a widely held opinion at the time in both Government and scientific circles,

Figure 8.1
Front elevation of one of the smaller
'sound mirrors' at Greatstone-on-Sea.

though thankfully for the sake of the generations that followed, it was not the belief of a few, determined men who felt sure that some technology, perhaps unknown at that time, could be brought to bear on the problem. They had decided some years before that some means of detecting enemy bombers must be found in order to give the home defences enough time to counter any offensive of this kind, and so many different projects were begun with varying degrees of urgency and practicality to this end.

The problem was that enemy bombers were so hard to detect. During the day and in fine weather the observer corps should see and report aircraft coming from several miles away, but at night or during foggy conditions (very prevalent at the time), no such method could be relied upon. The only defence, it was concluded, was to listen for the approaching aircraft and detect them by methods of sound amplification. During the First World War, several forms of sound detector had been used to locate enemy gun positions. These small dish-shaped detectors were now considered as possible detectors for aircraft and were to be scaled up accordingly.

Experiments were carried out in great secrecy along the cliff-tops near Dover with a variety of sound reflector mirrors that had been built. Through the late 1920s and into the early 1930s the experiments continued until 1931, when, it was proposed, a series of massive, long-range aircraft detector systems, should be constructed

Figure 8.2
Back elevation of the same sound
mirror. The dish is approximately 18
feet in diameter.

along the coast from Romney Marshes, near Hythe, towards Dungeness. Built of reinforced concrete, these enormous 'ears' would, it was hoped, be large enough to pick up the noise from an approaching, but still distant aircraft, and focus the sound by means of a probe placed at the centre of the dish to a central listening position. This probe was nothing more elaborate than a collecting horn for the amplified sound collected by the huge dish. From this, it was expected, a bearing could be determined.

There was an element of scientific basis in the design, it had been known for some time that sound detection was possible but, that it relied on very accurate hearing, excellent detection equipment and perhaps most telling of all, knowledge of which direction the aircraft was coming from. Regardless of the fact that an undertaking such as that proposed at Romney might bear little if any positive results, several of the massive concrete sound reflectors were actually completed, the largest of which was at Greatstone-on-Sea, near Dungeness.

This non-active location system did in fact work, with the listening position situated directly under the great dish of the reflector. Here, an operator would sit, usually at night when it was quietest, and plug themselves into the reflected sound picked up by the probe at the centre of the dish. The massive dish itself was nothing more than a huge amplifier, with the operator listening at the base of a periscope tube, down which two sound tubes fed directly into a stethoscope-like apparatus. The listener would then wait for the approaching aircraft, take a bearing and pass the information on to headquarters nearby. When two or more bearings were taken from several sound detectors along the coast, the direction of the aircraft could be calculated.

However, despite the fact that approaching aircraft could indeed be heard, the system was only effective when the aeroplane was less than a few miles away. This had not been much of a problem in the late 1920s when bombers were flying at around 80 mph but, ten years later, bombers were flying at nearer 280 mph which meant, in real terms, that the early warning advantage given by the sound detectors was perhaps only two or three minutes more than the observer corps. This in turn gave little if any warning to Fighter Command for possible counter measures to be launched.

There was another, completely unforeseen problem with the system. Not only did it pick up the sound of an approaching aircraft, there was nothing to stop the dish from picking up and amplifying the sound of everything else on Romney Marshes. This included sheep grazing, passing cars and seagulls flying overhead. In summer,

Figure 8.3
Some of the larger sound mirrors at Dungeness. The long 'wall' mirror on the left is over 200 feet long.

the 200-foot wall of one of the mirrors was often used as a giant wind-break by pic-
nickers; during one demonstration set-up for the Air Ministry, their officials were
almost deafened by the sounds coming from a horse-drawn milk float.

The project at Romney Marshes was scrapped, the full series of sound detectors
were never built, and the remaining concrete detectors abandoned. They stand to
this day, slowly decaying, sentinels to a bygone age of great suspicion and little scientific know-how. Incredibly, they are now scheduled as 'ancient monuments',
which puts them in the same category as Stonehenge or Salisbury Cathedral. While
they may not have worked in the manner for which they were originally constructed,
the sound mirrors inadvertently laid a very important foundation for the war years
to come. The operating system for the early detection of incoming aircraft, which
had been put in place around them, would now serve to help with the development
of aircraft detection by radio means, which the Air Ministry, rather embarrassed at
the failure of the sound mirrors, turned to next.

## The beginnings of radar

By the early 1930s, the British Government, among many others, were busy putting
together teams of scientists to look into the possibilities of producing 'death rays'
which could potentially be used against attacking aircraft. It must be remembered
that at this time there were real considerations that such a serious threat to the airspace of the British Empire could be brought about. Motion pictures of the period,
such as Fritz Lang's 'Metropolis', released in 1926, were full of images of the world
of the future in which, it was imagined, life would be dominated by aircraft. As far
fetched as it may seem now, the possibility of 'death rays' that could incinerate an
aircraft and its pilot as they flew overhead, were taken very seriously and had to be
investigated.

The Director of Scientific Research in the Air Ministry, Harry E. Wimperis, had, by
1934, enough concern about the possibility of attack from the air and the consequences thereof, that he sought the advice of A.V. Hill (the distinguished
Cambridge physiologist who had been a gunnery officer in World War One), about
possible means of destroying enemy aircraft. Following the conversations that
Wimperis had with A.V. Hill, a memorandum was sent to Charles Stewart Henry
Vane-Tempest-Stewart, Lord Londonderry, Secretary of State for Air, on
12 November 1934, advising him to set up a committee 'to consider how far recent
advances in scientific and technical knowledge can be used to strengthen the
present methods of defence against hostile aircraft'.

Wimperis went further by recommending to Lord Londonderry that the then
Chairman of the Aeronautical Research Committee, Henry T. Tizard, should chair
this new committee, which was to be called the Scientific Survey of Air Defence and
would be formally responsible to the Committee for Imperial Defence. The other
members Wimperis suggested should include A.V. Hill, Patrick Maynard Stuart
Blackett (who had been a Naval Officer before and during World War One, and who
had since proved himself to be one of the brightest young scientists of the day
through his work as a professor of physics at Birkbeck College), and of course
himself. Upon receiving the advice from Wimperis, Lord Londonderry wasted no
time, and forwarded the invitation to chair the new Air Defence Research
Committee (ADRC) to Tizard, and invitations to the others on 12 December 1934.

Figure 8.4
Robert Alexander Watson-Watt.

The new committee's first meeting was held in Room 724 of the Air Ministry in London, on the morning of 28 January 1935, in the presence of Tizard, Hill, Blackett and Wimperis along with Wimperis' assistant Albert Percival Rowe (subsequently to become head of the Air Ministry Research Establishment). The meeting lasted from 11.00 a.m. until 1.45 p.m. and among the topics discussed was the possibility of such a death ray existing. There had been disturbing reports supposedly leaked from places such as Soviet Russia about high frequency anti-aircraft rays of the type, it was thought, that could bring an aircraft down. Tizard wanted his team to look into whether such a ray could be feasible and, if so, could it really be used to knock down aircraft in some way. Despite it being somewhat far fetched, he and the others felt that further advice should be sought.

One of the first decisions taken by the Tizard committee (as it had now become colloquially known), would prove to be one of its most significant, as they turned to Robert Alexander Watson-Watt, Superintendent of the Radio Research Station, part of the National Physical Laboratory (NPL) at Datchet, near Slough, in Berkshire for advice. Tizard and Wimperis put the question to him: 'Could such a death ray be constructed and used against aircraft?'

Watson-Watt had at his disposal a team of highly dedicated scientists who had access to very sophisticated radio equipment for the time. They also had one of the very earliest oscilloscopes, so in actual fact, though they weren't to know it, they had all the equipment required to build a rudimentary radar system already in place. The question of whether a death ray could be built or not was put by Watson-Watt to one of the members of his staff at NPL, Arnold F. Wilkins, who was charged with calculating how much energy would be required to damage an aircraft, or its crew, and if such a ray was possible could it have already been produced?

Figure 8.5
Arnold F. Wilkins.

Wilkins investigated the possibility of such a ray and concluded that while in theory it was possible, the power required to make it effective would be so prohibitively high that he considered it quite impossible that such a device could be built. Arnold Wilkins reported his findings back to Robert Watson-Watt and the two men concluded that in effect this was what they had expected.

Watson-Watt reported back on 4 February 1935 to Tizard and Wimperis that there was no possibility of a high energy ray harming the aircraft or the aircrew. Once this had been circulated to the members of the committee, A.P. Rowe (as secretary of the committee) approached Watson-Watt on 6 February 1935, to see if he could help them with their inquiries further. Robert Watson-Watt then asked Arnold Wilkins how he felt they might be able, if at all, to help the Air Ministry with their task.

As it happens, some years before this, British Post Office engineers had noticed that an aircraft, flying through an experimental high frequency beam, had caused the beam to 'flutter' a noticeable degree. The matter had been reported in a Post Office report which, while it had not received wide circulation, had come to the attention of Arnold Wilkins who had been working at the time with Watson-Watt at the Radio Research Station.

Wilkins remembered this report and thought that it might be used as the basis of a system for detecting aircraft. Watson-Watt was sufficiently interested to ask him to calculate the possibility of using this and, after some work, Wilkins came back with the conclusion that it was quite possible to develop some form of aircraft detection system given the right equipment. It seemed that the probability of an aircraft, re-radiating a radio signal aimed at it back to the source, was very high.

Robert Watson-Watt drew up a document of the results of Arnold Wilkins' work to report back to Tizard, Wimperis and Rowe on Tuesday, 12 February 1935. The report was entitled 'The Detection and Location of Aircraft by Radio Means', and in it Watson-Watt wrote his now famous memorandum that 'Although it was impossible to destroy aircraft by means of radio waves, it should be possible to detect them by radio energy bouncing back from the aircraft's body.'

Tizard and Wimperis were very impressed by the report and its conclusions, so much so that they immediately arranged a meeting at RAF Farnborough with Robert Watson-Watt and Air Vice Marshall Hugh Dowding, who, at the time, was Air Ministry Member for Research and Development. At the meeting, on Wednesday, 13 February 1935, Dowding was at first rather unimpressed with the calculations which, after all only showed the vague possibility of such a detection system working. However, Air Vice Marshall Dowding was nothing if not far-sighted and, despite his initial misgivings, he felt that if a practical demonstration could be arranged, and the mathematics that Arnold Wilkins had concluded in the report proved in principle, it might sway his conclusions.

## First tests

The initial demonstration was to take place on Tuesday, 26 February 1935, just 13 days after the meeting at Farnborough, which indicates the urgency with which the Air Ministry was treating the matter. The RAF loaned a Handley Page Heyford bomber (K6902) for the experiment to act as a 'target' to be tracked. The Heyford was a slow and lumbering aircraft which, having first flown in 1930, was already some years out of date by 1935. As an aircraft design it was remarkable for few reasons, perhaps the most obvious of which was it was only capable of a maximum speed of 142 mph; it was also the last heavy bomber of the biplane design to serve with the RAF. For the purposes of the test however, with a wing span of 75 feet, it did provide quite a large object to 'aim at' in the sky.

The test was to be carried out in a field outside the little town of Weedon, near Daventry. The Heyford was instructed to fly on a path between Weedon and the BBC transmitter at Daventry. The detection equipment used consisted of a rather large receiver which was fitted with one of the early oscilloscopes, all of which had come from the laboratory at NPL, and tuned to the 49-meter wavelength that the BBC transmitter worked on. The equipment was loaded into the back of a small van and driven on the evening of Monday, 25 February 1935, to the field, where Arnold Wilkins and an assistant prepared, in the cold and the dark, for the test the next day.

The pilot of the Heyford, Flight Lieutenant R.S. Blucke, took off from Farnborough the next morning, and climbed to 6000 feet and started to fly the course on his flight plan. In the van Arnold Wilkins and his assistant tuned their radio receiver to the frequency of the BBC transmitter at Daventry. As the Heyford bomber flew overhead, the steady signal of the transmitter which was being received and displayed on the oscilloscope, began to move up and down, indicating that a measurable amount of radio energy was being reflected from the aircraft above. The Air Ministry men in the van watched as the signal indicated the aircraft in their vicinity and they were able to track it for nearly five minutes (which corresponded to a range of approximately eight miles). The experiment was a complete success and had proved conclusively that detection of aircraft with radio means was possible.

Figure 8.6
A Handley Page Heyford bomber
similar to this was used as a
'target' by Arnold Wilkins for the
first RDF tests on 26 February
1935.

From that first test, an initial sum of £10,000 was granted to continue the work, staff were drawn from the people at the Radio Research Station and these included Watson-Watt and Wilkins of course as well as L.H. Bainbridge-Bell who had been with them since the 1920s. In April 1935, several new recruits were added from a list Watson-Watt had drawn up of promising radio graduates. Among these were Edward G. Bowen who had an honours degree in physics from University of Swansea and who was destined to play a very important role in radar development later on.

A second demonstration was arranged to convince the members of the Tizard committee of the merits of the new unit which had been moved to the remote site of Orfordness on 13 May 1935, a small spit of shingle on the East Suffolk coast some 90 miles north east of London. The entire project had been classified as secret by now, and it was decided that a place was needed where the 75-foot towers for the project could be built away from prying eyes.

The initial test was arranged by Watson-Watt for Saturday afternoon, 15 June 1935, with a Vickers Valentia bomber flying short legs up and down the route between nearby RAF Martlesham Heath and the station at Orfordness. Unfortunately the test was a complete failure, much to the embarrassment of Watson-Watt and the entire Tizard committee as the aircraft, which they could see and hear, flew by several times, yet the radio detection equipment remained obstinately blank and silent.

The next day however, Sunday, 16 June 1935, another inadvertent demonstration was conducted when Watson-Watt noticed on the screen the distinct return of an aircraft heading in the direction of Felixstowe Air Station at a distance of about 17 miles. He called the commandant of the station and asked him if one of his aircraft had just returned. The rather surprised station commandant said that a Scapa flying boat had just returned from a routine patrol. Watson-Watt convinced the man to instruct the pilot of the Scapa to go up once more and re-fly the route, at which time the Tizard committee members all witnessed a satisfactory demonstration of the possibilities of radio detection finding.

At a later similar test to convince Air Ministry officials, Watson-Watt was able to show the equipment working this time using a Westland Wallace aircraft as the 'target'. The Wallace was a significantly smaller aircraft than the Scapa or Heyford bomber, having a wingspan of just 49 feet. The aircraft type had been derived from the Westland PV3, an aircraft with extremely good altitude potential. This was proved on 19 April 1933, when PV3 G-ACAZ piloted by Squadron Leader the Marquis of Douglas and Clydesdale, accompanied by L.V.S. Blacker in the cabin, became the first aircraft to fly across the peak of Mount Everest at 29,028 feet.

However, for the purposes of the radio location experiment for the Air Ministry such altitudes were not needed, and the Wallace flew at the more sedate height of 7000 feet over the test area. During the experiment, which was attended by several high ranking Air Ministry officials, three Hawker Hart aircraft inadvertently also flew into the test area being flown by the Wallace. They too became targets for the radar demonstration, and as the ground crew watched one of the Harts broke formation and flew away from the other two. The system had convincingly demonstrated the feasibility of multi-target detection, and needless to say the Air Ministry were satisfied that the system would work.

## Political pressures

In May and June 1935, Lord Londonderry came under extreme pressure from Winston Churchill who, while still vocal on the problems of Britain's air defence, was, at the time, in the twilight of his career and on the periphery of mainstream politics. Churchill had however recently been elected to the Imperial Defence Committee (which the Tizard Committee reported to); one of his provisos for joining was that the Tizard Committee be enlarged to include his friend Professor Frederick Alexander Lindemann (later Lord Cherwell), Professor of Experimental Philosophy at Oxford University. Lindemann and Churchill had been friends since the First World War when Lindemann had first spoken of his concerns over air defence. By 1934, both men had been voicing their concerns about the rise in power of the Nazis in Germany since Hitler had become Chancellor in 1933. Churchill and Lindemann both felt that there was an urgent need to awaken their countrymen to this threat, especially that of the German Air Force which, despite the Versailles Treaty, was being rebuilt and re-armed with alarming speed.

Months earlier, completely ignorant of Wimperis's conversation with Lord Londonderry, Churchill and Lindemann had gone to Prime Minister MacDonald and insisted that a committee be set up to investigate air defence. On 10 January 1935, MacDonald agreed with them that something should be done, and indeed had been. Churchill and Lindemann were informed of the setting up of the ADRC (Air Defence Research Committee) – without them. While both men were naturally more than a little disgruntled at having been 'overlooked', they accepted that the ADRC was at least a step in the right direction, and began lobbying MacDonald to set up a sub-committee on which both Churchill and Lindemann could sit.

In the meanwhile, Churchill contented himself with gathering all the information he could on the ADRC, even attended the fourth meeting on 25 July 1935, when Tizard had made his first report that the preliminary experiments in radio direction finding had justified further executive action. At that meeting the service departments had been invited to formulate plans for the setting up of a special organization based around Watson-Watt's radar group at Orfordness (who had moved there in May 1935), and the building of a chain of radar stations around Britain's coastline.

Now fully aware of the aims of the Tizard Committee, Churchill felt that he needed more than the regular advice on proceedings that a member of parliament would receive through the official channels; he felt that he needed to have somebody he could rely on as part of this body. He chose Lindemann because he was a man he could trust to deliver the scientific advice he wanted, and who had a firm scientific background.

Tizard and Lindemann had known each other for many years; they had originally been good friends, they were both good tennis players having been doubles partners at one point, with Lindemann even competing at Wimbledon. In 1920, Tizard had left his Fellowship at Oriel, to become the chairman of the Department of Scientific and Industrial Research (DSIR), and in 1926, at Tizard's instigation, Lindemann was appointed a member of the Council of the DSIR. Lindemann however made himself unpopular with the other members of the council through his critical attitude to the work being carried out; in fact the Aeronautical Research Committee refused to have him as a member despite Tizard's intervention.

Lindemann probably knew that Tizard had proposed him, but despite this it seems he blamed the rebuff directly on Tizard; from there on their relationship seems to have gone downhill with Lindemann especially showing signs of impatience with Tizard. In the years that followed, Lindemann became very vocal with his concerns about future air defence preparations; certainly he was not the only scientist that was anxious about new developments, but he did do more than most to bring scientific developments to public notice. Following Lord Londonderry's decision to set up the special committee to be chaired by Tizard, many felt, Churchill included, that Lindemann should have been a part of this committee, hence the pressure he placed on Lord Londonderry to include Professor Lindemann.

The proposal to include Lindemann as part of the ADRC was strongly opposed by the existing members but, eventually, under extreme political pressure from Churchill, they acquiesced and finally, begrudgingly invited Lindemann to join. Somewhat surprisingly Lindemann delayed his acceptance (almost certainly on the advice of Austen Chamberlain) pending a parliamentary debate of the matter – the Government had changed. James Ramsay MacDonald had lost control of the Nationalist Party at the General Election, and was to about to be replaced by Stanley Baldwin. This would also mean a change in the Secretary of State for Air, and Churchill, Lindemann and Chamberlain needed to have the house in agreement on the issue of air defence committees.

At the debate, which was held on 7 June 1935, Winston Churchill spoke in favour of his friend pointing out that he had had several conversations with Professor Lindemann the previous autumn in the presence of Baldwin, with regard to the need for a special committee on air defence working directly under the Committee of Imperial Defence (CID). Chamberlain now explained that it was in fact he who had been responsible for a 'misunderstanding' between the Government and Professor Lindemann which had 'aggravated the delay in inviting him to become a member of the committee'.

The next day the ministerial changes took place in Parliament and Stanley Baldwin became Prime Minister. Sir Philip Cunliffe-Lister, Lord Swinton as he became, who had been Secretary of State for the Colonies, now replaced Lord Londonderry as Secretary of State for Air. A couple of days later at an informal meeting in the smoking-room in the House of Commons, Baldwin asked Churchill if he would be willing to join the new Committee of Imperial Defence on Air Defence Research. This committee would become known as the 'Swinton Committee'.

Churchill agreed only if he could reserve his freedom of action and if Lindemann could join as a member of the Technical Sub-Committee as he depended on his scientific aid. Baldwin agreed, and in July 1935, Professor F.A. Lindemann finally became a member of the ADRC, but he never overcame the suspicion that he had been out-manoeuvred by Tizard, his erstwhile friend; it certainly tainted his behaviour as a member of the committee.

## Controversy

The Tizard committee was thereafter surrounded with controversy as meeting after meeting became long and disputed primarily over which of the projects they were considering should receive priority for research and development funding. Tizard argued that Lindemann had objected when priority was being given to radar, though Lindemann's personal papers on the matter would seem to suggest exactly the opposite to this. It would seem that in fact Lindemann's objections had been that too little use was being made of Robert Watson-Watt. He proposed that Watson-Watt should not only be put solely in charge of radar development, but also of the new communications system which would be necessary for its effective use.

It is true that Lindemann proposed that time and money be given for research into the detection of aircraft through infrared radiation (basically the heat given off from the aircraft engine emissions), something he had voiced his thoughts on as early as 1916. Though according to the notes of Dr (later Professor) R.V. Jones, who worked as a scientific advisor with both Tizard and Lindemann from January 1936 onwards, the infrared detection system was always thought to have 'slender prospects of a speedy success.'

Jones was in a good position to recall, as it had been his task to develop the original apparatus that was used for the detection of heat radiation from aircraft engines. He had constructed an electronic amplifier that caused a spot of light to broaden if there was even the slightest trace of infrared radiation around. On 24 February 1936, Watson-Watt had arrived at the Clarendon Laboratory, where Jones was working, to talk to Lindemann (who was head of the facility), about recruits for the new research group he was establishing. During his visit, he was shown the results of Jones' work, commenting that the device showed little promise of practical success. It would seem that all projects, regardless of how little progress was likely to be made using them, were discussed. P.M.S. Blackett wrote as part of his notes: 'The Committee investigated and assessed a large number of projects, good and bad: we visited many establishments and discussed all aspects of air defence problems with air marshals, with pilot officers, and with scientists in their laboratories'.

On another occasion Lindemann had proposed that funding be made available for a project to develop parachute-carrying bombs, which would be dropped in front of enemy night-bombers. On yet another, Tizard had argued that 'for technical reasons' he was not sure whether an airborne radar detection system would work at ranges of less than one mile. Lindemann therefore proposed in December 1935, that an airborne infrared detection system might overcome this problem; and so on and on it went, neither man giving ground to the other.

The situation grew gradually worse until, in July 1936, Lindemann insisted on submitting a minority report 'differing from the satisfaction that the rest of the committee felt regarding the progress of air defence research', this report had been borne out of his impatience with the committee to take either radar or Watson-Watt seriously. In August 1936, during a meeting when the level of abuse between the two men became intolerable, things finally came to a head.

In his notes, kept as secretary of the committee, A. P. Rowe recalled that during this meeting the deterioration in the relationship between Tizard and Lindemann became so acute that: 'The secretaries of the committee had to be sent out of the room to keep the squabble as quiet and as private as possible'. Soon after that meeting, it was decided that the situation had in fact grown so bad that A.V. Hill and A.P. Rowe concluded 'the committee could not function satisfactorily under such conditions'. Both resigned without giving an emollient excuse for doing so.

Lindemann would not be moved. The remaining members of the committee, resentful of his connection to Churchill, and exasperated by his ideas which they felt he would always press indefatigably, regardless of the majority position on the matter (even when he was defeated by a majority in the committee, Lindemann took his arguments straight to Churchill, who then raised them at the Air Defence Research Sub-Committee, responsible to the CID, which had subsequently been set up by Prime Minister Ramsay MacDonald in response to the original Churchill/Lindemann lobby). Under such circumstances the remaining members of the Tizard committee felt they had no option but ask Lindemann to resign his position with the ADRC, or for them to reform without him.

Certainly, the friction between Lindemann and Tizard and the other members of the committee contributed to the breaking up of the group. Lindemann's insistence on submitting his minority report couldn't have helped matters. There was however another reason for the tension between the men. Lindemann had introduced Watson-Watt to Churchill behind the backs of the Tizard committee and its associated civil servants. The result was Winston Churchill asking yet more searching questions of the Air Defence Research Sub-Committee. With Churchill pressing for more action, and Lindemann now running for parliament (on the now pressing issue of air defence), the natural resentment at such actions left the members of the Tizard committee with little other course of action than to reconvene without Lindemann.

The re-constituted committee met again for the first time on 8 October 1936, this time without Lindemann. Instead they had invited Professor Edward Victor Appleton to replace him. Later still on 3 January 1939, Professor T.R. Merton also joined. It was during one of these meetings of this new committee that Tizard would predict that once hostilities commenced, the German Luftwaffe would probably start daylight bombing on England. Tizard felt sure that the RAF, with the integration that was underway to develop Fighter Command's response, would overcome this but under such circumstances, the Germans would then resort to night-time bombing raids. This then made the need for an air defence early warning system all the more acute. Little did he know at the time how prophetic his statements would prove in the summer of 1940, during the Battle of Britain.

Lindemann and Tizard never fully reconciled their differences, though over the years that followed several attempts were made by their colleagues to do so. In 1938, Lindemann was appointed to the CID Sub-Committee where he again met Tizard, only this time on equal terms. In the late autumn of 1936, Tizard had become the chairman of the new Committee for the Scientific Survey of Air Defence (CSSAO), with Blackett becoming a member shortly before the outbreak of war. In October 1939, the CSSAO amalgamated with the original Tizard committee to form the Committee for the Scientific Survey of Air Warfare (CSSAW).

By September 1939, Lindemann had joined Winston Churchill at the Admiralty (where Churchill had returned earlier that year). He became Churchill's advisor on scientific and economic matters. After Churchill became Prime Minister, Tizard found his position at Whitehall increasingly uncomfortable as Lindemann effectively became scientific advisor in Downing Street. At this time, the combined efforts of Churchill and Lindemann rendered the unique position of influence of the CSSAW almost negligible and the committee basically died a natural death. Hill went out as attaché to Washington in May 1940, Blackett divided his time between various committees and the Royal Aircraft Establishment (RAE) and Tizard, who had earlier taken full-time appointment as Scientific Advisor to the Chief of Air Staff, now headed a scientific mission to the United States, which in fact gave many of the British radar development secrets to the Americans, and yet he ended up playing an ever decreasing role in operational matters.

## 'Chain Home'

Luckily for radio detection finding (RDF), as Watson-Watt had now called the system he and Arnold Wilkins had developed, the decision to go ahead with the development of that particular project was agreed upon by all parties. Following the success of the initial February 1935 demonstration, the team of early RDF developers had now moved to the remote site of Orfordness, on the Suffolk coastline, 90 miles north east of London. Here Robert Watson-Watt, Arnold Wilkins, Edward G. Bowen and Sidney Jefferson began the erection of the first experimental radar system. In mid-June 1935, they had succeeded in detecting radar echoes from the Scapa flying boat at a range of 17 miles. By September the range for the detection of aircraft had increased to 40 miles, and by the end of 1935, they had increased the range so that they could detect aircraft as far away as 80 miles. It was a remarkable achievement blanketed for many, many years in total secrecy.

During this time Watson-Watt put together the basis of the patent for his radio detection system which became British Patent No. 593,017 applied for on 17 September 1935. The completed specification was in fact fully accepted on 31 May 1937, though because of the delicate nature of the work, its publication was withheld under section 30 of the Patents and Designs Act, 1907 to 1932, not being published in full until 17 July 1947.

The success of the experiments carried out at Orfordness led the Air Ministry staff to ask the Government for money sufficient to build five radar stations to cover the approaches to the Thames estuary. The Treasury now allocated, in total secrecy, a sum of £10 million for this work to be completed as soon as was feasibly possible. Despite the very fierce rivalry between Lindemann and Tizard, both men agreed in principle that the development of radio detection finding should continue. In his notes, P.M.S. Blackett wrote: 'The full backing of the Committee became an effective way of getting high priority given to a project. Without such mutual trust the scientific development of radar would not have forged ahead at such speed, nor would the tens of millions of pounds have been provided so rapidly and secretly by the Treasury to build the radar chain'.

Between the end of 1935 and June of 1936, many more tests were carried out, as the Air Ministry rapidly set up a radar research establishment at the Aircraft and Armament Experimental Establishment (A&AEE) based at RAF Martlesham, near Woodbridge in Suffolk. In March 1936, the Orfordness group were moved to Bawdsey Manor a little further down on the Suffolk coast and during this time plans were put into action to construct enormous radar detection aerials all around the eastern coastline of England and Scotland. The first of these was built between June 1936 and June 1937, and became part of the system that would eventually become known as the Air Ministry Experimental Stations, Type One, or 'Chain Home' (CH).

Robert Watson-Watt was made Superintendent of the Bawdsey Research Station, which although initially still run by NPL and DSIR, was transferred to the Air Ministry on 1 August 1936. The team was also enlarged with the addition of Gerald Touch (whom Watson-Watt had recruited straight after finishing his doctorate under Lindemann at Clarendon, and who arrived on 8 August 1936), Robert Hanbury Brown, Keith Wood and Perc Hibberd.

Early in 1938, Tizard had asked J.D. Cockcroft of the Cavendish Laboratory to join him for lunch at the Athenaeum club in London with a view to asking him how, in the event of war, the physicists at Cavendish would help with radar development. This put Tizard in a very tricky position, as he had to explain the details of the then still very secret radar detection finding system to Cockcroft, who in turn would have to tell his staff at Cavendish.

Once Cockcroft had the developments of radar explained to him by Tizard, the two men decided that yes, Cavendish Laboratory could indeed be of help to future developments in the event of war, but Tizard was not in a position to obtain the necessary authority to release the radar secrets to the staff at Cavendish just yet. On 13 September 1938, Tizard wrote to W.L. Bragg, the Cavendish professor of physics, informing him that a decision about the introduction of the Cavendish scientists to the secrets of radar development would have to be put off until the following autumn. Instead, it was decided that a list of people who might be prepared to commit themselves should be prepared. It was the eventual breakdown of the political situation in Europe that would dictate matters however, and things now began to move apace.

In May 1938, Watson-Watt was appointed Director of Communications Department (DCD) at the Air Ministry with A.P. Rowe succeeding him as Superintendent at Bawdsey. By February 1939, the need for reservations about secrecy from the staff at Cavendish had disappeared, and the list of scientists became a matter of some urgency.

Watson-Watt visited Cambridge in February 1939, for a meeting chaired by W.L. Bragg. At the meeting were four senior members of the Cavendish staff, including Professor P.I. Dee and W.B. Lewis, both of whom would play increasingly important roles in the future, Dee as Chief Superintendent, and Lewis as Deputy Chief Superintendent of the soon to be formed Telecommunications Research Establishment. They now drew up a list of scientists and divided them into groups, each of which to be allocated to the various CH stations. Among these scientists from the various universities all over Britain on that initial list were: Bernard Lovell, John D. Cockcroft, J.A. Ratcliffe, M.V. Wilkes, N. Feather, John T. Randall (who was destined to play an enormous role in the development of centimetric radar), C.E. Wyn Williams, H.W.B. Skinner, G.E.F. Fertel and L.G.H. Huxley – all of whom would eventually be called up to join with A.P Rowe at Bawdsey on 14 August 1939.

The CH system was extended by Easter 1939 to cover the east and south coast of Britain from Ventnor on the Isle of Wight, to the Firth of Tay in Scotland. A 24-hour watch closely integrated with the headquarters of Fighter Command had commenced. All along the coasts of England and Scotland, gigantic radar detection aerials, often over 300 feet high, had appeared much to the bemusement, and eventually chagrin, of those who had to live near them.

A typical CH station would consist of four steel towers, usually around 360 feet high, which carried the transmitter aerials; these were six half-wave dipoles, stacked vertically with a tuned reflector curtain. This produced a polar pattern with a horizontal beam width of about 60°, effectively floodlighting the area in front of the station. Early on during the conception of CH it was foreseen that enemy jamming of the system was possible, therefore each of the four transmitter towers carried a different spot-frequency array within the 20–50 MHz range. This allowed any one of the four spot radio frequencies to be used thereby diminishing by a factor of four the possibility of enemy jamming proving effective. Most of the static CH stations operated on one or two frequencies within the 22–30 MHz range (approx. 12 meters wavelength), with the mobile radio units (MRUs), being used temporarily while the main stations were being constructed, working at 45 MHz (6.67 meters wavelength).

The receiving aerials were usually smaller, often 250 feet high, four wooden towers this time with two or more pairs of crossed dipoles at different heights. The dipoles in each pair were mounted at right angles to each other, being perpendicular to the axis of the transmitted beam. This was done in order that signals could be received with horizontal or vertical polarization equally well. These were then tuned to the corresponding frequency of the transmitter towers. In this way, whichever polarization was used for the transmitter, it would be reflected by the target in either plane or a mixture of both, depending upon the dimensions, shape and orientation of the target.

By comparing the strength of the radar echo in the two aerials it was possible to obtain an estimation of the approaching aircraft's bearing; by comparing the echo strengths from the dipoles at different heights you were able to achieve an approximation of the height of the aircraft. The CH operator had a range CRT display which gave range ($x$ axis), calibrated in miles up to 200 miles, and signal strength ($-y$ axis). Near to the ground station, the transmission pulse reflections were very strong and echoes could not be determined in what was known as the 'ground ray'. Further away however, echoes would be displayed along the calibrated range line dependent on how far from the CH station they were.

On 20 June 1939, Winston Churchill, under Tizard's guidance, visited the Bawdsey and Martlesham research sites to see and evaluate the potential of the new radio direction finding systems on which so much time, effort and money was being spent. A.P. Rowe had set his staff the task of organizing a full demonstration for Churchill that included an interception of an aircraft with CH as well as discussions of future plans for the system.

Figure 8.7 Map showing the Chain Home cover of Britain at the outbreak of war in September 1939.

Churchill was undoubtedly impressed with what he saw as Sir Edward Fennessy, CBE (Bawdsey Research Station and Staff 60 Group), recalled:

'Following the lunch in the mess, Churchill addressed us ... I can still recall the essence of his speech: "Today has been one of the most exciting days of my life, for you have shown me the weapon with which we shall defeat the Nazis. But gentlemen, you still have one problem to solve ... I am a German pilot ... I cross the English coast and I am a very happy pilot. Why is that? Because I have flown from the twentieth century into the early stone age. And that, Gentlemen, is the problem you must solve."

Churchill was not telling us anything we did not know. CH tracking was excellent until the target crossed the coast, but once inland, accuracy fell off badly, and Observer Corps watchers peering out from hilltops like primitive man had to try and pick up the track. Churchill's graphic illustration brought it vividly home to us and encouraged our further work.'

Churchill left Bawdsey for London, stopping at Martlesham on the way where 'Taffy' Bowen and his team were to demonstrate the work that had been achieved with airborne interception (AI) radar systems. When he arrived it was too late for a practical airborne demonstration to be arranged, so Churchill climbed into the rear of the cockpit of one of the two Fairey Battle aircraft (K9208) that had been allocated to the unit for use on the experiments. In the cramped space where the experimental AI system had been set up, Churchill watched on the CRT as Robert Hanbury Brown 'targeted' the other Fairey Battle (K9207), which was flying tight circuits above the airfield at 1500 feet. The demonstration proved that airborne detection of aircraft could work and, evidently, Churchill climbed out of the aircraft after a very satisfactory day indeed. He was invited by the Commanding Officer for some refreshments before he left for London, and was offered a cup of tea by Bowen who recalled the reaction:

'"Tea? Tea!?" exclaimed Churchill, "Fetch me a brandy – a big one!", he demanded (perhaps to celebrate what he had just seen), and this was apparently supplied by the mess after a brief search. Then Churchill, surrounded by the pick of Britain's best test pilots and research engineers, each holding a cup of tea, watched as the great man downed his brandy with some relish while holding a newly lit cigar in his other hand. He then got up and promptly gave a rousing 15-minute speech, congratulating the team on their work.'

In his own recollections of the day in his enormous work *The Second World War*, Churchill remembered of Bawdsey and Martlesham: 'A weak point in this wonderful development is of course when the raider crosses the coast it leaves the RDF, and we become dependent upon the Observer Corps. This would seem transition from the middle of the twentieth century to the early stone age. Although I hear that good results are obtained from the Observer Corps, we must regard following the raider inland by some application of RDF as most urgently needed. It will be some time before the RDF stations can look back inland, and then only upon a crowded and confused air theatre.'

Chain Home it seemed couldn't have been completed at more important moment in time. Just weeks after Churchill visited the sites at Bawdsey and Martlesham so did the Germans, but in a rather different manner. On Thursday, 3 August 1939, exactly one month before war finally broke out, the radar operators at the CH stations along the coastline of Britain saw the largest echo they had ever witnessed appear on their screens. The enormous return, moving extremely slowly (approximately 60 mph), was travelling up and down the British coastline from the Thames up to Scotland, roughly 8 miles out to sea, for most of the rest of that day and into the early evening.

The echo the operators had seen, which one had reported back to Headquarters Fighter Command as 'between 50 and 100 aircraft', was in fact the enormous hulk of the German airship LZ130, the *Graf Zeppelin*. General Martini, the Director-General of Signals in the Luftwaffe, had arranged for the ship to be packed with special listening devices to discover the extent of British radar development. The *Graf Zeppelin* was on a spying mission.

That the Germans left, having concluded from the mission that the British radar system was at best primitive and almost certainly ineffective, is one of the major mysteries of the Second World War; but that is exactly what happened. Trying to unravel the reason why the Germans felt so confident about this is hard to explain, though it seems that the scientific team aboard the great airship were confused by a form of interference throughout the flight which may have been derived from a number of different sources.

One possible source of this 'interference' came from the way in which CH operated: instead of producing a narrow radar beam in the same way as the German systems then under development, Chain Home worked at the comparatively long wavelength of 12 meters. This wavelength literally 'flooded' the area in front of the station with 'illumination'. Because the huge Zeppelin flew along the coast, its flight path was continuously within the overlapping beams from the CH stations; consequently, instead of seeing many narrow beams the Germans only picked up a continuous irradiation level.

Why should they have imagined this was some form of interference? It seems that the 25-cycle pulse rate of the CH systems, which was derived from the 50-cycle mains supply frequency, was interpreted by the scientific team as 'arcing and sparking from the British electricity distribution system'. Moreover, because the airship flew so close to the transmission aerials of the CH stations, the equipment they were using to detect radio emissions would almost certainly have been flooded with signal, probably overloading the input circuits of the sensitive devices. It may well be that the enormous structure of the very airship itself distorted the radiation levels that the scientific teams aboard were measuring.

So while the ship flew gracefully along the coastline seemingly hearing and seeing nothing of British radar, the entire flight from beginning to end had been plotted and recorded from as far away as 100 miles out to sea. All in all it was to prove a stroke of enormous luck for Britain; as the zeppelin flew back to Germany that evening to report that the British radar system was ineffectual, it might not be overstating the situation to say that Goering had lost the Battle of Britain before it had even started.

## Other measures

So, the Chain Home system worked, and it worked well, but not everybody was convinced by CH, or the plans for air defence that were being based around it. Many influential people, both in parliament and in the armed forces, believed that any active warning system such as Chain Home, which was based on radio techniques, would simply be jammed by the Germans once hostilities began. Or worse, the fragile radio aerials, which were now appearing on the cliff-tops all along the coastline of Britain, would be destroyed in the first bombing waves. Indeed, some voiced the opinion that it was only the non-active warning systems such as the original sound detectors like those built at Romney Marshes, or the infrared system originally put forward by Lindemann, that would provide the only early means of a warning against an air attack. Chain Home, they argued, would be rendered inoperable almost immediately.

This attitude, and the other factors that were involved, led to the once discarded sound detectors being used again, though not in the way that they had been at Romney. The attitude of the Air Ministry towards the Air Defence of Great Britain (ADGB), who were in theory responsible for all forms of defence against air attack on the British Isles (which consisted mostly of anti-aircraft gun batteries), was such that all air defence applications, including the new RDF system, should be controlled exclusively by Air Ministry staff only.

This position naturally caused some considerable friction between the forces, this in turn meant that the supply of RDF sets to the British Army by the Air Ministry was a matter of high politics by 1938, when, the Army withdrew from all active development of RDF systems for its Anti-Aircraft (AA) gunnery batteries which had direct contact with Air Ministry staff. Instead, they decided to proceed with the work through their own 'Army Cell' based at Bawdsey. This ridiculous situation had come about simply because it was understood by the Army that, in the event of hostilities, the Air Ministry would supply the RDF systems for the AA early warning batteries. The Air Ministry were also responsible for 'putting-on' the precision optical rangefinders/predictors (as they were called) which were used to help range the guns.

Under these circumstances the War Office, who controlled Army equipment development, elected to place a higher emphasis on sound detection for searchlight direction at night than active RDF development, and for the putting-on of gunnery predictors they elected to work closely with EMI (who had won the contract to develop these in 1938), to improve detection range, bearing and elevation measurements.

Thus at the beginning of 1939, there were in fact two quite separate forms of early warning anti-aircraft system being deployed around the coastline of Britain; the active Chain Home network was by now almost complete, and the extensive tests being carried out by A&AEE were proving that the CH system was worth the time, effort and not inconsiderable amount of money that been spent on it. There were also non-active systems such as the AA sound detector sets, which usually had six (though sometimes eight) large, four foot diameter, collection horns mounted on a transportable, fully rotational and directional platform. An AA battery operator would sit in a location, central to the sound locator horns, and listen intently as the entire apparatus was pointed skyward, much as it had been employed in the First World War.

Having foreseen the problems that would be encountered as early as 1936, were hostilities to break out, Tizard recognized that even with all the air defence measures that were now under development there were still serious problems for the RAF. Even under good weather conditions, a night fighter could not be expected to intercept an attacking bomber at a range of less than 1000 feet. Worse still, the ground based CH system couldn't possibly be relied upon to provide positional information that accurate. It was now that the committee pressed Watson-Watt to begin the rapid development of a miniaturized radar set which could be carried aboard an aircraft to allow a night fighter to home in on a hostile from a distance of about four miles. There were other measures that had been looked into by this time also. The infrared detection system that Lindemann had championed for so long was still being considered at the time, but it seemed that the logical solution was to look into the possibility of an airborne radar detection system. This could be carried aloft by a series of aircraft that would scan the skies looking for inbound hostiles.

In the late autumn of 1936, an RAF Heyford bomber, piloted by Flight Sergeant Shippobotham and with Bowen as observer (operating the rudimentary equipment), was used for tests to experiment with the feasibility of such an airborne radar detection system being fitted. This would later be known as Airborne Interception (AI). Flying from Martlesham Heath and making circuits at around 2000 to 3000 feet over the Bawdsey transmitter, Bowen managed to obtain echoes from the aircraft that happened to be in the area at a range of about 8 to 10 miles. The tests proved conclusively that there was enough potential in an airborne radar detection system for development to proceed, and even though the initial trials had not given reliable indications from other aircraft (primarily because of the receiver's poor sensitivity), echoes from the coastline, harbour installations and more importantly ships at sea, were all observed while the aircraft was in the vicinity of Harwich.

It was considered that suitable valves, which would improve the receiver, would soon be available, and so, all efforts were to be concentrated on the development of an airborne installation on a shorter wavelength. By the summer of 1937, this had been built to operate on a wavelength of 1¼ meters. The new system consisted of a pulse transmitter and modulator, a new receiver, which had come from EMI (and had the required sensitivity needed), a CRT indicator and the various associated power supply units. There was also a separate single dipole aerial system, which was provided for transmitting and receiving.

During 1936, EMI had produced the first high-definition television receivers that were now inadvertently to play an enormously important role in the development of the world's first airborne radar detection system. EMI had offered the receivers for sale to the public in time for the opening of the first tests of the BBC high-definition transmitters during that autumn. A number of the first batch of high-definition receivers found their way to the laboratory of Robert Watson-Watt at Bawdsey Manor, where Bowen and his team which included Gerald Touch, Robert Hanbury Brown, Keith Wood and Perc Hibberd, were given the urgent task of developing an airborne radar detection system for night fighters. The object of the airborne system was to make up for errors in detecting the positions of the attacking aircraft from the RDF-1 ground stations.

The EMI receiver had been designed to receive high-definition television pictures (which forced operating frequencies up to 45 MHz), this required the equipment to receive the wide bandwidths necessary to accurately produce a full television signal (which was equally true of radar). At the same time, a CRT display, incorporating a time base, and fed by the new-high frequency receiver, was essential to both television and radar. The new EMI television receiver was ideal for the radar development team, however, they had to adjust it first to work on a wavelength of 7.2 meters (the same frequency as the sound transmitter at Alexandra Palace on 41.5 MHz) before it could be installed in the Handley Page Heyford trials bomber on loan from RAE Farnborough. Once the installation of all the bulky equipment was completed, the aerial was slung between the wheels of the aircraft. A second Heyford bomber acted as a target for the first, and was 'illuminated' by a transmitter from the ground in such a way that reflections from the target bomber would be intercepted and picked up by the Heyford carrying the EMI receiver. In this way the receiver would actually be picking up two signals, one from the transmitter and the other reflected from the target Heyford bomber.

Because the relative positions of the signal from the Heyford would be constantly changing, a deep fading effect was produced when, at times, the two signals complemented one another whilst at others they opposed each other. The net result of all of this on the display was a variable frequency pulsation that indicated the level of transmission power being reflected from the target. The system would undoubtedly work.

Efforts were hurried to produce a lightweight transmitter that could be carried in the aircraft. The team came up with a heavily modified version of the EMI receiver and a 7-inch display along with a 1 kW portable petrol power generator unit and the new breadboard transmitter. In June of 1937, the Air Ministry gave A&AEE two Avro Anson Mk.1 aircraft (K6260 & K8758) for more test flights to improve the ability of the radar system and, in July 1937, K6260 became the first aircraft to be fitted with equipment that could attain the then very short wavelength of 1¼ meters.

There were still problems to overcome. The aircraft needed substantial screening, which involved the perfection of the aircraft bonding by joining electrically all metal parts of the aircraft, and the fitting of an engine ignition screening harness. These problems had been negligible when trials had been carried out in the mostly canvas-covered wooden Heyford; but the Anson, which was primarily an all-metal aircraft, posed entirely new problems.

This experimental installation thus became in August 1937, the world's first complete airborne radar (which was subsequently coded RDF-2 to distinguish it from the ground stations), with K6260 equipped with the radar gear and K8758 as an early 'target'. Because of the early indications that had been received from shipping, the first trial flight made from Martlesham Heath with the new radar was concentrated on locating echo returns from ships, and on 17 August 1937, Wood and Touch reported clear echoes from vessels two to three miles out to sea off Felixstowe. Later the radar would pick up clear and strong signals from a 2000-ton freighter at a distance of five miles and, realising the potential for radar exploration for ships at sea, further tests were planned for September.

By pure chance a combined exercise of the Home Fleet and aircraft of Coastal Command had been arranged for 4/5 September 1937, the object being for 48 aircraft of Coastal Command to look for them as they zigzagged their way between the Straights of Dover on their way to a planned rendezvous point. On the evening of 3 September 1937, the day before the exercise was to take place, Bowen and Wood took off from Martlesham Heath in K6260 with the object of testing the radar equipment by trying to find the fleet as it made its way to the exercise area. The battleship *Rodney*, the cruiser *Southampton* and the aircraft carrier *Courageous* along with four destroyers were intercepted and located 10 miles off Beachy Head. The Anson flew over the fleet several times at a height of 1500 feet and Bowen reported later that clear and unmistakable echoes were received from the vessels at a maximum distance of four miles.

The following day with the weather rapidly deteriorating, another flight was made by K6260. In conditions of extremely poor visibility all the other aircraft of Coastal Command, about half of which were Ansons while the remainder were mostly flying boats, were recalled by wireless telegraphy (W/T) and instructed to remain inactive for 48 hours while the weather passed. Anson K6260, again with Bowen and Wood on board, was not however equipped with W/T, and therefore did not hear the recall message; it continued on its reconnaissance mission.

Once again, HMS *Southampton*, HMS *Courageous* and the four destroyers were found though the Anson did not locate HMS *Rodney*, which was presumed to have sailed further to the north. The naval force commander, knowing that all aircraft had been recalled thought that his fleet was under attack and in fact ordered the 15 Fairey Swordfish aircraft that were aboard the carrier to give chase. By then however thick

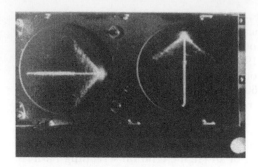

Figure 8.8
AI Mk.1 tubes indicating elevation and
azimuth of the target aircraft.
(*Courtesy of EMI*)

cloud had extended from sea level to 12,000 feet and K6260 had returned to base.

The mission had proved a complete success and would become the basis of an air to surface vessel radar detection system later known as ASV. The trial flights of K6260 were greeted with keen interest from the members of the Scientific Survey of Air Defence, in particular Professor Appleton, but apart from the small team of scientists working with Bowen under Watson-Watt at Bawdsey Manor and NPL at Datchet, knowledge of the work still remained shrouded in secrecy for several years to come. This was a key period in radar development and also the beginning of direct involvement from two future key players, the EMI Company and Alan Blumlein, though it is fair to say that both were only playing incidental roles at this stage.

By now, Chain Home could detect high flying aircraft from as far away as 120 miles, but was limited to distances just beyond this because of the curvature of the earth. Nevertheless, it was calculated that this would potentially give an early warning of aircraft approaching of between 20 and 30 minutes. The CH system was not without its problems though, and one in particular had perplexed the scientists at Bawdsey for some time – CH could not detect low flying aircraft very well.

The Army Cell at Bawdsey, which had continued their research into gun-laying radar throughout 1938 and 1939, had by late-1939, begun work with a new form of experimental short-wave radar called 'GL' or gun-laying radar. Elements of this system were designed by Alan Blumlein and then manufactured by the Works Designs department of EMI. Bawdsey were now using an adapted version of this to detect vessels out at sea. This could be used to give the shore-based guns target bearings and ranges in darkness or during poor visibility. Working at a frequency of 60 MHz and at wavelengths around 1½ meters (as opposed to the 12 meter wavelengths of CH), this detection system would eventually be designated Coastal Defence (CD) and Coastal Artillery (CA) by the Army, and consisted of two separately mounted rotating transmitting and receiving aerials. The transmitting aerial illuminated the target area in front of it with a narrow beam roughly 25° wide. The receiving aerial was also trained upon the target, but while its natural beam width was of the same order, greater bearing accuracy was achieved by splitting the beam into a pair of overlapping beams. The strengths from these overlapping beams was then compared (the two responses only being the same over a much smaller arc), the differences would then indicate a much more accurate target bearing.

The system worked so well that it was adopted for surface vessels used in the detection of submarines with results as good as 25 miles on a surface based vessel, and up to 70 miles for an aircraft. Having found that the system had a far superior range than that of the CH system for the detection of low-flying aircraft, the Air Ministry now adopted the CD and CA short-wavelength system, redeveloping some of the equipment for use with the RAF. A second CRT, called a plan position indicator (PPI), was needed to aid the monitoring of fast moving low-altitude aircraft. The PPI was positioned next to the ranging CRT, and usually set at about 50 to 60 miles as required (the ranging CRT usually indicated 180 to 200 miles), the intensity of the rotating PPI trace was set to be just visible with echoes painted up as brighter arcs.

Because the system relied on the bearings being taken from the angular position of the PPI trace, the aerials needed to be positioned to take advantage of the maximum downward elevation. This meant that many of the new type of radar stations were sited on high cliff tops with the rotating aerials angled down to illuminate out to sea. Where no high terrain was available the aerials were sited on their own towers usually about 200 feet above the ground, or even on one of the cantilevered platforms atop the existing CH towers. The Air Ministry now renamed the new system Air Ministry Experimental Stations, Type Two or 'Chain Home Low' (CHL).

By December 1938, fourteen of the planned CH stations were operational with temporary equipment. By the outbreak of war in September 1939, Britain had all 21 CH stations set up at intervals all along the coast with overlapping sectors of coverage. With stations running from Netherbutton in Orkney to Ventnor on the Isle of Wight, the CH project had cost in excess of a million pounds but by the time of the Battle of Britain in August/September 1940, with the now 54 CH and CHL stations in operation 24 hours a day, they would prove to be well worth the expense and effort.

## EMI, Alan Blumlein and radar development

Though none of the EMI scientists had a hand in the invention of the first radar system, they were to become very influential in the various versions that were developed for operational use. In particular Alan Blumlein, Eric White, Maurice Harker, Felix Trott, Cecil Browne and Frank Blythen would all play major roles in the contribution to improvements to the original specifications, and EMI in general had watched the development of radar from 1937 onwards with keen interest. Throughout the early development of the radio direction finding system, EMI seem to have played a gradual, but increasingly important role, beginning with the introduction of the high-definition television system for the BBC London Television Station at Alexandra Palace. The company had played their first real part in the development of radar in supplying the Radar Research Development team under Watson-Watt with the receiver types required for airborne radar detection.

Since however, the work being carried out at Bawdsey was secret, EMI had no knowledge of the details of their work. The fading of television signals due to the reflection from aircraft was however, a well known phenomenon, and echoes from large objects such as gasometers were strong enough to be a nuisance, causing 'ghosts' on television receivers. Eric White, while at Cambridge, had built a pulse transmitter and receiver for studying the ionosphere, thus it was not difficult for people working with wide-band pulse-type systems, such as television, to guess that work on what would eventually be called 'radar' was going on.

There were of course other companies that the Government had on their list of candidates for radar development contracts but EMI had its Research Department and the Gramophone Company departments, as well as Blumlein and his team of course, all working on part of the component make-up of radar development which had sprung from the work that they had been carrying out developing the high-definition television system. EMI was chosen for this work because of their track record in pre-war CRT development, originally intended for television applications. Naturally, they, like most British companies at the time, were keen to pick up lucrative war contracts that could prove decisive not only to victory for the nation, but also to the long-term success of the company as a whole after the conflict was over. Curiously then, the initial pressure for work did not come in the form of a Government request or contract, quite the opposite in fact; EMI had to tender a proposal for an aircraft detection system in the context of a private venture.

Under the direction of Alan Blumlein, the research and development team at EMI had continued their work that had begun in the early 1930s for the proposed public television service. By 1938, they were at the forefront of the science, having proposed 405-line television and eventually having it adopted by the BBC as the standard broadcast system. EMI had also improved designs for television cameras, aerial transmission systems, amplification systems, and cabling methods for carrying video signals to and from the main transmitter at Alexandra Palace to the BBC's Broadcasting House at the top of Regent Street.

The development team were in fact capable of putting together just about all the elements of television circuit and transmission design, and had the war not interrupted this work would have undoubtedly led to a vastly quicker television service than the one that Britain eventually got. With television transmission based on high frequencies, it was natural then that many of this team working with Alan Blumlein would have turned their attention to the new possibilities of radar; but this had probably not been fully appreciated by EMI until the beginning of the building of the Chain Home stations in June 1936. Radio detection would have been a natural development from television as the two systems employ many of the same pieces of equipment and similar frequency ranges (though later these were raised much higher in radar).

The path for EMI from television development and consumer receiver manufacturer, to the much higher priority radar development and manufacture, was a difficult one. First, EMI were not considered by any of the armed forces to be capable of producing military specification equipment. Second, all radar development was, at the time, shrouded in the highest levels of secrecy and EMI, despite being aware that elements of their equipment were being modified and used for radar system development, were powerless to become involved to any greater degree until they could prove their worth to the military hierarchy.

## VIE – visual indicating equipment

The first steps for EMI towards becoming part of the radar business came when, in June 1938, George Condliffe had suggested the idea of using part of Blumlein's binaural system to indicate the direction of a sound source. There followed a series of tests that proved the theory worked, and EMI saw their chance – this was an opportunity of getting into the business of military contracts, and so they ensured that the scheme took some priority. Blumlein proceeded to correct the circuits' deficiencies, and the system was demonstrated to the Air Ministry. Not surprisingly, the authorities were impressed by what they saw and contracted EMI, who, under direction from Blumlein and Clark, were contracted to manufacture a series of units consisting of a pair of microphones in a single parabolic reflector. The design was specified to be an add-on unit for the existing sound detector locators already being used.

EMI had been interested in sound location and detection systems for some time, and in their extensive archives are several reference documents on such devices, including one in German, dated 15 August 1936. Originating from a Dr Edgar Kutzscher in Berlin, this document, entitled 'Physical and technical principles of locators for sound from aeroplanes', had been translated from German into English and directed to Blumlein for evaluation (it remains, to this day, in one of Blumlein's many files in the archive). The document explains, in some detail, the use of sound locator systems which had been originally designed for use in the First World War and, in the post-war period, had been used to locate aircraft with the aid of sound waves produced by the propeller and airframe.

Several of the requirements of a sound locator are outlined, including the use of binaural sound (though in this case, the German scientist probably refers to two-ear sound, rather than an electromagnetic system such as that which Blumlein had patented only a few years earlier), there are detailed plans for the construction of several sound locator devices, including the 'Hornbostel' type, used by Germany in World War One. There is also reference to a new type of sound locator being constructed by the German firm, Electro-acoustic, which had a combination of four horns arranged in a ring. The report concludes that the technical design rendered the set considerably more suitable than all other known sets of the period. Blumlein would undoubtedly have studied this document, though at the time, August 1936, he would have been extremely busy with the last stages of the testing of the high-definition television equipment for the soon-to-be-operational Alexandra Palace, London Television Station.

When, in 1938, Blumlein had the opportunity to apply his knowledge of binaural sound to the construction of sound location devices, he wrote a five-page internal EMI report entitled 'Principles of EMI system of sound location'. It is dated 10 October 1938, and was work undoubtedly originated as a direct result of the experiments that Condliffe had instigated four months previously. In the report, Blumlein states that:

> The location of the direction of a source of sound is usually effected by means of two listening devices (human ears, microphones, etc.), the direction of the incident sound being determined by the relative sound intensity at the two listening devices. The manner in which two relative sound intensities vary at the two listening devices may be varied by altering the location, or spacing between the two devices, as for example, the two devices may be put far apart to give great angular sensitivity as in the present sound locators. The EMI system is based on the fact that the criterion of the direction of incident sound at the listening devices is the difference of sound intensities at these devices as compared with the average intensity. The sound intensities are therefore observed by two microphones and the outputs from these microphones are at once com-

bined to give two further representatives of the sum (average) and the difference of the original outputs.

The report was supplemented with a diagram of three possible circuits, one based on the principle of aural detection using electrically enhanced base length (effectively two microphones with a pair of headphones at the end of the audio chain), a second system similar to the first, but with a cathode ray tube replacing the headphones to give a visual representation of the object's location; and a third circuit, in which the principle of combining the visual and aural observation using two microphones and one reflector is suggested.

There are two vital factors contained in this one diagram that would eventually lead both Blumlein and EMI into the realms of radar development. First, the use of a cathode ray tube, shown in the second and third circuit diagrams, is evidence that Blumlein was already thinking in terms of using a display of some kind to visually represent an object. It must be stressed that all information regarding radar development at this time was considered 'Top Secret', and while the principles of radio detection had probably been known of by the EMI scientists for some time, recent developments of the Bawdsey Research Station would have been totally unknown to Blumlein. Here then, is the first direct indication that Blumlein had envisaged an object location system, in this case a binaural sound detector, which, while it would work (the initial tests carried out in July 1938, had proved that), was undoubtedly a stop-gap measure in the search for a far more elaborate, CRT-based system of detection.

The second important factor in the report was that the combination of both the visual and aural observation of an aircraft, using two microphones and one reflector, would become the basis of a system that EMI would soon construct called VIE or Visual Indicating Equipment. The demonstrations given to the Air Ministry in October 1938 would lead to contracts being placed upon EMI to construct VIE sets, with EMI making the cathode ray tubes and Standard Telephones & Cables manufacturing the large quantity of microphones. These were, in turn, to become part of

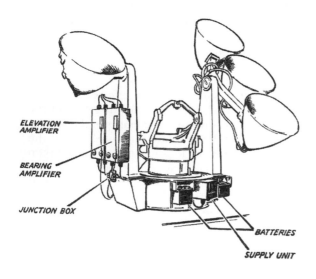

ELEVATION
AMPLIFIER

BEARING
AMPLIFIER

JUNCTION BOX

BATTERIES

SUPPLY UNIT

Figure 8.9
VIE (visual indicating
equipment) front elevation.
(*Courtesy of EMI*)

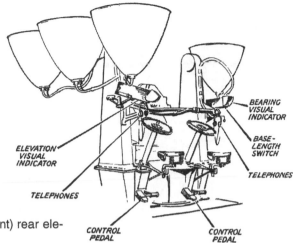

Figure 8.10
VIE (visual indicating equipment) rear elevation. (*Courtesy of EMI*)

adapted, existing sound locator units. By the time war was imminent in late August 1939, discussions had started on the manufacture of the first six prototype VIE sets by the Research Department at EMI, with further manufacture to be carried out upon receipt of contracts by the EMI factory.

The microphones for the first six prototypes were loaned to EMI by Biggin Hill, these were the Standard Telephones & Cables type 4021C, which were similar to the commercially available type 4021A. This microphone had been modified electrically to military standard and had been chosen by the Air Ministry for sound locator applications (this is why EMI did not provide their own microphones for the VIE sets). Testing began by Blumlein, Clark and Westlake in September 1939, just after war had been declared, and while a great deal of trouble was taken to find a reliable cathode ray tube which eventually lead to EMI manufacturing their own, the six prototypes were ready by late-December 1939.

EMI were, by the end of December 1939, deeply involved in negotiations with the Air Ministry for many different manufacturing contracts for all forms of radar equipment. Many of their rival manufacturers had been sceptical of EMI's ability to fulfil the task by the agreed completion date – the sound locator detectors, with their VIE sets installed, had been required by mid-January 1940. Some of these rivals had been most vocal in their opposition to EMI taking on vital military specification work; indeed, the Air Ministry itself had doubts if EMI could manage the task. The VIE sets did however, get built on time, with the first prototypes installed at Horsfield, near Bristol in mid-January 1940, by Westlake. Despite the cruel winter of 1940, one of the worst this century, and which caused Westlake to report difficulties in reading the bearing scale on CRT screen because of the thick layer of ice, the VIE units worked well. Having proved their worth, more contracts were won and by the end of the year, EMI would manufacture over 400 sound locator sets.

Some of the later experimental and ground staff training work with VIE was done in Gunnersbury Park, not far from where Blumlein was living at the time. In the evenings, when the system was being used most (as German aircraft over-flew London – the Blitz had yet to begin at this time), Blumlein would wander down to

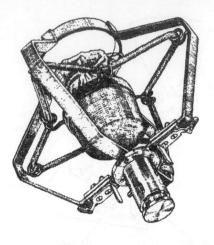

Figure 8.11
VIE (visual indicating equipment),
Standard Telephones & Cables micro-
phone type 4021C and mounting.
(*Courtesy of EMI*)

Gunnersbury Park and talk with the men who were operating the sound locator VIE
sets to see how they were getting on with them. This is typical of Blumlein. With
complete disregard of any risk to himself, he wanted to see how an application of
his invention was working, in order that he might learn from the men using it, ways
of improving the VIE set.

## Gun laying – the '60 MHz job'

The work on the visual indicating equipment had planted the spark of an idea into
Blumlein's fertile mind; now, working with Eric White, he began the almost simultaneous development of a radio detection system that could indicate the relative
position of an object, for example an aircraft, or a surface vessel or submarine, on
a cathode ray tube. The system would operate at a frequency of 60 MHz, and would
be able to calculate the altitude, distance and heading of an object in the vicinity of
the radar. The design, which EMI called 'GL' (gun-laying) radar was based around
an array of three receiving antennas, which were to be placed in a triangular pattern
some distance apart from each other, with the central transmitting aerial somewhat
taller than the receiving ones. The central aerial would then transmit a signal in the
direction of the target and the multiple receiving aerials would detect the reflected
echo back from the object at multiple distances. In this way the bearing of the object
could be determined from the time differences taken for the reflected echoes to
arrive at the various receiving aerials.

From this, the height and speed of the object could be calculated, as well as the
direction in which it was travelling. Blumlein's first patents in the field of radar stem
from this work with specification No. 543,602, dated 21 June 1939, being for a
system of indicating two versions of the same signal using a dynamometer as a mul-
tiplying device, when these two primary complex signals have effectively zero relative time delay. The dynamometer produced two auxiliary complex waves, which
could then be compared to the primary source waves. It was, in effect, sum and difference once again, but this time for a quite different purpose.

With the height, range and bearing of the aircraft determined there remained the problem of identifying friendly aircraft from the enemy ones. This was solved with an ingenious system known as IFF (identification, friend or foe), that had been invented by Arnold Wilkins and R.H. Carter at Bawdsey Manor. When, in the summer of 1939, a new member of the team arrived, Frederic Calland Williams, he too was put onto the IFF project and given the task of perfecting it. The way IFF worked relied on a transponder, which was carried in allied aircraft which responded to a received radar pulsed signal and in turn 'replied' with a modified transmission which could be received on the ground. In this way IFF allowed British fighters to identify themselves to the CH operators through these intermittent series of pulses transmitted within a very specific frequency range (157–187 MHz), which showed up on the CRT as a larger and intermittent response to that of enemy aircraft. Each aeroplane carried a small IFF transmitter known as the 'interrogator', which was picked up by a receiver at the CH station known as the 'responder'. The responder output was fed directly to the main radar display and aligned with the radar echoes to which the aircraft's identifying intermittent pulses referred.

Frederic Williams set to work to improve the system which, in 1939, was in an early stage of development. The transponder interrogator contained a super-regenerative receiver, which was manually adjusted so that oscillations in the radio frequency oscillator did not normally build up in the absence of a radar pulse signal. Only the arrival of a signal would therefore cause the oscillations to be initiated. The oscillator output was rectified and the resulting waveform was amplified and fed back through an RC wave-shaping network to the oscillator without changing its polarity. This caused the oscillation's amplitude to increase, which in turn produced a powerful burst of oscillation, radiated at the same frequency as that of the initiating radar signal. On the observer's screen in the CH station, this had the effect of intensifying and widening the echo on their CRT and was thus identifiable as a friendly aircraft.

The contract for the manufacture of the Mk.I IFF sets was given to Ferranti Ltd, and Williams would become heavily involved in IFF revisions over the next year or so. IFF Mk.II sets worked on the same principle as the Mk.I sets except that three separate wavebands were covered, and these pre-production models were tested by Ferranti at AMRE (the Air Ministry Research Establishment) in Dundee in November 1939. However, as new radar wavebands came into use, Williams began to think that a universal IFF, which would work in a separate band reserved for IFF transmissions, would be the way to go. Such a system would require additional interrogators and responders at each CH station and it was this design that eventually became IFF Mk.III. There were two significant features of IFF Mk.III: first, the use of a cathode follower in the positive feedback loop which had been widely used in the laboratories at EMI by Blumlein as part of the Marconi-EMI high definition television system; and second, the inclusion of automatic gain stabilization (AGS).

Williams used the cathode follower to drive the grid of the oscillator bias so that the set would operate on a received radar pulse but not on noise. A reduction of the bias allowed the oscillator to produce continuous waves that caused interference on radar. He used AGS to rectify the output of the oscillator with a mean rectifier, and the resultant DC signal, after amplification, was then fed back with polarity, which reduced the tendency to oscillate. The super-regenerative detector was thus maintained automatically in its most sensitive state. With IFF Mk.III first installed in

February or March 1940, it became the first beacon operating on the 1½ meter waveband. This system, much modified of course, is, in effect, still in operation today, though modern returns from aircraft are given in alphanumeric form.

At about the same time as Williams had been perfecting IFF Mk.I and Mk.II, Alan Blumlein and Eric White had been working on the experimental 60 MHz project, subsequent to the provisional patent application made on 20 July 1939, which eventually became Patent No. 581,920 ('Improvements in or relating to Methods and Apparatus for Determining the Direction or Position of Sources or Reflectors of Radiant Energy'). It is worth pointing out that Blumlein and White never referred to the project as 'GL' in the laboratory, they always described the project as the '60 MHz job'. This described how by comparing the time of arrival of the reflected modulations of a radio signal, directional indications of the source of the reflected signal can be determined. On 3 November 1939, certain small amendments were made to the 60 MHz job and these resulted in Patent No. 585,906. A series of demonstrations was planned for various scientists working for the British Government to show the capabilities of the system, and so a collection of these men were gathered at Lake Farm, adjacent to the EMI research labs at Hayes in Middlesex. The demonstration was quite a success, and convinced EMI, as well as many of those present, that the project had merit and should be pursued further.

Eric White, who worked extensively with Blumlein on the new system recalled the events:

'He had an idea for a radar for getting not only the direction of the aircraft, but its height and this was done by having a single transmitting aerial and three receiving aerials spaced around it at distances of about 50 yards in a triangular formation. We gave a demonstration of this in early December 1939, and got various people from what became TRE at Malvern to come along and see it. I think probably the fact that we had been doing this work and had obviously shown we could produce plenty of ideas, led the Ministry of Defence to then give us some work that they wanted doing. At Lake Farm, three receiving aerials were employed, at the corners of an equilateral triangle of about 70 meters each side, and the transmitting aerial was centrally disposed. The carrier frequency was of course 60 MHz, modulated by 0.5 µs pulses at a pulse rate frequency of 5000 per second. The whole system was driven from continuously running oscillators, so that the pulse period was constant, and the carrier in successive pulses was coherent. The pulses were plain and nominally rectangular. The rectified echoes from each receiver were displayed in turn, following three successive transmitted pulses, and two adjustable signal delays were set to show coincidence of the pulses. The delay values allowed bearing and elevation to be calculated. The maximum delay required was 0.23 µs. Clearly it would be very difficult to align pulses with sufficient accuracy when displayed on the full range time-base of 200 µs, so a portion of the scan containing the echo was selected and expanded.'

From the initial tests, a working system was developed (from which the CA and CD gun laying systems mentioned earlier were developed), and Blumlein improved his original specification further in Patent No. 585,908, dated 4 December 1939, which reproduced the repetitive waveform at a lower frequency by comparing the time recurrence of the signals in successive cycles, and selecting these signals so as to increase the minimum ratio of their amplitudes. At the same time, he produced a method of improving the reduction of interference that these rapidly recurring signals tended to produce. Although the claims of Patent No. 581,920 covered the EMI experimental equipment, this differed in several respects from the examples given. The form of modulation used was a simple 0.5 μs pulse, at a pulse rate frequency of 500 per second. Time differences of arrival of the echoes at the three receivers were measured by aligning enlarged echoes on a CRT A-display by inserting adjustable delays in the signal processing circuits. By using a master oscillator to control the frequency it was possible to use coherent integration of successive received echoes at RF or IF, thus improving the signal-to-noise ratio even when the signal was less than that of the noise.

This was the basis of Patent No. 585,907, dated 1 December 1939, which selected the recurrent signals, but this time different parts of the repetition rate were retained, these being integrated with the original signal, thereby cancelling much of the unwanted interference. Blumlein had realized that subsequent integration improved the signal-to-noise ratio, but would only really be effective if the individual signals were at least equal to the noise. He therefore introduced the integration before rectification to remove this limiting factor, and ended up with a system that has become known as a 'coherent radar'. A strobe pulse was adjusted in timing to select the desired echo, and any additional signals or noise would be rejected before or after it. Since the transmitted pulses were coherent in phase, integrating the echoes at RF or IF, i.e. before demodulation by rectifying, would result in the signal amplitude being proportional to the number of successive integrations, but the noise, being random, integrates in power, not amplitude. The resultant noise amplitude would therefore increase as $\sqrt{n}$, and hence signal-to-noise ratio increases as $n/\sqrt{n} = \sqrt{n}$. When the target range is altering, the Doppler effect causes successive changes in phase of the echo. This puts a practical limit on $n$ depending on wavelength, pulse rate frequency, and target radial velocity.

The purpose of Patent No. 585,908 is to 'slow down' the signal waveform, rather like optical strobing, by taking samples a fraction of the duration of the radar pulses so as to cover the whole echo in say 10 samples, stretch each sample by the pulse repetition period, so that the stretched samples are now contiguous, and display them (suitably smoothed) on a time-base as long as the 10 or so repetition periods. Thus, an echo pulse is displayed which is much enlarged in width. If Patent No. 585,907 is also used, the signal-to-noise ratio is improved, so that the echo is fairly 'clean'. Eric White's patent, No. 585,909, extended Patent No. 585,908 to provide a way of measuring the relative time delays to the three or more aerials, and hence the bearing and azimuth angles, by showing the enlarged echoes from the three aerials in succession. This was done at a rate big enough to utilize persistence for integration, etc. from each aerial. In this way the three echoes were observed apparently superimposed. The relative timings were adjusted manually, e.g. by tapped delay lines, to give coincidence, the settings of the adjusting devices then giving the delays and hence the bearing and azimuth. To facilitate this adjustment, the three echoes were suitably marked for identification, e.g. by different phases of on/off modulation, or by different small vertical displacements of each echo.

Two further patents, No. 589,228 'Improvements in or relating to Apparatus for the Control of the Timing of Recurrent Signals', dated 13 December 1939, and Patent

No. 589,229 'Improvements in or relating to Radio Receivers and Applications thereof', dated 10 January 1940, made up the remainder of those for the 60 MHz job. Patent No. 589,228 was a method of adjusting the time differences of pulses with an accuracy that was extremely high relative to the repetition rate, by generating a train of pulses at an integral multiple of the repetition rate and adjusting with moderate accuracy and gating the two trains together. Patent No. 589,229 was a homodyne receiver which was insensitive to the phase of the receiver carrier. It used two local signals in phase quadrature, and then combined the resultant signals after separate detection of the target was made.

After this latest system had been perfected, it was decided that the development team should organize a demonstration of the system to the Air Ministry which was set for mid-December 1939, and once again, EMI proved the effectiveness of the system. A further demonstration followed for W.B. Lewis, now Deputy Chief Superintendent of the Air Ministry Research Establishment (AMRE) in May 1940, and, as a result of these tests, a report on the possibilities of gun laying radar being developed by EMI was written by A.P. Rowe, Superintendent of AMRE to the deputy director of coastal defence in which he outlined the full potential benefits of the system. He concluded in his report that EMI should be encouraged to continue with their work and that the scientific team should be instructed to try to find other applications for their apparatus that would help the war effort.

At about the same time as this report was drawn up, another, written by J.D. Cockcroft at Cavendish, was pointing out the positive features of GL, but also its drawbacks, which Cockcroft went on to describe in some detail. He concluded that while GL had potential, especially if it was adapted for the detection of low-flying aircraft (which of course it eventually was, becoming an integral part of the CHL or Chain Home Low system), he felt that it would be more beneficial if the team at EMI were directed towards other radar detection finding systems currently under development and with a much higher priority than gun laying. In the end, the 60 MHz job was considered somewhat impractical, it was hardly portable and not really suited to multi-target situations, and this too was reflected by Cockcroft in his report. The form of experimental equipment that had been put together in the laboratory at EMI ended up being the only one ever built.

## Airborne interception – tracking the minimum distance to target

As keen as EMI were to get involved with Government contracts for the radar work, there were just as many people in influential positions who were to question whether an industrial company, particularly one not previously considered a 'secure' supplier of radar equipment, could be trusted to manufacture and come up with such equipment. Most vocal among these voices was the Air Ministry itself which questioned whether EMI should be considered in the same light as Pye Radio or A.C. Cossor Ltd of Highbury, for example, who were producing the majority of the airborne interception radar sets. As it turned out, other events would eventually force the hand of the Air Ministry, brought about by an ingenious invention from Alan Blumlein which undoubtedly proved that EMI not only had the resources to produce the equipment for radar production, but that they had the genius to ensure its further development.

Despite the early enthusiasm with the RDF-2 airborne radar detection system it did suffer from one serious problem. The minimum distance which a target could be tracked by the intercepting fighter needed to be such that the hostile bomber had to come well within the range of the aircraft's guns so that it could be destroyed. The problem was that maximum target range that could be displayed was limited by ground reflections to the height being flown by the RDF-carrying interceptor aircraft. As the earliest Airborne Interceptor sets only allowed an operational window of about 1½ to 2 miles in which to detect and visually identify the hostile bomber, and bring it within range of the attacking aircraft's guns, the task proved impossible to all but the most skilful of RAF pilots and RDF operators.

By the early summer of 1939, Alan Blumlein had secretly been made aware of this problem through EMI's association with GEC Hirst Laboratory in Wembley. The two companies had been working closely together for some time, to develop special high frequency valves (and other equipment), for use in a range of smaller RDF sets for land, airborne and seaborne use. Because of the work that GEC were carrying out (they had been decreed a higher security rating than EMI) EMI had been informed of the problem, and they in turn naturally turned to Alan Blumlein informing him of the nature of their joint development work. EMI were supplied with details of an experimental AI set being used by Pye Radio and A.C. Cossor to try to solve the problem and these were given to Blumlein to work on in secret.

Blumlein realized straight away that what was required was a new technique for generating very short, high-energy pulses in the RDF transmitter with a constant amplitude directly into the aerial. He also felt certain that he had come up with a solution to the problem. At the time, the AI sets being used were basically miniature versions of the entire CH ground transmitter with a low power modulator carrying the sound waveform which was then applied to a low power RF carrier oscillator output. The resultant modulated signal was then passed through an output driver power amplifier into an output high power amplifier and then to the aerial. Two displays gave the AI operator (usually the navigator/gunner) an azimuth and an elevation indication, though on these early sets much of the detail was lost in the ground clutter returns.

How had Blumlein achieved this solution so soon? The key to his new technique of high power modulation was the invention of a new artificial delay line in which a high powered sharp step of voltage was launched into the delay line. This would begin the generation of the high power RF oscillations which, when they reached the end of the delay line, were reflected back to the launch point. Once at the launch point, the oscillations had the effect of cancelling the input step voltage thereby removing the power from the oscillator stage. The net result was the desired very steep-sided narrow pulse of RF energy, whose pulse length could be determined exactly by the length of the delay line, and which was totally independent of all power supply variations. In effect, by applying a high power RF signal from the oscillator, the modulator would become a simple on/off switch that controlled the generation of the high power carrier. This in turn would switch on and off the high-tension voltage in the oscillator.

Working on his own from May, until August 1939 (it should be added that this was done in conjunction with the work on the 60 MHz job), Blumlein's designs would eventually become the basis for the world's first true radar pulse transmitter, and has been used as the basis for all radar designs until relatively recently. By the end of

1939, the new 'short pulse modulator' idea (as Blumlein originally called it) which he had been working on in secret was ready to be fitted to one of the new AI Mk.III sets, where it could be demonstrated to the Air Ministry. In early January 1940, Blumlein and Eric White travelled to Dundee, where the Bawdsey team had been moved, in order to see the work that had been carried out on their experimental AI Mk.III set also built by Pye Radio and A.C. Cossor Ltd. Blumlein and White were particularly interested in the efforts that had been made to tackle the problem of improving the detail on the screen and of course the problem of the minimum distance to the target. Quite against all security protocol, they took one of the AMRE AI Mk.III sets with them back to Hayes on the night sleeper train, in order to modify it at EMI in accordance with the series of ingenious ideas that Blumlein had been working on.

Once back at EMI, they began applying the short pulse modulator and began testing it in the laboratory. By placing 'taps' on the delay line, Blumlein, working mostly with Eric White and Maurice Harker, now managed to get the RDF set to provide a number of pulse lengths for either short range high resolution during the final phase of an attack, or for long range target detection. The first result of this work was Patent No. 579,725; applied for on 27 January 1940, it would eventually become better known as the 'break modulator'. The original design for the AI Mk.III set used a simple 'squegging' push-pull self-oscillator for the output stage, where the grid coil was coupled to the anode coil to cause self-oscillation, and its centre point was returned to ground via a capacitor and grid leak in parallel. Oscillation then drove grids positive, each in one half-cycle, and the resulting grid current charged the capacitor until the negative bias produced was sufficient to stop the oscillation. Thus the size of the capacitor controlled the duration of the pulse, and the time-constant (the capacitor times the grid leak) controlled the interval before the charge had leaked away enough for oscillation to start again, i.e. it controlled the pulse rate frequency.

As has been explained, Blumlein's background had been in telephone engineering, and he was therefore well acquainted with transmission line theory and its extension to lumped impedances, i.e. low-pass filters, which act as delay lines up to the cut-off frequency. He introduced his team to such delay lines, and the first application was to television to compensate for the delay of controlling pulses and scanning waveforms applied to television cameras. These would often be connected to multi-core cables of variable lengths up to 300 meters long, plus similar delays to the picture signal returning. Compensation was necessary to allow the signals from several cameras to be mixed (faded) satisfactorily, when using different lengths of cable. The 'transversal filter' (Patent No. 517,516), is another of Blumlein's patents in which he, Heinz Erwin Kallman and William Spencer Percival, produced a shaped waveform by passing it along a delay line and combining outputs from a number of taps, each with a designed 'weight', some of which may have been negative.

The break modulator originally used a single inductor, which was charged with current through a pentode from a modest supply of 1000 volts during a period of 20 μs immediately before the transmitter pulse was required. On sharply switching off the pentode, the current continued to flow in the inductor and thence into the stray capacitance, giving rise to a transient oscillation rising to a peak of several kilovolts. This was applied to the self-oscillating transmitter valves, presenting a load which damped the supply oscillation to allow about one half-cycle, with a duration of about 2 μs. This worked much better than the squegging arrangement of the original AI Mk.III unit, but was still not good enough until the idea of driving the grids positive after the 2 μs pulse, in order to damp the residual oscillation, was introduced. This idea would subsequently become Patent No. 581,561 a couple of months later in June 1940. At the time of the original application for the break modulator in January 1940, Blumlein had only contemplated using a simple inductor to give a squarer shaped pulse. However, by the time the patent was completed a year later, in January 1941, he had resorted to the use of a delay line for this operation. Politics however, once again, played a major role in delaying the application of this revolutionary new radar type as Air Ministry officials refused to believe that non-contracted industrial companies, such as EMI and GEC, were capable of coming up with a better solution to the problem of target ranging than their own contracted companies Pye Radio and A.C. Cossor Ltd.

EMI suggested to the Air Ministry that perhaps if a comparison test were carried out this might resolve the problem; this suggestion was passed on to A.P. Rowe, by now superintendent of AMRE (and who had also been trying to solve the target ranging problem since late 1938). AMRE had been concentrating their efforts in solving the problem of reducing the time wasted between paralysis of the RDF sets receiver (by the outward RF high-powered pulse), and the subsequent recovery of the receiver to the point when signals from the target aircraft could be displayed. This recovery time represented the range between the intercepting fighter and the attacking bomber, so it was felt that by reducing this time lag, the range to target could be decreased giving the interceptor a much higher chance of a kill.

The scientists at AMRE had, by early 1939, using their modification of the original AI Mk.III set, been able to demonstrate a minimum range to target of between 1000 and 1500 feet. However, the maximum effective range of the Browning .303 guns carried on most night fighters at the time was just 600 feet – this included the Bristol Blenheim, which had been chosen as the night fighter of choice for the RAF. This left the pilot with up to 1000 feet of visual range tracking, which was usually done by looking for the engine flare from the attacking bomber. The problem with this was that most German aircraft had simple exhaust flame shields fitted, which rendered a visual trace impossible.

AMRE, like the Air Ministry, refused to believe that a company such as EMI could have come up with the solution to the problem and so quickly. A. P. Rowe went as far as to firmly denounce any comparison tests as a waste of time, and so a stalemate situation ensued. It was solved by Air Marshall Hugh Dowding, who, as Head of Fighter Command, had supported the principle of AI radar as an essential part of the same ADGB policy that had produced the CH radar stations. It was Dowding who had sanctioned the use of the Bristol Blenheim aircraft as night fighters, and he was now being informed by his Group Commanders that the results they were achieving were so poor that they were totally unacceptable. In December 1939, after reviewing what had become a critical situation, Dowding concluded that a comparison test should be carried out with EMI and GEC, inviting them to demonstrate Blumlein's theory to see if could work. AMRE were also to trial their latest version of a modified AI Mk.III set to see which could range the target closest.

The test flights were carried out in February 1940, and while the modified AMRE AI radar reduced the range to target from the original 1500 feet to just under 800 feet,

Figure 8.12 Bristol Beaufighter with AI Mk.IV aerials fitted to the nose of the aircraft and on the leading edge of the wings. (*Courtesy of EMI*)

when the EMI modified version of the AI Mk.III set was tested, it was found that Blumlein's short pulse modulator enabled the target range to be reduced to just 300 feet. This not only almost ensured visual contact, but also brought the target aircraft well within the range of the interceptor's guns. To their credit, neither AMRE or Rowe disputed the results and, within days of the trial flights being carried out, the modifications were being integrated into the final version of the AI Mk.III RDF sets. These would now become the AI Mk.IV, the principal Airborne Interceptor radar used by the RAF during World War Two. On 2 July 1940, a Bristol Blenheim became the first RAF aircraft using AI Mk.IV to shoot down a German aircraft, a Messerschmitt 110.

## A proving time

The year 1940 was something of a proving time for EMI as the company moved more and more towards radar development and manufacture. Alan Blumlein was responsible for four other patent applications during 1940, all of which were by that time associated with work for radar of various types. For his Patent No. 581,161, of Wednesday, 10 January 1940, Blumlein returned once again to his AC bridge invention. In this version, a self-balancing variation of the bridge was used to design a low level altimeter which was based upon the capacitance of the earth itself. By measuring the change in capacity of two electrodes on the under surface of an aircraft (they could be attached to the undersides of the wing for example), the mean distance to the conducting surface could be measured as it altered (it could be the earth, but later trials proved the device worked just as well, if not better, at sea). This would give an accurate indication of the relative position of the aircraft, which could then be displayed to the pilot. Though methods of measuring the changes in capacitance using bridges had been known for some time, it had also been observed that difficulties could arise in the measurement of small capacities since stray capacities could interfere with the degree of accuracy. This was not so much of a problem if the accuracy was not of paramount importance, but to a pilot this could mean the difference between safety and danger above a surface.

Blumlein therefore arranged his bridge with the first source of AC voltage connected to the common point of the arms with an adjustable known capacity between the source and one arm. A second source of the same frequency but of a greater voltage was connected to a common point to both arms with a relatively small and unknown capacity connected between the second AC source and the arms, but away from the common point of one arm. In this way, AC voltage was injected across the outer ends of the bridge arms and the two sources of AC would be replaced by impedances across which AC voltage could be measured. Blumlein was evidently preoccupied while working on this invention (possibly with the break modulator) as its application was made on two further occasions, 11 March and 17 June 1940 (twice), due to the various minor adjustments made to the design.

As a direct result of this work Blumlein applied for another patent, No. 581,164 also on 10 January 1940. This specification, using two different voltages tapped from an autotransformer, meant that variances in impedance could be measured and compared. Blumlein pointed out that the invention could be used not only for accurately measuring such things as distance above a surface, but also that the measurement of small capacities such as inter-electrode capacities between valves and the like could also be made.

January 1940 seems to have been a particularly busy and inventive month for Alan Blumlein with no less than five patents applied for. It would seem that he was working on a whole host of ideas practically at the same time and does much to intensify his image as an engineer capable of working with many people at once. Yet, as has been seen, he could work without ever losing track of each project and its current state of development. On 26 January 1940, Patent No. 567,227 was applied for co-written with Cecil Browne. The invention was for multistud switches for tapping impedances or delay lines, and worked by an arrangement of tapping where a fine-step switch was made to interpolate accurately into steps of a coarse-step switch. On 29 March 1940, Blumlein applied for Patent No. 579,154 in which he describes the basis of a radar detection system similar to that which had been embodied as VIE. The system had a continuous transmission switched between two frequencies where the returning echo of frequency A was heterodyned by frequency B at the receiver over a period which depended upon the delay time of the echo.

Yet another variation of the closely coupled inductor ratio-arm bridge would become Patent No. 587,878 (derived in part from a combination of Patent Nos 581,161 and 581,164). Applied for on 17 June 1940, Patent No. 587,878 was designed to prevent unwanted coupling of the two inductances due to common resistance which could affect the bridge balance. This was done with the provision of separate earthing leads for the generator and the common point of the ratio arms for the two inductances. In this way, if the direct connection between one terminal and the generator and the common point of the two inductances was broken, and that terminal returned separately to earth (preferably near the electrode of a known capacitance to which the other terminal of the generator was also connected), the current flowing from the generator through that capacitance would no longer pass through the resistance and any unwanted coupling would be eliminated. Once again, Blumlein would later rework this patent, making a second application for it in a modified form on 10 February 1941.

Two additional patents were also applied for on a rather busy 17 June 1940. The first of these, No. 541,942 is yet another variation of the Blumlein bridge, here used to compare low resistance with separate current and voltage terminals. The second patent, No. 563,464 is far more important and useful, and can be considered another of Blumlein's classic circuits in which he displays his dislike of dependence on uncertain parameters. The specification describes a simple method of stabilizing the amplitude of oscillators without relying on grid current. It employed an automatic amplitude control of a conventional oscillator circuit by using a valve with an extra diode.

As has been mentioned, during the development of Patent No. 579,725 in January 1940, the break modulator used a single inductor, which was charged a current of 1000 volts for a period of 20 µs immediately before the transmitter pulse was required. It had been discovered that while this method worked better than the squegging arrangement of the original AI Mk.III unit, it was still not good enough and so the idea of driving the grids positive after the 2µs pulse, in order to damp the residual oscillation, was introduced. This idea became Blumlein's final patent of 1940, Patent No. 581,561 applied for on 17 June 1940 (and again, a second modified version on 23 January 1941). 'Improvements in Methods and Apparatus for the Measurement of Small Capacities and Applications thereof'. Improvements had been made to increase the damping of the oscillator after the applied voltage had passed its peak amplitude thus preventing or at least reducing any sustained oscillation on the CRT which would appear as a 'tail' on the screen. This was achieved by introducing the damping after the delay network which had been made greater than the maximum width of the pulse. In effect, this meant that any radio frequency output of the signal from an approaching aircraft would reach a pre-determined maximum value, causing the voltage applied to the grids of two valves to swing positive and damp the circuit.

## In the radar business

With AI Mk.IV, not only had EMI and Blumlein produced an extraordinary piece of apparatus considered by some to be well beyond their capabilities, the manufacturing division of EMI even managed to deliver them on time, quite a feat on its own at this point of the war. This was evidently much to the delight of Isaac Shoenberg, who, in his personal files (still kept in the EMI archives), retained a letter dated 22 November 1940, from the Ministry of Aircraft Production, which reads:

> I would like to congratulate and thank you and your staff for the magnificent achievement in delivering the whole of the AI Mk.IV Modulator Units by the original date forecast by you. In view of the fact that the delivery date and final completion date forecast were considered to be an almost impossible target, I look upon this as a particularly creditable achievement.

The letter is signed by the Minister of Aircraft Production. There would be no further officials who would question or refuse to believe the achievements of a 'non-contracted industrial manufacturer'.

EMI had definitely made their mark, and so had Alan Blumlein. They were both now, officially, in the radar business.

# Chapter 9
# H2S – the coming of centimetric radar

## Life at home – the Blitz

Though Blumlein was often working night and day at Hayes on various projects to do with radar from the outbreak of war, there is a marvellous insight into his life at home from both Doreen Blumlein and his lifelong friend, J.B. Kaye. When war came, Blumlein decided to send his family away from London, so Doreen took Simon and David down to her family in Cornwall, leaving Blumlein and Kaye to make some ingenious warning and prevention devices against the bombs which were falling all around Park Royal at the time:

'I was still in touch occasionally with Blumlein, but I knew that he was working on some very secret work, and it was most inadvisable for us to discuss work of any description. We did meet occasionally, but rarely, because he was very busy, and it was a time in the phoney war when we were pretty busy. On the Sunday, 15 September 1940, that was the day, well it's known as Battle of Britain day, I had a phone call in the morning and it was Blumlein speaking from his house on Hanger Hill, and he said, "Look here, things have been a bit hot, and they seem a bit lively now, Doreen's gone with the boys to Cornwall", which was the wisest thing to do, "Would Millicent and I like to use his house?" And the upshot was we said "Yes, very kind." This was typical of him because the house was just a normal size house, but it cannot compare with the flat that we were living in. He immediately came over on that Sunday morning, with machine guns firing over head, and all sorts of things, they were really hard at it, the RAF and the Luftwaffe then.

He was having to work funny hours at EMI, and come over just before dusk to get to the gun pits at Gunnersbury Park where he was testing out the acoustic direction finding system for directing anti-aircraft fire. He and H.M. Clark, and Nind, had come, Millicent (Kaye) was looking after things with the cooking, and then in the morning they would come in looking pretty unshaven and bleary eyed and ready for breakfast. They'd have breakfast and back they'd go to EMI.

We all had to have certain arrangements made. First of all, in the porch there was a bucket of sand, a stirrup pump and a spade; we arranged that, if there was an incendiary, mainly we had to think about incendiaries, because if you've got an HE (high explosive), there's no good mucking about with it with a bucket of sand. We arranged that Blumlein would carry the stirrup pump. Millicent and I would carry the bucket between us, and one of us would carry the spade. We didn't sleep, because the nights then were very noisy because of the bombing, practically all of the night, into Park Royal, and all around us. I think I'm right in saying that none of us slept at all from that September, until the night blitz more or less had eased off. I can't remember the dates now, I've got no records, but we had one or two interesting incidents.

We were sleeping in a room at the back and there was a blast wall on the garden side of it to stop the blast from blowing the glass in, which was one of the things that damaged people rather badly even if they were some distance from the bomb itself. We knew about some of these blasts, and that it was the suction that would suck the windows outwards. They had a latch, and we got an ordinary nail used in woodwork, bent at right angles, and instead of latching the latch lever into a slot in the woodwork of the window frame, we latched it behind this nail, so that the head of the nail, which was obviously quite a thin bit of metal, would hold the latch in position. So, if the window was pulled open or struck from one side, it would shear the head off the nail and open, they were not metal-framed windows, because I remember we hammered them into the wood. In order to prevent the things from being blown open or pulled open by winds and things, we used elastic bands, from the office or some place, that was the only precaution we had downstairs.

Upstairs, we threaded sewing cotton in lengths across the loft of the house, just under the roof. One end was attached to a tobacco tin, the thread was run over a small pulley of the type used for hanging curtains. The other cotton was then taken across the loft to the far end, and it was attached to the opened end of a very large safety pin, which we got at Woolworth's. The other leg of the safety pin was rigidly screwed onto the roof beam. Then the cotton was attached to the end that was free to a spring, and the tobacco tin at the other end was filled with sand to the extent that it would keep that pin leg under tension, so it pulled it away from the other side. In between, we had a bit of brass strip, and we wired the safety pin to one pole, and to the brass strip at the other to a battery and bell circuit, and paralleled up a whole network of these things in the loft. So, if an incendiary came through, it might hit the cotton, or if it burst and was burning it would burn the cotton, the bell would ring, and then we would roar into life.

Anyway, what happened was, it was a very nasty night, a Saturday night, stuff was rattling around the place, and we were sitting in the back room waiting for the action, Blumlein, Millicent and I. There was an almighty clatter which I said, "I wonder what that was? It sounds like a nose cap or a piece of shell," and suddenly Blumlein said, "I know what it was, it was an incendiary, those were slates and things." So we rushed upstairs and found that we were intact. He said "Its next door." Right. So we charged in there and Blumlein and I went haring up the stairs, opening bedroom doors and rushing around generally. No sign of anything burning or anything, so we said it must be another house, but it sounded terribly like this one, and just as we were about to come down stairs, one of us opened the only door we hadn't looked in – the lavatory door, at the top of the stairs; and lo and behold, there was the incendiary in the lavatory pan. It hadn't burst, it hadn't gone off, fins sticking out, all looking very symmetrical and decorative.'

Blumlein had wanted to join one of the armed forces when war had broken out in September 1939. His first choice was of course the Royal Air Force; he was, after all a qualified pilot at a time when such men were in high demand, but as Doreen recalled, there was never any doubt that the Government would let him go:

'They wouldn't let him join the services. I think he would've like to have joined the RAF, but there was no question of it, they wouldn't let him go. Of course, he was a pilot, but they wouldn't let him go, there was no question of them letting him go in any case. Well, it was a waste anyway, in any case they wouldn't take him as a pilot, for one thing he was older you see, he was over thirty.'

Determined to do something (it would seem that, to Alan Blumlein, his work was not enough as it was), he decided to become an auxiliary fireman: 'That's why he took this extra wardens chemical course so he should know more about it, because they wouldn't take him in the services,' Doreen explained. Of course, living where they did right next to Park Royal and the industrial units there, once the Blitz began, it wasn't long before Blumlein had a chance to put his training to the test, as J.B. Kaye recounts:

'We were getting showered with them around that time, more incendiaries up the road, bright, burning and in those days they got the fire out quick so that there was no indication to the bombers to home onto the fires, which of course they did – they certainly did over Park Royal.

Well, this one time there was no doubt about it – a bomb had fallen near by. We were normally dressed in flannel trousers, wellingtons standing by, and a sort of sweater or something, and this time we went leaping out, and acting according to the prescribed routine, off goes Blumlein up the road with a stirrup pump. Millicent and I set off with the bucket of sand and the spade, charging up the road. By this time the sky was pretty bright with the lights of fires, and we passed a figure standing in the middle of the road looking like a very celebrated piece of sculpture. I can't remember who did it; it was somebody wrestling with a serpent, but this one happened to be Blumlein wrestling with a stirrup pump, he had put it over his shoulder, and then the coils did the rest. They descended. And he was in one heck of a mess, so we left him to it really.

I said, "Come on, sort yourself out we must deal with this quick." We'd got up to the far end at just about the same time as Blumlein had detached himself and had joined us. There was one or two of these things cooking away, because they had a nasty way of clinging to their sticks that they were attached to. They were supposed to burst in the air and scatter, but sometimes they burst on the ground, and sometimes they had explosive charges in them just to brighten the party up. I had seized the shovel, but had no sand. So, I hastily shoved up some earth and flung it across, or as I thought, at the bonfire. It passed straight through and hit the civil defence boys on the far side, who were also trying to put this thing out.

Blumlein got busy with the stirrup pump, and Millicent arrived playing merry hell with me for having being left behind, and I was accused of "desertion under fire". Blumlein got the stirrup pump going and not having had any practice before, he managed to get the jet right across the flaming things, and saturated some of the people on the far side. By that time, the civil defence boys had put it out because they'd had some practice, but there was this pantomime going on all the time with us.'

On yet another evening during the Blitz, J.B. Kaye recalled how Blumlein had decided that he wanted a piece of shrapnel to study for some reason:

'Another night, for some reason Blumlein wanted a souvenir, and there were a couple of flats on fire, not far from Park Royal tube station, I think it would be, down at the bottom of the hillside. There was quite a nice display going on, and I suddenly missed Blumlein. I had a look, but couldn't see Blumlein anywhere, then, way down in the firelight and glow, there was a figure looking rather as if he was enjoying Guy Fawkes day. These darn things were popping away and fizzing about, and this figure was dancing around this fire! The fire was put out, and eventually back comes Blumlein with a broad grin on his face, carrying an incendiary bomb. I think I'm right by saying that we kept it on the table in the hall.

Now, when the bombing was on in the evening, and Blumlein was back for the night, in other words some nights he wouldn't be going out testing this beastly gear of his, the sonic gear I mean; we would be there, and he would start work after his meal putting a slide rule and calculating and he never showed any sign of nerves at all. Millicent and I would be dozing, but not sleeping, we had Parker Knoll settees in this place. We couldn't sleep with what was going on, and anyway, you'd hear a whistle and when you hear a whistling bomb you knew it was somewhere in the vicinity, and if it sounded rather near, Blumlein would come hurtling into this back room. On one occasion, there was obviously something big whistling down, and that really got him off his feet. He shot through the door into this room, jumped in the air, hung on one of the metal scaffold poles which jacked the ceiling up, so, it meant that if there was a collapse of the building, we had some chance of not having a lot of stuff on our heads. He managed to turn upside down, and hang with his feet – all the contents of money from his pockets were scattered all over the floor, and he was hooting with laughter, and that was a typical evening at the Blumlein house on Hanger Hill in the Blitz.'

The Blitz on London lasted for months, night after night the German bombers came back to drop their high explosive and incendiary bombs on the city. Though it is probably not the case that the people of London ever actually 'got used to the bombing', people did begin to recognize patterns in the attacks such as how many bombs would fall in a 'stick', and so they would count them as they fell. It was also a widely held belief that the bomb that 'fell on you', was always the one that you didn't hear. One night in 1941, as the Blitz began to quieten down a little, this principle was put the test very vividly in the Blumlein house as J.B. Kaye remembered:

'Millicent and I were by ourselves in the front room, then suddenly we heard the old noise of the engines, the drumming sound of the bombers overhead and suddenly a whistle, and a crash, and then another crash, then another whistle which sounded absolutely on us. Funny sort of feeling, because we were once told that you don't hear the whistling approach of a bomb that gets you. We certainly heard this one. And we moved while the whistle was on, from the front room to the room at the back. And there was an almighty crash. Sufficient to press one's eardrums in to an extent that prevented them vibrating. In other words, you felt the crash rather than heard it, and also you felt the suction of air being drawn out of your lungs.

It didn't hurt either of us, we were perfectly all right. The windows were flung open due to suction of air going upwards. The nail heads were sheared beautifully according to the plan, the curtains, hanging down by the windows, went straight up in the air outside. They were not wrenched off the rails, but it looked so funny to see the windows open, the curtains go upwards, and, as I say, it was an almighty crash. Well we looked out, it was a moonlight night, of course the bombers like the moonlight, they prefer to come over on such nights so they could see what they were going for, and looking out of the back window having decided that we were perfectly all right. I was looking out of the window and said to Millicent, "What the hell's that great lump in the garden, has it always been there? And look at where the garden fence is." The garden fence, in the middle, was lifted up in the air and appeared to be resting on nothing, and there was what looked like a miniature volcano in the middle of the garden.

Then I saw white helmets bobbing about, and these were the ARP people. And the distance, which was measured afterwards, I think was 12 yards between where we were standing in the room to where this bomb dropped. I let out a shout, to the ARP man, and he turned looking very pale and said, "Good God. Are you alive?" I said, "Yes, as far as we know." They were completely staggered, but fortunately I did have a bottle of whisky handy, so they all came in and we sat down and had a drink, celebrating the occasion. What had happened was that this garden in the back, I think it was a new house, it had been built fairly recently before the war, the garden had not had time to be made up and there was clay underneath the soil. The bomb had gone straight down, drilling a hole in the clay, and then it burst, just as if it had burst in the bottom of a gun barrel. The force of the explosion blew everything out of the top end, forcing the ground up around it, and the blast was projected upwards creating a vacuum that opened the windows, took the curtains out, and more or less drew our breath out.'

## Time to move AMRE

As the war began, A.P. Rowe felt that the vulnerability of the Aircraft and Armament Experimental Establishment (A&AEE) made it too tempting and important a target for the Luftwaffe to risk staying where they were so, utilizing some of Watson-Watt's Scottish connections, the vice-chancellor of Dundee University was contacted with a view to arranging for the group to use some of their facilities. The Bawdsey Manor group were moved to Scotland on 3 September 1939, the day that war was officially declared. Here the A&AEE became the Air Ministry Research Establishment (AMRE). Bowen's airborne radar group, which had until then been using the nearby RAF aerodrome of Martlesham Heath, were told that they would be re-established at RAF Scone, near Perth in Scotland. Unfortunately RAF Scone was a very small facility with just two hangars, one of which was already being used for the vital training flights that were going on there. Bowen had managed to procure the other for some of the laboratory equipment, but it meant working in very cramped conditions with much of the fitting of the radar equipment into the first Blenheims being carried out in the open.

One of the original recruits to the Dundee AMRE team was Bernard Lovell, who arrived there on 29 September; he was quickly informed by Sidney Jefferson, one of the original Bawdsey group, that he had in fact been allocated to the airborne unit which was based at RAF Scone. It would appear that Blackett had been directly responsible for the change in arrangements writing, as he did, directly to Lewis on 19 September, suggesting that Lovell be sent to work with Bowen: 'I would be rather pleased' he wrote 'As it would give him a personal contact with the branch of RDF in which I am particularly interested in connection with the possibility of blind bombing from above.' A day later Lewis replied welcoming Lovell to Bowen's group at Scone.

In November 1939, it was decided to move Bowen's group once again, this time to the vast RAF St Athan complex, near Cardiff in Wales where a series of large hangars and a number of runways were put at their disposal, despite not being completed when they arrived. St Athan at the time was one of the Air Ministry's huge training and maintenance facilities with nearly 4000 staff already on the site. The actual reasons for this latest move (which placed the airborne unit some 300 miles from its parent group) has never been fully discovered. Bowen's group became attached to No. 32 Maintenance Group based at St Athan, who were primarily airframe and engine maintenance engineers. They were now given, so it would seem, the additional task of installing airborne radar equipment. Needless to say, with the main body remaining in Dundee, this made life a little tricky.

In April 1940, it was decided that AMRE should finally be moved away from Scotland, and that a suitable location had been found at a site just beyond the village of Worth Matravers, near Swanage, on the Dorset coast of England. Operational base for the airborne group was set up at the tiny RAF aerodrome at Christchurch nearby, which was about as different to the vast hangars and long runways of St Athan as you could get.

RAF Christchurch was little more than a field with a single small wooden hut containing light aircraft, a cafe that was closed, and a flying club room. The entire site was surrounded by private housing so there was no prospect of expanding the location, and a rather baffled Bernard Lovell, who had been sent ahead to look at the facility, found himself wandering how an experimental aircraft facility was expected to operate under such conditions. None the less, in May 1940, Bowen's airborne group moved to RAF Christchurch and Rowe's AMRE moved into their accommodation at Worth Matravers. This was the time just before the Battle of Britain however, and the work was to suffer greatly from the constant air battles that would take place above southern England during the autumn of 1940.

There were other worries as well. On 23 August 1940, a Heinkel HEIII flew over the site at 1000 feet. AMRE was an easy target, but German intelligence had not succeeded in penetrating the veil of secrecy that surrounded the place. Nevertheless it was warning enough that it was time to leave. On 24 August, the very next day, the orders came through to evacuate the Worth Matravers site and move to a school called Leeson House in Langton Matravers, only 2½ miles away, but about halfway between Swanage and Worth Matravers. The site was to prove perfect with views overlooking Swanage and across Swanage Bay to the Isle of Wight, ideal for the radar tests they were carrying out; the move took place during September 1940. On the evening of 11 September, Churchill told the nation to expect an invasion by Germany and on 22 September, the group at AMRE received the alert for the invasion which never came. At this time, the airborne group became known as the

Ministry of Aircraft Production Research Establishment (MAPRE). The predictions that Tizard had first made in 1936 were now becoming realities. The German Luftwaffe had been defeated by day and they would now turn to night-time bombing. The great Blitz on London had begun.

## Life at MAPRE

In August 1941, with the threat of invasion considered less likely MAPRE, the airborne group moved from RAF Christchurch and set up base at RAF Hurn, near Bournemouth on the south coast of England. Despite the fact that the Blitz had ended and the Luftwaffe attacks abated somewhat, there was still a near paranoia that surrounded the secret radar development groups, it was this which accounted for much of their continual moving from location to location. The Government was always acutely aware of the high priority that German attackers would place on knocking out the British radar development programme.

Then on the night of 27/28 February 1942, British forces attacked the German radar station at Bruneval, north of Le Havre on the French coast. British intelligence had spotted that a new form of German radar, known as a Würtzberg, was sitting on top of a 400-foot high cliff next to a house just beyond the radar station on Bruneval itself. The Würtzberg radar had been photographed by RAF photo reconnaissance unit pilot Tony Hill, and British Intelligence under the direction of R.V. Jones felt that even though the radar station was at the top of the 400-foot high cliff, conveniently, it sloped down to small beach some 400 yards away. Jones thought that it was definitely worth a try and with the backing of W.B. Lewis at TRE, the plan was put to Churchill, who gave it the go ahead.

A full moon and slight winds had been a pre-requisite for the raid, and despite the night of 27 February being one day after the full moon, it was the first occasion when all elements of the weather were in place. The British raiding force was made up of a total of around 120 men, who took off from RAF Thruxton in twelve Whitley bombers at around midnight to raid the installation. They had a single radar operator with them, Flight Sergeant C.W.H. Cox, whose job it was to dismantle the radar equipment that they hoped to find and retrieve. To help Cox, Jones sent Derek Garrard from the Intelligence Division as technical advisor, and had been briefed by Jones before the raid that in the event of capture, he would most likely suffer worst. TRE sent a man also, D.H. Priest, who had been one of the original Bawdsey group to help with the evaluation of the equipment once they arrived as it was expected that little more than half an hour would be available before the force would have to pull out. Cox, despite the fact that he had never even been in a ship before the war (having been a cinema projectionist), much less an aeroplane, never the less trained for this hazardous jump over enemy territory with 'C' Company, 2 'Para', realizing that he was the only one capable of examining and removing the radar equipment they expected to find.

The raid was a complete success and Cox removed elements of the Würtzberg radar and, together with a captured German operator, he and the Commandos were evacuated by the Royal Navy. The amazing success of the raid finally brought home to the Germans the massive implications of the 'electronic war'. They had, up to this point, not really considered just how vulnerable this new phase of the war was. The

Germans had coined a new phrase 'Hochfrequenzkreig' or high frequency war, and it was now generally accepted that some form of aggression against British radar facilities was to be expected. In the following months, British intelligence discovered that formations of crack German paratroopers were being sent to the Cherbourg peninsular. The conclusion was drawn that a raid was imminent and that AMRE and MAPRE could be, and probably were, among the targets for a raid (though there was never any positive evidence for this). On the morning of 26 April, two Messerschmitt ME109s attacked the railway station at Swanage and dropped bombs, which destroyed the local grocer's shop. They also machine-gunned the site at Worth Matravers presumably unaware that AMRE had moved the previous autumn to Leeson House.

Under the circumstances then, with Leeson House and Hurn so near to the coast, Churchill felt that a reprisal attack was very likely. He instructed, in mid-March, that efforts should be made immediately to move the entire unit, airborne division and all, to a safer haven away from the prying eyes of the Luftwaffe and far enough away to preclude the possibility of a German paratroop attack. All this was to be done before the next full moon and with the utmost secrecy of course. When A.P. Rowe learned of the decision to move away from the south coast, he began a frantic search inland for a suitable site for the unit. After much frustration looking at totally unsuitable locations, he was made aware by Vivian Bowden (later Lord Bowden of Chesterfield), that Malvern College in Worcestershire might be suitable as he thought it was empty at the time. When Rowe visited Malvern on 5 April, he found that it was indeed empty and in fact ideal for his needs but, the college, as it turned out, was only empty during Rowe's visit because it was Easter and the pupils were away on holiday. The Government sent their inspectors to survey the place on 25 April.

Malvern College had in fact already been evacuated once before for urgent war needs. Two and a half years before Rowe's visit, the site had been taken over by the Admiralty in case the evacuation of London had been necessary. Now, with Government men looking around his college again, the Headmaster, H.C.A. Gaunt, was naturally disturbed at the possibility of it all having to happen again. Unfortunately for Headmaster Gaunt and the pupils, they were informed that: 'Owing to an unexpected development in the course of the war, a certain Government Department had been compelled to move their establishment elsewhere immediately; that this was a War Cabinet decision and that Malvern College had been selected as the most suitable premises in the country.' The staff and pupils of Malvern College moved to Harrow to make way for Rowe and his unit who, it was decided would move there between 25/26 May 1942.

With the decision made to move there now remained the not inconsiderable problem of where the 1000+ members of staff were going to be accommodated. Many of the members of AMRE and MAPRE were living in Swanage and the surrounding areas with their families in relative comfort, none of which could be guaranteed yet at the new site in Worcestershire. Rowe, Dee, Cockburn and Bernard Lovell (who had joined Rowe in August 1939, when he succeeded Watson-Watt as superintendent at Bawdsey), were very concerned about the situation and contacted the Controller of Telecommunication Equipment (CTE) to request an urgent meeting to discuss the matter. On 19 May, just one week before the scheduled move was to take place, the men from AMRE/MAPRE convinced the CTE of the urgency of the situation; with the intervention of the Minister of Aircraft Production, Malvern Winter Gardens was commandeered and quickly converted into a huge canteen for the staff, and the search for accommodation for everybody given the highest priority.

As can be imagined the local population, who were now called upon to do 'their bit for the war effort', were not exactly overjoyed at the prospect of being forced to billet complete strangers. The situation was compounded by the fact that, because of the total secrecy surrounding the unit, none of the people of Great Malvern were told why their quiet little town was being turned totally upside-down. Some reacted angrily, placing beds for their billetees in garages or on the drives to their houses. In addition, of course it rained, making everything and everyone miserable and wet. Eventually however, despite the enormous queues for food at the Malvern Winter Gardens Pavilion, which Lovell later described as: 'An incredible organization staffed by the Women's Volunteer Service and serving 1000 people of all sorts ... they gave food, and plenty of cheese and butter. After this we felt better,' the mood had changed, and life began to settle down.

The final stage in the group's development and re-location then, came in May 1942, with the creation of the Radar Research and Development Establishment (RRDE) at Malvern. MAPRE now became the Telecommunications Research Establishment (TRE), and its flying wing was now called the Telecommunications Flying Unit (TFU). All previous work that had been carried out as the original A&AEE was now to be disbanded, regardless of where it was being carried out, and the entire secret radar research operational unit was to be moved to Great Malvern in Worcestershire. At the same time, the airborne group that had been carrying out flight tests from RAF Hurn was now to be re-located to the new purpose-built installations based at RAF Defford. This move also took place between 25/26 May 1942.

The layout of RAF Defford, which was the closest airbase to RRDE at Malvern, was typical of Royal Air Force aerodrome design of the Second World War. The facility had three runways of between 1200 and 1500 feet in length, enough for Lancaster, Halifax or Stirling heavy bombers to land and take off. RAF Defford had six hangars on the aerodrome site, four were for TRE and TFU and two for the RAF SIU (Special Installation Unit). In addition, there were 67 parking plans for large aircraft of the heavy bomber type, six of which were fitted with dual anti-blast bunkers. Of the remainder, 23 were widely dispersed to avoid losses during an air raid. Several 'Blister' hangars were also situated on the site with purpose-laid hard standings for aircraft. As well as all this, the site had facilities, workshops and quarters for the radar testing equipment and the military and civilian staff in attendance.

The RAF SIU responsible for the manufacture and fitting of radar equipment to aircraft would now be able to work alongside the TRE and its civilian manufacturing and installation section, the TFU. However as the SIU was a military operation and the TRE/TFU was strictly manned by civilian scientists certain responsibilities of design and research were kept apart. About 2000 personnel in all were based at Defford. The work that was now to be carried out at RAF Defford and at TRE/TFU Malvern, during the darkest days of the war, was later said to be 'The most important of the war', by none other than Winston Churchill. Many aspects of the enormous effort that was required to get such a vast organization in place and working, under extreme pressures from all angles, but, which eventually would lead to the much improved airborne radar systems that Bomber

Command would use, were as a direct result of pressure applied by the Prime Minister himself.

## Further progress – other developments in British and German wartime radar

The AI Mk.IV set, the circuitry for which had been designed extensively by Alan Blumlein, was now readied for operational use by having extensive flight tests mostly carried out by Maurice Harker (based initially at Christchurch then Hurn and later Defford). Though AI Mk.IV was a vast improvement on the AI Mk.III unit, development to improve airborne interception did not stop there. Frederic Williams, who had been responsible for improving the IFF system so much, turned his attention to AI in the Autumn of 1940. His designs for manual strobe circuits and switching circuits for the amplified strobe output would eventually become the AI Mk.V unit. The importance of this work was that much of it was applicable to automatic tracking; this was required because of the introduction of the newer types of aircraft such as the Bristol Blenheim and later Bristol Beaufighter. Originally the night fighter of choice for the RAF had been the Boulton Paul Defiant (which had been withdrawn from daylight operations because of an extraordinary and lamentable feature of the aircraft being it could not fire forwards), the air gunner could not operate the manual AI probably because there was insufficient room in the turret which was mounted aft of the pilot's seat. Work was begun on automatic strobe circuits which would enable the AI set to search out from zero range to the ground returns and lock automatically on to an echo. If no echo was found, the strobe ran on until it encountered the wide ground echo, which caused it to return to minimum range and restart its search cycle. As it happened, Blumlein was also working on this problem at EMI in the autumn of 1940, and it was here that Williams and Blumlein met for the first time.

Blumlein made a great impression upon Williams, and the latter was said to have never lost his admiration for him. Williams was particularly moved by Blumlein's approach to engineering and circuitry at EMI, and recognized with greater clarity than he had ever done so before that with the right approach circuits could be designed. Blumlein sought to achieve designability through the use of long-tailed pairs: the current in a pair of triode valves was defined by the value of the common cathode resistance (the tail resistance), and the voltage to which it was returned. The voltage across the tail resistance was made large enough to render variations in the triode grid bases of negligible consequence. This is the defined current approach. For non-linear circuits, voltage amplitudes were determined by catching diodes and/or by allowing valves to operate either bottomed or cut off, so that variations in valve characteristics had no effect on the performance of the circuit. This is the defined voltage amplitude approach. For linear circuits negative feedback was used, so that again variations in valve characteristics were of secondary importance. Following this meeting with Blumlein, Williams' approach was quite changed and he too adopted designability as the driving force behind his work.

A later version, AI Mk.VI, designed mainly by Eric White, with assistance from Maurice Harker and Ivan James, greatly improved the system by automating it and simplifying the presentation so that the pilot could operate it. Until this time AI sets had to be operated by a second member of the flight crew, often the navigator/gunner. With the installation of AI Mk.VI, the pilot had a CRT placed in the instrument panel in front of him, which gave indication of location, bearing and distance of the target aircraft. Harker recalled how Blumlein would work alongside the development team during this period:

'Blumlein always managed to keep in touch with every facet of the work that each of the development teams was working on. He had the most extraordinary ability to pick up a conversation where he had left it off, regardless of when he had spoken to you last. He might be deep in conversation with you one minute, and then be called away for some reason the next. A week might go by before he caught up with you again, and damned if he didn't just pick up the conversation exactly where he had stopped before. Woe betide the engineer who didn't remember what they had been talking about to him, as I never knew Blumlein to forget, it was the most uncanny thing.

There were many occasions during the war, after a full day at Hayes, when he would leave his office and come and join us in the laboratory. Blumlein would take off his jacket, pull on a lab coat and often end up working until far too late to go home, so he would sleep in one of the beds that we had put in. On such occasions, we had an arrangement with the Hayes canteen that a dinner would be prepared for us at 6 p.m. though, if we didn't make it in time, they would leave it somewhere for us. Well, one evening I do remember, we had worked very late and suddenly became hungry, so off we went over to get some food only to discover that we were barred from entry by a six foot high gate leading to the works' canteen. Quick as a flash, Blumlein shinned up this gate and was over. These were the days when EMI was protected by guards at night, I dread to think what would have happened if we'd been caught climbing into a locked area.'

Not long after AMRE had moved to Worth Matravers in May 1940, a new team of scientists began to form with the object of developing an AI system with a short enough wavelength so that the ground return clutter could be avoided. Despite the fact that AI was now working quite well on the 1½-meter wavelength, it was really quite hard to distinguish the actual target town from the ground interference from other objects which were being reflected back up to the aircraft. This problem had in fact become so acute that the team at AMRE were seriously concerned whether even experienced operators would be able to use the equipment in combat situations; worse still, many of the pilots who were expected to fly these night fighter missions were often only trained just beyond the basic level required. This meant that they would have precious little in the way of skill when it came to operating and discriminating with the AI set in their aircraft during a battle.

What was needed was an operational system that would reproduce clear signals reflected from the ground that a pilot could be trained to recognize quickly and use to navigate to and target an inbound hostile aircraft. This however, required a much shorter wavelength than that being used. Luckily the Government had had the foresight to contract companies such as EMI and GEC to produce valves which could operate at high frequencies with sufficient power which equated to detection range. Both companies had used their skills during the time when television was being perfected just before the war, to develop miniature valves with a short electron transmit time inside the valve (which in turn means that the number of oscillations between the filament to the grid could increase, hence a higher frequency).

These valves were now actually capable of working in reasonably high power situations at the 30 centimetre wavelength range, though they were very fragile and unlikely to work well in the kind of conditions expected under combat. This indeed proved to be the case, with test after test having to be cut short because of a failed valve in the AI set. It was GEC who eventually came up with further advances. In March 1940, they introduced the VT90 valve to the tests, which could operate as a pair of valves, giving 10 kilowatts of pulsed power. Later variants of this valve were even to yield several kilowatts of pulsed power at wavelengths as short as 25 centimetres, but the developments that were being made in thermionic valve design were about to be cut short as the introduction of the klystron and the magnetron valves took place.

The secret war of radar development offensive and counter offensive had been going on since 1940, and, whilst the Germans had been using an operational radar system long before the British, the scientific research teams that had been organized by Tizard and Cockcroft as early as 1938 had now caught up; in many respects they had even surpassed the early German development advantage. The entire history of the radio war up to 1942 had been one of measure and counter measure as the intelligence and scientific bodies of the warring nations tried to out-wit the other with newer systems for aircraft navigation to and from their targets.

The Germans had produced an accurate navigation and targeting system code named 'Knickebein' (which literally translates as 'crooked leg'). This was followed by the X-Gerät and Y-Gerät (Wotan) systems from May 1940, onwards, all of which were totally misjudged by British Intelligence. Knickebein, X-Gerät and Y-Gerät used high frequency beams based on the Lorcnz blind landing system that had been used extensively since 1937. This was a method intended to guide an aircraft, without the use of any visual aid, to a safe landing using a series of electronic pulses on a specific radio frequency.

The pilot or navigator of the aircraft simply had to identify the frequency of the beam, usually between 28 and 35 MHz, and then fly along it listening to the pulses which would indicate if the aircraft was to the right or left of the target. If the aircraft was flown on the centre line of the beam then the pilot knew they would be directly in line with the target itself. In later versions, such as X-Gerät, multiple beams were sent from Kleves and Stollberg in Germany, and later (after June 1940 and the fall of France) from Calais and Cherbourg. These gave the navigating crew accuracy over the target as high as half a mile from 30,000 feet by triangulating their position. The problem with all these beam systems was that they could be easily jammed using the same frequency as that which they were being transmitted on, and indeed, this is exactly what British Intelligence did. Later, with the Y-Gerät system, which only used one beam, the redundant BBC television transmitter at Alexandra Palace was used to jam the beam as it produced a frequency of 45 MHz, which the Germans had rather poorly chosen as their transmission frequency.

Originally, the earliest British navigation system, code-named 'GEE', had used a grid to direct aircraft to their targets. GEE had first been proposed at Bawdsey Manor by its designer R.J. Dippy in 1938, but was rejected by the Air Ministry and so did not become fully developed by AMRE until July 1941. It projected a dense grid of radar pulses over Western Europe, which made navigation up to 260 miles from Britain relatively accurate, but over Germany GEE was only as good as ±5 miles. GEE would be used throughout the war, and by 1945, had gone through several improvements resulting in GEE Mk.III used in the last raids over Germany in March and April 1945. But the limitations in accuracy that GEE suffered meant that the Air Ministry was always looking for a more precise system.

The successor to GEE was code named 'OBOE'. Following GEE, a new system code named OBOE was devised. It was designed by A.H. Reeves and F.E. Jones in 1941, and first demonstrated operationally on board a Wellington bomber on the night of 20/21 December 1941, during a raid against the Lutterade Power Station in Holland. OBOE consisted of two transmitter stations in England based at Dover in Kent and Cromer in Norfolk. These were given the code names 'Cat' and 'Mouse'. The two stations sent out pulses, which were in turn re-transmitted back by the aircraft flying to the target. By measuring the time it took for the pulses to be returned a very accurate fix could be made on the aircraft. The 'Cat' station, at Dover, was the master tracking station, and locked on to the lead aircraft in the formation. The 'Mouse' station, at Cromer, was the bomb release station, and gave the coded command to the aircraft. The bombardier could in this way be radioed from England with a bomb release signal when it was computed that the aircraft was directly over the target. The accuracy of OBOE could be measured in tens of yards, and was probably the most accurate of all the radar systems employed during the war.

OBOE worked on a quite low 1½-meter wavelength, the same wavelength as the Germans had been using, but like GEE it relied totally on the ground stations zeroing in a single aircraft to the target, and both systems had limited range. As only one aircraft at a time could be controlled the first squadrons to receive OBOE were naturally the Pathfinder outfits. Both GEE and OBOE (which was very like X-Gerät in operation), suffered from the problem of being easily jammed once the transmission frequency was known, although strangely the Germans left OBOE unjammed for several months after its introduction. What was needed was an AI system that would be independent of ground transmitter stations.

As has been seen from the earlier airborne radar developments, made as long ago as 1937, such a system did exist and in fact was already in active service. It had become known as ASV or air to surface vessel. As its name suggests, it was used by Coastal Command and the Navy primarily to detect shipping and especially U-boats off the coasts of Britain. The ASV system also worked on the quite low 1½-meter wavelength, and so had limited results over the ground. However, on a flat sea, carried by a Fairey Swordfish torpedo interceptor, it was a very effective weapon, as the U-boats found out very quickly. ASV was kept very secret and initially the Germans had no idea of its existence. U-boat crews were so aware of its potential that the mere sight of a distant aircraft would cause them to crash dive straight away. The principal advantage of the metric wavelength ASV system was its total independence; it was limited in range only by the aircraft carrying it. It was this principle that would be applied to the new centimetric radar systems under development at RAF Hurn; they could work entirely on their own, and with the higher frequency would now be effective over the ground too, without the assistance of transmitter tracking stations.

Eventually the secret of ASV was determined by German Intelligence and a counter measure developed, but by that time, the new centimetric radar was being used. When centimetric ASV was first used against the U-boats, Coastal Command enjoyed a quite prolonged period of success as the German captains found their metric ASV warning systems redundant. By 1945, with the war at sea all but lost, and the development of newer faster U-boats cut short by concentrated allied bombing of the pens and docks where they were being built, Hitler (and after him Admiral Doenitz, the last fuhrer of the Third Reich) would attribute 74 per cent of U-boats lost to ASV. In fact, the figures were even more decisive than could have been realized at the time, from a total fleet of 832 U-boats built during World War Two, a staggering 791 of them – 93 per cent – were sunk. Of 39,000 men, 29,000 were killed and another 5000 captured, the highest casualties of any service, from any nation, during World War Two.

Despite all these developments in radar type, the single most important development of them all was about to take place through a series of interconnected events which, as so often happens in history, all occurred just at the right time. What was needed by Bomber Command was a radar system that could image unseen objects on the ground on a CRT at a much higher resolution. With better imaging, they could manage more accurate bombing and this would dramatically cut down the losses that they were suffering. By 1942, those losses were approaching the intolerable – something needed to be done. In order to produce better ground imaging a much shorter wavelength would be required; the klystron and the magnetron were about to be introduced and centimetric radar was about to be invented.

## Centimetric radar – the klystron and the magnetron

It was only with the invention of the new centimetric radar systems that the possibility of greater resolution of ground reflected objects, in order to accurately navigate to and from targets, finally came about. There are many intricately woven factors which brought about the final conclusive development of centimetric radar (i.e. with a wavelength of less than a meter), but it was undoubtedly with the invention of two main components that everything became possible – the klystron and the magnetron.

Towards the end of 1940, and throughout 1941, there was growing recognition in the Bomber Command (and somewhat in the Government), that the night bombing raids being carried out by the RAF were in fact having little or no effect on the German war machine, and were actually rather expensive, both in machinery and manpower, little more than morale boosting exercises. None of this was made public of course as, following the evacuation from Dunkirk at the beginning of June 1940, the primary job for the Royal Air Force Bomber Command had been to restore some of that morale which was (not unsurprisingly) at an all time low. These were the darkest days of the war for the British people and Churchill wanted every bomb possible dropped on Germany after the Battle of Britain to atone for the Blitz during the winter of 1940/41.

The problem was that, statistically, only one bomber in ten was actually getting anywhere near its intended target area. An analysis of photographs taken from a series of night operations between 2 June 1941 and 25 July 1941 was made in the hope that the Air Ministry could gain some kind of understanding of just how accurate the claims being made by Bomber Command actually were; the object was simple, could a way be found to improve the effectiveness of the raids. Some 650 photographs were investigated relating to 28 targets over 48 nights and perhaps 100 separate raids. Of these pictures, nearly half had been taken independently of actual bombing, but in every instance, the target in the photograph believed to have been bombed was named. The conclusions that were evaluated from the study of these pictures made very worrying reading indeed. Of those aircraft that were reported as having actually attacked their targets, only one in three got within 5 miles of the actual target. Against French targets, the proportion was one in two, against German targets it was one in four. Against targets in the Ruhr, the heart of German industry, it was only one in ten that even got close to the actual target.

During the full moon periods, naturally the results became a little better, though, overall the proportion was two in five aircraft. During new moon periods, it dropped to just one in fifteen. The 650 photographs amounted to approximately 10 per cent of all sorties carried out during the evaluation period, and it is true that the months of June and July 1941 were particularly bad weather wise. None the less, the overall conclusion that had to be drawn from this investigation was that radical improvements were needed in both navigation and target location methods.

Churchill was informed of the results and, appreciating the seriousness of the matter, he immediately demanded that action be taken to improve the situation. On 15 September 1941, he told the Chief of the Air Staff, Marshall of the Royal Air Force Sir Charles Portal: 'It is an awful thought that perhaps three-quarters of our bombs go astray. If we could make it half and half we should virtually have doubled our bombing power.' The Chief of the Air Staff was equally moved replying that: 'Much more must be done to improve the accuracy of our night bombing. I regard this as perhaps the greatest of all operational problems confronting us at the present time.'

Some rumours began to trickle back to TRE that the results that had been claimed by Bomber Command were not exactly accurate, and that on the whole the situation was pretty poor. This came as a surprise to most of the staff who worked at TRE at the time because, like most, they believed the propaganda that was placed in front of them daily by the War Office through the newspapers.

There had been no appeals from Bomber Command to TRE for radar aids such as those that had come from Fighter Command for the AI sets and Coastal Command for the ASV system. In fact at the time of the operational bombing raids of 1941, the only such aid that was in development was Gee, which was still going through its testing stage, and in any case was not a bombing aid as such, but more of a navigational one.

Professor Lindemann, Lord Cherwell, was enthused by the possibility that TRE could perhaps help in the situation and shortly after the communication between the Prime Minister and the Chief of Air Staff in September, Cherwell appealed to Rowe for help. On 26 October 1941, at one of the regular Sunday meetings that Rowe held with high-ranking RAF and Government officials (known as 'Sunday Soviets') Cherwell insisted that Bomber Command must have a self-contained radar bombing aid in their aircraft to help with the bombing of unseen targets and as soon as possible. He said that because of the great ranges of the bombing operations that were taking place, the navigational aid Gee and the blind-bombing aid Oboe were both ruled out as they relied too heavily on transmissions from England. Although Rowe later recalled that 'The day ended sadly, for I recall that we went to our homes tired and without an idea', Dee later recalled that it had stimulated him to reflect on the possibilities inherent in the centimetric radar systems that he had been developing for AI and ASV at the time, which had already yielded quite promising results.

## The origins of centimetric radar

Immediately after the AMRE team had arrived at Worth Matravers in May 1940, they had begun their work on an advanced version of the AI system using centimetric thermionic valves which could hopefully withstand the power demands that were to be made upon them, as well as produce pulsed bursts at centimetric wavelengths. At first, the team had been working with the developments in valve design and miniaturization which were being produced by GEC in Wembley and EMI in Hayes. Even though these companies, whose combined efforts had come about from their pre-war work in television development, were now able to get thermionic valves to work at 25 centimetre wavelengths at sufficient power output for the type of AI set that was needed, other events were about to take the stage. This was the time of the great Blitz on British towns and cities.

By the end of 1940, the Air Officer Commanding-in-Chief, Air Marshall Sir Philip Joubert de la Ferté, had been complaining bitterly to the Telecommunications Research Establishment that the progress of centimetric airborne interception was taking far too long. The lack of success of the Army anti-aircraft batteries and the poor results of the RAF night fighters meant that precious little was being done to prevent the German night bomber squadrons from freely bombing British cities night after night. The 1½-meter AI systems, which had been installed in Bristol Blenheim night fighters, had seen little success, though by the end of 1940, the more powerful Bristol Beaufighter aircraft began to come into service. The Germans were however, suffering very few losses, perhaps only 1 per cent, and most of these were due to the anti-aircraft guns rather than intercepting night fighters. Something had to be done to improve airborne interception.

### The klystron

During the early months of the war, the Committee of Valve Development (CVD) had placed development orders on a number of institutions for the transmission and receiving valves of 10 centimetre wavelengths. One of these had gone to the University of Birmingham, whose research group was lead by Professor Mark L.E. Oliphant (later Sir Mark Oliphant), professor of physics, and who had been looking at the klystron (which Oliphant had first seen during a visit to America where it was invented, in 1938). At Stanford University in America, W.W. Hanson had been investigating the properties of cavity resonators since 1936. A cavity resonator is essentially a stream of electrons projected across a cavity by means of a thermionic valve, which produces an intense alternating field of high frequencies. A series of electromagnets reverses this stream of electrons every half turn in the cavity, so that the stream will return and re-return through the cavity several times before it escapes. This device is known as a rhumbatron.

The brothers R.H. and S.F. Varian took up Hanson's ideas and combined two rhumbatrons into one tube. Electrons leaving the first cavity, called the 'buncher', then passed through the second, called the 'catcher'. The passage of the bunches of electrons caused the 'catcher' to resonate and RF power could be drawn from it via a coupling coil. Such a device is an amplifier, and it is called a klystron. By connecting together the coupling loops in the buncher and catcher, the device becomes a self-oscillator (such as the 10 kW pulsed klystron that would be originally tried in the EMI H2S system). Alternatively, a single klystron can be caused to oscillate by reflecting the beam of electrons after being bunched by one passage through it, back through the rhumbatron again, by means of a negatively biased reflecting electrode. Such reflex klystrons were developed as low-power local oscillators for superheterodyne receivers, and manufactured by EMI.

By the end of 1937, klystron oscillators were being constructed that were giving constant energy at a wavelength of 12 centimetres and, in 1938, shortly before the findings were published in the *Journal of Applied Physics*, Oliphant visited Stanford to see for himself the progress that was being made by Varians and Hansen with the device. Upon returning to England and the University of Birmingham, he set about forming a group which would improve the results obtainable from a klystron; by the end of 1939, they had achieved an oscillator which could produce about 400 watts of continual power output on a wavelength of 10 centimetres. Following this development a pulsed oscillator was developed, but there were serious problems to be overcome when the possibility of an airborne centimetric klystron system was considered, not the least of these was how to isolate the high powered electron beam inside an aircraft.

As the development of this equipment progressed, a second possible candidate for the production of a centimetric radar system emerged – the 'magnetron' oscillator.

### The magnetron

Simple split-anode forms of what would eventually become known as the magnetron had been around since the 1920s when GEC in Britain, as well as laboratories in Germany, America, France, Russia and Japan, had all built low-power units which could oscillate at frequencies that produced a wavelength of around 10 centimetres. These early split-anode magnetrons were the only form of producing 'microwaves' other than the sparks that were originally used in early radio transmission. By 1939, with the physics department of Birmingham University buzzing with talk of microwaves, Professor Oliphant decided to put two of his researchers to work to investigate the possibility of using a 'Barkhausen–Kurz tube' as a possible microwave detector; they were John Randall and Harry Boot.

In 1920, Barkhausen and Kurz had applied a high positive potential to a grid of a triode and a small negative potential to the anode so that electrons were accelerated through the grid but were returned to the filament without actually reaching the anode. From this they deduced that the frequency produced was determined by the time it took the electrons to travel from the filament, through the grid, and back to the filament again. Randall and Boot set to work to see if the Barkhausen–Kurz tube could be used in the way Oliphant had instructed, but as Randall later pointed out: 'We were unsuccessful and disenchanted with the task ... and were naturally more interested in the main problem, the generation of microwaves.' Aware of the problems that were dogging progress with the klystron oscillator, Randall proposed that another form of microwave resonator be tried. In July 1939, while on a short holiday, he had bought an English translation of Heinrich Hertz's book *Electric Waves* by H.M. Macdonald, published in 1902. In the book Hertz describes how, in 1889, he used a resonator as a detector in his experiments that consisted of a simple piece of wire, bent into a circle with a small gap as the detector. The resonator was in fact a loop and not a cavity as was the case in the klystron.

Randall and Boot took the idea of the loop and extended it to a cylinder with a slot for the gap. Several of these slotted cylinders could be constructed to fit symmetrically around a cathode of a conventional magnetron. It was hoped that by machining such a device with its cylindrical segments from solid copper, it would allow good heat transfer and hopefully high power dissipation. As there were no theoretical calculations of the wavelength that could be expected from a resonator of the cylindrical-slot type, Randall and Boot now turned to Hertz once again who had shown in the same book which Randall had purchased on his holiday (and which had been subsequently confirmed theoretically by Macdonald in 1902), that the wavelength from a simple Hertzian dipole was $\lambda = 7.94d$, where $d$ is the diameter of the ring. Using this as the only basis for their calculations, six resonators, 1.2 cm in diameter, with slots $0.1 \times 0.1$ cm, were used with the length of the resonator cylinder set at 4 cm. Randall and Boot calculated if a potential of 16,000 volts were applied within a magnetic field of approximately 1000 oersteds, it was hoped that this configuration would produce a wavelength, in an anode hole of the same diameter, of around 10 centimetres. Randall and Boot called their device a cavity magnetron.

The cavity magnetron had its first demonstration in their laboratory at Birmingham University on Wednesday, 21 February 1940. Even though the device had a fairly feeble power output of only 400 watts, it did produce a wavelength of 9.8 centimetres, which was measured by means of Lechter wires against a known scale. The success of the project was now shrouded in complete secrecy by the CVD. Only Rowe and Lewis were told at AMRE and Dee was informed upon his arrival in May. On 22 May 1940, Dee visited Randall and Boot at Birmingham University to see the cavity magnetron working; he wrote in his diary at the time: 'Oliphant at Birmingham has developed a klystron giving a kilowatt at 8 cm and Randall, also at Birmingham, has a magnetron on the same wavelength.'

## Testing unknown quantities

As the magnetron and klystron oscillators both represented unknown quantities, it was decided that both should be produced for tests to be carried out at Worth Matravers as soon as suitable examples could be constructed. The examples of the centimetric cavity magnetron were to be made by GEC research laboratories, who essentially copied the original Birmingham laboratory version piece by piece. The first GEC magnetron model was known as the E1188, but this was followed only a couple of weeks later by the E1189 which could operate at 8.5 kilovolts in a magnetic field of 1500 oersteds.

The centimetric klystron valve was delivered first, arriving on 2 July 1940. The magnetron valve arrived a couple of weeks later, on 19 July. Tests now needed to be carried out using a spun parabola that Lovell had been perfecting at AMRE following the discovery that it created a polar pattern up to ten times the size of the parabolic dish itself. Once these had been manufactured and the magnetron system assembled the tests could begin.

On 12 August 1940, during the Battle of Britain which was raging in the skies above AMRE, the first 10 cm echoes were received from an unidentified aircraft flying along the coast a few miles away using an E1189 magnetron from GEC. The next day more echoes were recorded. Dee had brought Watson-Watt and Rowc to see these historic events for themselves and, in the afternoon, one of the junior assistants at AMRE, Reg Batt, took his bicycle and a large tin sheet and rode along the cliff top in front of the experiment set-up. As the parabolic dish was swung to follow the bicycle, the tin sheet produced a strong echo on the CRT. Lovell noted at the time that it was 'amazing considering it [the tin sheet] should have been right in the ground returns'.

Various improvements were made to the cavity magnetron by GEC which included increasing the number of resonator cylinders, lengthening and shortening the cylinders before an improved unit (called the CV38, but known as the E1198 by GEC), producing wavelengths of between 9.1 cm, with a peak power of 15 kW, a pulse rate of 50 per second and a pulse length of 30 microseconds. It was this unit that was considered suitable at last for the first airborne flight trials. It was a GEC E1198 that Tizard had taken with him to America in August 1940. Under instruction from Churchill, with a team of British scientists, Tizard was to explain to the Americans all the knowledge obtained by the British up to that point from klystron and magnetron development. An example of the cavity magnetron was also to be given to them to investigate and trial themselves.

By November 1940, the TRE team (AMRE became TRE in November 1940), had been increased with several specialists in the fields related to the development of the radar system. Among these was Dr Sam Curran, who would later head the EMI/TRE liaison team. At this point Curran, who was a specialist in pulse amplifiers at Cavendish, set to work on the problem of producing 50 kW of pulse power, with the scanner rotating at 2000 rpm, which had hitherto been considered an unworkable proposition. Several other delays had resulted from the problems associated with the scanner: the shape and size of the parabolic dish, the angle at which it was to be elevated and the motion pattern which would give the best results had been Lovell's task for some months by this point. Two types of scanner had been built and tried: one used a spiral pattern to scan the region in front of the aircraft – this would later become known as AIS – while the other used a helical pattern to scan what would become known as AIH.

The problems with the scanner had finally been resolved in February, and the choice of the spiral pattern had been made, as it proved more responsive in the flight tests. With all the necessary elements in place, the first centimetric AIS flight tests took place in March 1941, in a Blenheim aircraft (N3522), which demonstrated positive effects from the new system. Ranges of three miles were soon achieved on target aircraft and ground returns from the main beam showed up as an artificial horizon, which rose and fell with the varying pitch of the spiral that the parabola took. It was determined that this could prove very useful in the interception of a target.

There were however, unwanted features. A large ring of ground returns caused by scattering in the centre of the display, relative to the position directly below the aircraft, would dog the progress for many months while it was worked out how to reduce this unwanted ring. Up to this point the new centimetric radar system had been called provisionally 'TR' for Transmit-Receive. The metric AI system in use at the time with the Beaufighter night fighter had been slowly improved to the point where it had reached the Mark VI stage. The new centimetric system would now became AI Mark VII.

It was decided that four Beaufighters were to be equipped with the TR system, and twelve sets of scanners were ordered from Nash & Thompson with the associated electronics coming from GEC. The Bristol Aircraft Company, which produced the Beaufighter, had to be convinced to produce the aircraft with Perspex noses as it had been discovered by now that the aircraft's profile contributed to the degree to which the unwanted ring of returns was seen.

## 'The turning point of the war'

It must be remembered that until Rowe's 'Sunday Soviet' of 26 October 1941, the urgency with which centimetric radar had been developed was driven by the pace of the scientists working to develop what they saw as a logical improvement on the existing airborne interception radar system. Centimetric AI had been driven forward because of the poor results that the RAF night fighters were having against German night bombers during the Blitz. The knowledge of the poor Bomber Command results had not been commonly known at TRE, nor had the subsequent investigations into what was causing them or how they could be improved. Bomber Command, as mentioned earlier, had never approached TRE for any targeting or navigational bombing aids as other services had done. Because it was assumed that the bombing results over France and Germany were acceptable, even morale boosting to the British public, nothing had been done to speed up the development of centimetric radar in relation to bombing aids – though it could undoubtedly help. All that was about to change however.

Dee now wanted to organize a specific test to formulate the possibility of centimetric radar being used to scan the ground for echo returns. By modifying the rotation of the parabolic scanner to a fixed angle of depression to the horizontal, he felt certain that objects as large as a town would show up despite the ground scatter. Dee had installed, in August 1941, a double paraboloid helical scanner in another Bristol Blenheim Mk.4 aircraft (V6000), for flight trials on the AIH type of scanner. He was so certain that this method would work for the type of tests that would detect ground-based objects that, in the week following the Sunday meeting with Rowe, he asked Dr Bernard J. O'Kane (who was a member of the GEC staff attached to TRE, and the project leader for V6000), and Geoffrey S. Hensby (the 23-year old radar operator), both of whom were actually carrying out the tests, to make the necessary modifications.

Hensby had been an engineering apprentice at the Royal Aircraft Establishment, Farnborough, from 1934 to 1938, during which time he achieved a BSc and MSc as an external student at Birkbeck College. He was appointed 'Technical Assistant III' in July 1938, but chose instead to take up a DSIR studentship at Birkbeck in September, studying cosmic rays under H.J.J. Braddick. At the outbreak of war,

Figure 9.1
Bernard J. O'Kane. (*Courtesy of Sir Bernard Lovell*)

Hensby had returned to RAE, and taken up the position of Junior Scientific Officer working on proximity fuses when they were in their infancy. Hensby joined TRE in November 1940, along with several other scientific officers from RAE, all needed to assist the urgent development of radar. It was at TRE in November 1940, that Hensby first met Sam Curran, who was immediately impressed with the young man, and asked him to come and work with him on solving the problem of driving the magnetron, which they did by introducing triggered sparkgaps. Hensby then turned his attentions to the problems of AI systems for night fighters, becoming instrumental in the installation of, and later flight-testing of aircraft fitted with experimental centimetric radar from Hurn airfield, near Christchurch. For a while, Hensby had also been aboard the test aircraft as observer during many of the AIS trials. As some of these flights were conducted over enemy-held territory, Hensby was given the protection of an RAF honorary commission and uniform.

On 27 October, O'Kane and Hensby visited GEC in Wembley, in order to obtain the agreement needed to interrupt the AIH tests, and to make the modifications to the scanner that Dee had thought necessary. This done, the TRE workshops then modified the scanner (which had been taken there on 28 October). The first flight test was carried out in V6000 on Saturday 1 November 1941, and soon after take-off, at about 5000 feet near the Solent, radar operator Hensby reported to O'Kane that he could clearly see the large echo from the city of Southampton below. He noted especially that because of the absence of an echo return from the sea the coastline was shown in sharp relief. Encouraged by the success of the initial flight, O'Kane and Hensby made subsequent flights at 7000 feet over Salisbury Plain with photographic equipment this time to record the results. The photographs taken directly from the CRT clearly show the towns of Salisbury and Warminster, as well as the military

Figure 9.2
Geoffrey Hensby.

encampments on the flat ground between them. When Dee was shown these photographs, allegedly before they were even dry, he apparently rushed them over to Rowe's office. With the wet photographs in front of him Rowe exclaimed, 'This is the turning point of the war.'

## Naming the system 'H2S'

There are several versions of the story of how the new radar system came to get its name of 'H2S'. When the device had first been brought to the attention of R.V. Jones, by now Assistant Director of Intelligence and a close contact of Lord Cherwell, he noticed that it had been given the code name 'TF' which, he felt inappropriate as it would be interpreted as 'Town Finding'. This was not specifically what the system did nor was intended to do, and Jones decided to speak to Lindemann about it. According to Jones, Lindemann apparently remarked that it was 'stinking that TRE hadn't come up with the idea for the device years earlier, and that an appropriate name for it would be H2S'. H2S is the chemical symbol for hydrogen sulphide – a colourless, poisonous gas having the odour of rotten eggs!

There is another story of how the system got its name: Rowe said that Lindemann had chosen the name 'Home Sweet Home' because the system enabled one to locate and home in straight onto the target. That in turn became shortened to the name 'H2S' by TRE, though it is uncertain who actually abbreviated the new name first. It may have been Lindemann looking to find a simpler formula for the device, though more than likely it was the actual bomber crews first using the device, who shortened 'Home Sweet Home' to 'H2S'. In any case the name stuck. Rowe informed Churchill of the results of the flights made by O'Kane and Hensby on

20 November 1941. He was cautious enough to point out that a decision to use centimetric radar should only be made after serious consideration had been given to the possibility of losing the equipment over enemy territory. Churchill was very enthusiastic about the potential for the new 'BN' system (Blind Navigation) as it had now been called (before H2S had stuck).

Over the coming two months, through November and December 1941, many more test flights were carried out in V6000 to determine which altitudes gave the best results. At this time Southampton, Bournemouth and Wolverhampton were all detected at ranges of up to 35 miles with the angle of approach immaterial, even the hangars of aerodromes could be detected with the radar. The tests were classified as 'Most Secret' by now, and the first murmurings about the consequences of losing a magnetron over enemy territory were voiced. This would be a subject that returned time and again as the possibility of losing the new oscillator and its secrets being discovered plagued those who had to decide whether or not to use it.

Because of the potential disaster that could come from losing a magnetron, the tests were continued on the basis that the klystron valve might have to be used instead. They were subsequently carried out by O'Kane and Hensby using a magnetron to simulate the results that they expected to obtain from a klystron valve. Crucially however, instead of actually using a klystron during these tests, they actually reduced the overall power of the magnetron, which at the time showed 'No sufficient reduction in range that would impair the usefulness of the apparatus'. This observation would be destined to cause much trouble later on.

The next phase of the development of BN would be to evaluate its use at higher altitudes. The Blenheim that had been used up to now could not fly above 10,000 feet because the experimental equipment would flash over. A Beaufighter could fly at 20,000 feet for high altitude tests, but what was really needed were evaluation tests in the new series of bombers that were currently in development for Bomber Command. The four-engined Halifax, Stirling and Lancaster bombers were to replace the two-engined Wellington, Hampden and Whitleys as soon as they were available.

On 23 December 1941, the Secretary of State for Air, Sir Archibald Sinclair (who had succeeded Lord Swinton as Air Minister with the appointment of Churchill as Prime Minister on 10 May 1940), called a meeting with Lord Cherwell and directed that TRE proceed with the development of H2S for the new breed of bombers.

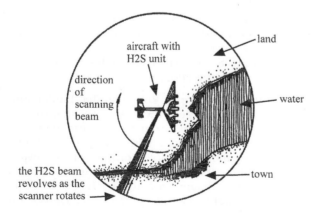

Figure 9.3
The H2S scanner beam.
(*Courtesy of EMI*)

Figure 9.4
The H2S plan position
indicator screen as the
operator would have had
to interpret it.
(*Courtesy of EMI*)

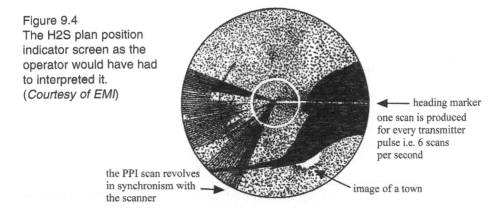

◄── heading marker
one scan is produced
for every transmitter
pulse i.e. 6 scans
per second

the PPI scan revolves
in synchronism with ──►
the scanner

image of a town

Sinclair also instructed that H2S should be equally evaluated, in a simplified form, as a ground-mapping method for the identification of a target, rather than the range/azimuth indicator that the current TRE radar projects all used. Cherwell pointed out that such a system would be too complicated and take too long to develop. He suggested a compromise in which a split aerial system with a left/right display could be used to home in on a target. This he assured Sinclair could be developed much easier and would be adequate. Finally, the men decided that despite the promising results from the magnetron-based system, the klystron and the magnetron would need to be tested to evaluate which would be most suitable for the device. The Ministry of Aircraft Production had been instructed several months before that any project involving the klystron should be considered an additional project, and that nothing was to impair the production of the magnetron valve at any cost.

However, since there remained some doubts about whether the response from H2S were unambiguous related to specific targets, Cherwell insisted that 'Further experimental flights were to be made immediately to determine whether signals obtained on the device could be definitely associated with specific ground objects.' Of the other matters that were discussed, the issue of when the new device might be ready for active service was brought up. Cherwell and Sinclair both concluded that even

height marker
set to the level
of the first
ground
returns

range marker

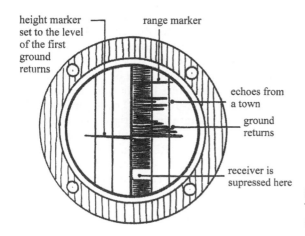

echoes from
a town

ground
returns

receiver is
supressed here

Figure 9.5
The H2S height tube indicator.
(*Courtesy of EMI*)

with a joint development programme involving TRE, EMI and GEC, H2S would not be available before the autumn of 1942, but, once available, it was more than likely to be successful.

Immediately after this meeting, Dee, O'Kane and Hensby visited the Aircraft and Armament Experimental Establishment at Boscombe Down to evaluate the three types of new aircraft to see which would be best suited for initial instalment of the BN equipment. After looking over the Lancaster and Stirling, they decided that the Halifax offered more scope and alternative positions for the installation of the scanner than the other two aircraft. Therefore, it was decided that the Halifax bomber, then only just coming into service, would become the first aircraft to trial with H2S.

Rowe now needed somebody to head the BN development programme for TRE. He chose Bernard Lovell as director, because of his close association with the development of centimetric radar, not least of which had been his work perfecting the parabolic scanner. Lovell, at first, was none too keen on the idea. In his book *Echoes of War*, he recalls how, at the time, he was busy working on a new feature of the improved AI system which would lock onto and follow an enemy aircraft (known as AIF):

'On 29 December 1941, Rowe had summoned me to his office and told me I was to cease working on the lock-follow AI and take charge of the development of a new device, then known as BN (Blind Navigation) to help Bomber Command. I responded that I did not want to do this because I was anxious to get the AIF system into the Beaufighter. Dee was absent but when he returned on 31 December, he said I had to do this and insisted on driving me to the new aerodrome at Hurn, hoping thereby to lessen my resistance. Early the next morning I was again taken to Rowe's office whose patience was rapidly evaporating as he abruptly terminated my further objections by the terse statement that "there was no alternative". In that manner I was ordered rather than "given charge of" a task that I did not want to do and knew little about.'

One of Lovell's first tasks was to explain to Handley Page that the two brand new Halifax bombers that were to be delivered to TRE would need to have a Perspex cupola mounted on the underside of the aircraft fuselage, in order to house the parabolic scanner. Lovell had to face Mr Handley Page, G.R. Volkert, his chief designer, and his design team on 4 January 1942, in the company of Bob King, also from TRE, who would be responsible for the fitting of the equipment into the aircraft, and A.W. Whitaker from Nash & Thompson who would fit the scanner. The three men had to explain that the modification to the Halifax needed to be made, regardless of the fact that Handley Page protested it would affect his aircraft's performance which would be 'ruined', and the bomb load severely reduced. Lovell remarked that it would surely be better to drop some of the bombs in the right place, than the entire load over the open countryside but, of course, he could not say anything about the nature of the work that they were doing at TRE, even to Handley Page at this stage. Other than give the specifications of the Perspex cupola – 8 feet long, 4 feet wide, 18 inches deep – there was little else Lovell could do to pacify Handley Page, who would just have to get on with the task.

It is remarkable today to think how quickly the aircraft were in fact delivered to

TRE. Halifax V9977, the first of the two, arrived at RAF Hurn on 27 March 1942, just two and a half months after Lovell's visit to the Handley Page works, with the second aircraft, R9490, arriving on 12 April, surely a testimony to the importance now placed on the task of perfecting H2S and the direct pressure being applied by Churchill on TRE.

## EMI are brought in to the picture

Following the meeting of Sinclair and Cherwell on 23 December 1941, it was decided that an immediate contract for 50 units of electronics for the H2S system would be placed on EMI. O'Kane was to bring the team of EMI scientists led by Alan Blumlein up to speed on the developments that had been made to that point with H2S; Lovell remarked at the time that he 'did not, at first, anticipate any serious problems'.

TRE were informed on 6 January 1942 that a system capable of detecting a town at a range of 15 miles from a height of 15,000 feet would be acceptable and in order to avoid delay, an order to produce 1000 units would be placed upon the relative manufacturers. The device was not to be produced as a navigational system, but should be considered solely as a means of identifying and homing to a target. The klystron should be considered as a vital part of the system as the potential disaster associated with the loss of a magnetron-based system was considered too great at this stage. Magnetron development was to be continued however, with the highest priority in order to assess its potential further.

By 21 January, the Chief of the Air Staff, Marshall of the Royal Air Force, Sir Charles Portal, had informed Churchill that the progress on the magnetron system had been so satisfactory that a full development plan for H2S could now be formulated. Portal proposed that the existing contract on EMI for the electronics, on GEC for the valves and on Nash & Thompson in Surbiton and Metropolitan Vickers in Manchester for the scanners (both of whom had been originally sourced by Lovell when developing the scanner), should now be increased. The Ministry of Aircraft Production should draw up a programme with dates, for the provision of 1500 H2S sets, to be manufactured by EMI, with a view to introducing the system as soon as possible. At the same time it was decided that Dr Sam Curran should be the liaison officer in charge of all communications between TRE and EMI. The RAF would also have their own liaison team comprising an experienced captain and navigator; this team was headed by Wing Commander G.P.L. Saye with Group Captain W.E. Theak and Flight Lieutenant E.J. Dickie. They would act as liaison officers between the RAF and TRE/TFU.

As the evaluation of the two valves continued it became obvious that the tests should be split into two separate groups, one working on the klystron while the other concentrated on the magnetron. At a meeting at TRE on 26 January, a decision to conduct a series of experiments to determine which valve should be used for the H2S system was called for. The experiments, which, it was hoped, would take two to three weeks to complete, should determine finally which of the two should have the undivided attention of the manufacturing facilities. EMI, it was decided, should conduct the tests into the potential of using the klystron and, should it prove to be the more favourable of the two types, they would be contracted further to manufacture the electronic components. They had of course also been involved with other elements of radar development, not the least of which was the GL project, also led by Alan Blumlein. The EMI team was assembled with Blumlein in charge, working with C.O. Browne and F. Blythen at Hayes, while F.R. Trott and M.G. Harker were to go to TRE and work on-site testing the equipment as it was manufactured and delivered. E.L.C. White would also spend some considerable time on site working with TRE. Because of the work that EMI had already carried out on GL and later AI, Blumlein's team had been working with klystron valves for some time and, bear in mind that O'Kane and Hensby had initially reported that there should not be any reason to consider the use of a klystron with H2S as a reason to impair the device.

The major problem with the use of the klystron in the H2S device was that O'Kane and Hensby had made an error in judgement when concluding that the klystron would work as suitably as the magnetron valve. This was clearly not the case. Their deduction had come from reducing the power on a magnetron during a test to 'simulate how a klystron might respond'. Magnetrons however, working at much higher power outputs, could detect targets at 35–50 miles. With the klystron valve installed,

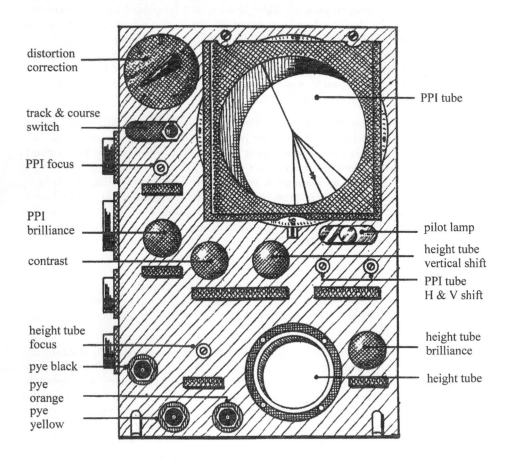

Figure 9.6 H2S type 184a indicator unit. (*Courtesy of EMI*)

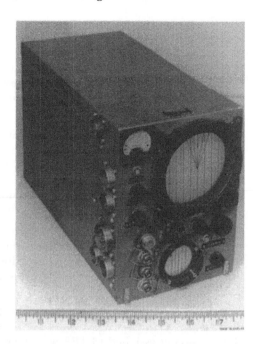

Figure 9.7
H2S indicator unit front and side
detail. (*Courtesy of EMI*)

initial results would yield results that only obtained targets at ranges of 5 miles or less. A single magnetron valve was given to EMI in order to assess the variances in their work with the klystron, though the majority of the evaluation tests on the magnetron would be carried out by O'Kane, Hensby and Lovell's team based at TRE. At the same time, work was also to continue towards the introduction of Gee which, by February 1942, had been developed to a stage where it was almost ready to be installed in aircraft in active service.

## Pre-production tests

When Churchill had given his express orders for the two Halifax bombers to be made immediately available to TRE/TFU, they were literally coaxed away from Bomber Command at a time when the air war was going very badly for the Allies. Aircraft losses in those first few months of 1942 were increasingly high, and it was felt by many at the Air Ministry that few if any could be spared for research that might potentially risk the aircraft and crew, with no guarantee that any development might benefit the RAF. In March 1942, during the many raids that were carried out that month, nearly 200 aircraft were lost to enemy action. That figure, apart from the enormous loss of men and materials that it represented to Bomber Command, was also very near the maximum for aircraft production and crew training programmes to keep pace. Despite the very definite opposition voiced by some to the huge effort, not to mention considerable cost that was now flowing into the radar development programme, Churchill continued to champion its cause. That was why the two Halifax aircraft were given to TRE with such a priority attached to them.

At EMI Blumlein was busy assembling the team that would work on the klystron development of H2S and bringing everybody up to speed on the new system and how it worked. The scientific team that he chose would consist of himself, Frank Blythen, Felix Trott and Cecil Browne. The engineering team, who would actually be responsible for the building of the first H2S units, comprised H.L. Oura and W.H. Cox. The work itself was carried out on the first floor of the Works Designs Department, Building No. 1 at Hayes, in Middlesex. Curran liaised directly with Blumlein who in turn reported to Lovell all aspects of the development progress.

At GEC the klystron and magnetron valves were already under contract to be manufactured although progress was proving painfully slow. Nash & Thompson were manufacturing the hydraulic scanner apparatus and Metropolitan Vickers were building an electrically driven version of the same. The assembly of the various elements of the system would obviously take some time, during which flight trials on the original magnetron prototype would have to continue. While the two Halifax bombers were being readied, another Blenheim Mk.4 (V6385) at RAF Hurn was used.

During one of the 20 or so flights that were carried out, the radar operator recorded that strong echoes were received from the aerodrome hangars at RAF Yelverton from a range of 30 miles. Although the visibility was so poor that the pilot didn't see the hangar with his eyes until only 2 miles away, with the heading he was given he could fly straight toward it.

At the English Electric works in Preston (where the first aircraft was being assembled) and at Handley Page (where the second aircraft was being assembled), the Halifax aircraft assigned to TRE were being fitted with their Perspex pods which were slung under the belly of the aircraft. This 'cupola', as Lovell preferred to call it, protected the delicate rotating scanner, but enabled the high frequency energy to penetrate and scan the ground without affecting the transmitted or received signals. During the time the development work on H2S had been progressing, centimetric ASV had also reached a stage where a provisional manufacturing contract had been placed with Ferranti in March 1942. It seemed obvious that many of the component parts of the new H2S system would be the same as a centimetric variant of ASV (which would eventually be designated ASV Mk. 3), and that in order to save duplication, there should be some form of agreement where one installation could cover the needs of both systems.

This posed a problem because centimetric ASV had been designed entirely around the use of the magnetron, whereas H2S at this point in time still had development basis on the klystron, which was not considered effective enough for ASV use. The continuing debate over which of the two valves to use would now have to be resolved in order for centimetric ASV and H2S to proceed jointly manufactured. The problem facing the Air Ministry and the Government was simple. Which of the two valves performed the best? If it was the magnetron, could it be risked over enemy territory? The answer to the first question was becoming clearer by the day. Even though neither Halifax had been fitted yet with the development equipment, the flights being made by O'Kane and Hensby in V6385 were now definitely showing that their first observations regarding the results of the klystron were in fact in error. Not only did the klystron not prove anywhere near as effective as the magnetron results, the ranges they were obtaining and the altitudes at which the device detected towns was well outside the Air Ministry's minimum of 15 miles at 15,000 feet. It was becoming obvious that the magnetron was winning out over the klystron. Because this new system would effectively make any aircraft fitted with it independent of any ground control station for target identification, it would be limited only by the range of the aircraft itself. Exactly what Bomber Command needed. This then, left the dilemma of whether such a secret device should be risked in an operational aircraft over enemy territory. What if an aircraft was captured and the equipment fell into enemy hands?

The last thing the security conscious authorities wanted was to give away any advantage that might have been gained to the Germans. For this reason they favoured the klystron system as it was considered technology that the Germans already knew of, having been published in detail as far back as 1939. True, it was not as good as the magnetron system, and it was significantly lower in power output although it worked in roughly the same frequency range, but could it be risked? Could a captured magnetron system be dismantled and copied by the Germans? If this were the case, then the German night fighter defence system would be able to match Bomber Command during their night raids. The potential losses in aircraft and crew would be enormous.

It was decided to see if the device could be destroyed with the use of a detonation charge that the crew could activate in the event of being shot down or imminent capture. However, even this had its problems. The main magnetron core was a solid copper block with eight bore holes around a central cavity through which the agitated current was passed. It was nearly indestructible. Explosives experts at Farnborough devoted some considerable time and effort to the task of fitting a suitable destructor to the magnetron block and its connector systems. Even with the amount of explosive charge that it was considered safe to carry onboard an aircraft, a magnetron was extremely difficult to camouflage. During the destruction trials, a 2-ounce explosive charge was used (which blew a 10-foot hole out of the side of a captured German Junkers JU88 aircraft used for the test), the magnetron core was still identifiable, and worse, the fragments left over would have been, it was deduced, able to be studied to find out how it worked.

Clearly unsatisfactory the Air Ministry tried a method of explosive ejection and destruction in mid-air. But this also proved too dangerous for the aircrew whose responsibility it was to confirm complete destruction near the aircraft, but risked endangering the aeroplane from the recoil of the explosion. The idea of resorting back to the klystron oscillator to power the H2S system was reconsidered, but in the end it was events in Bomber Command that would really decide the outcome for the authorities.

Figure 9.8
An H2S indicator unit with the outer casing removed showing interior detail and construction.
(*Courtesy of EMI*)

By mid-1942, with losses over Germany getting close to the limit of 200 aircraft a month that had been set by Churchill, Dowding and Harris, the use of H2S, and the risk associated with its loss over enemy territory was considered justified – if it would considerably reduce the numbers of aircraft lost.

On 15 March 1942, the Ministry of Aircraft Production agreed that a proposal for one company to develop jointly the electronic components for centimetric ASV and H2S should be EMI, and that this development should concentrate on sourcing as many components common to both as possible. At this point they held short of finalizing which of the two valves should be used as it was decided that this should be resolved with a short series of test flights in the new Halifax bombers which were now almost ready. TRE and EMI drew up a component list of interchangeable parts for centimetric ASV and H2S. The cupola, magslip, scanner, and cabling could be made common by changing the aerials of the H2S system which was already being made by EMI anyway; the power supply unit, receiver and timing circuits, control box and indicator unit could also be easily modified to work in both roles. In fact the only major differences at that time were three for centimetric ASV: the magnetron, a modulator and another power supply. For H2S, the differences were a klystron, and a modulator.

The Gramophone Company division of EMI were now given the task of putting this joint development programme into action which was known as H2S/ASV. There were to be 15 pre-production models completed by EMI as soon as possible, and the original manufacturing order placed on the company at the end of January for 1500 H2S units, was now changed to 1500 H2S/ASV units. The Air Ministry, TRE and EMI now decided that the final decision as to which valve to use with H2S should be resolved by a short series of demonstration flights. This would determine the merits of both systems and then the authorities could make their minds up based on hard facts. By now the two Halifaxes had had the modifications that Lovell, King and Whitaker had asked for and were ready to be delivered to RAF Hurn.

## Delivery

On Friday 27 March 1942, Halifax bomber V9977, a Mk.2 built by the English Electric Company in Preston, was delivered to RAF Hurn. It was followed, on Sunday 12 April 1942, by Halifax R9490, also a Mk.2 but built by Handley Page. During the following 10 weeks, these two aircraft were to prove that Churchill's dogged persistence with radar development was worth all the effort expended.

Halifax V9977 was designated as the number one demonstration aircraft and was duly fitted with the H2S radar system which incorporated the magnetron oscillator being tested by the TRE personnel. The radar equipment was linked to a Nash & Thompson hydraulic scanner, essentially a parabolic aerial, suspended in the pod under the aircraft. The ground reflections were to be displayed on a circular CRT mounted inside the aircraft, positioned at the navigator's station towards the front of the aircraft and immediately below the cockpit. From here the radar operator, peering into the CRT, could distinguish from the echoes received between flat countryside and taller built-up areas with varying degrees of eminent light. The operator would view a sweeping clockwise rotary image with the echoes appearing as areas of light and dark sketches tracing out the countryside just like a map.

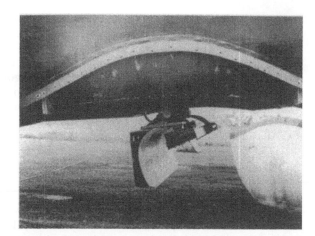

Figure 9.9
The Metropolitan Vickers
electrically driven parabolic
aerial underneath Halifax
R9490 in 1942.
(*Courtesy of EMI*)

Halifax R9490 would also be fitted with prototype H2S radar equipment, but this time using the klystron oscillator system, being tested by the EMI team, and which was linked to a Metropolitan Vickers electrically driven parabolic aerial. The two aircraft were now ready to complete their flight tests and trials that would conclude which of the two systems worked the best and which would eventually be fitted to the RAF night bombers of Bomber Command.

From its arrival on 27 March, until 15 April 1942, V9977 was in the hangar at Hurn having the magnetron H2S system installed. The first trial flight was to have taken place on 16 April, but there was an electrical problem with power to the H2S system supplied by an 80-volt alternator driven by one of the engines. The first test flight actually took place on 17 April, by which time R9490 had arrived at Hurn also, and was in the hangar being fitted out. During the flight, which was made by Lovell, Hensby, and O'Kane with Pilot Officer C.E. Vincent (who was a member of the TRE group responsible to Lovell), at the controls, an altitude of 8000 feet was achieved at which point several good ground returns were seen on the CRT, but, disappointingly, they were only able to detect these at a range of four or five miles.

Other flights took place in V9977, similar to that carried out on 17 April, but most of these also produced poor results compared to those that had been achieved in the Blenheim. This was mainly because they were carried out at higher altitudes which led to the conclusion by Lovell and Hensby that the scanning apparatus needed to be redesigned and repositioned.

As if the scientists at EMI and TRE didn't already have enough to worry about with the poor results they were obtaining, the numbers of bombers that were now rolling off the production lines and reaching Bomber Command meant that the possibility of a serious bombing offensive in the autumn now looked like becoming a reality. That offensive was reliant upon H2S being ready in time. As things stood at the end of April 1942, EMI didn't feel they could deliver more than 15 completed H2S units by Christmas – assuming they could get it to work properly.

Blumlein and the EMI team had been working non-stop during this period with Oura and Cox at Hayes, constructing the klystron version of the system. The work had taken longer than they had anticipated because of supply problems with certain components, having finished the work however they managed to assemble the apparatus in a coach and get it working ready for the drive down from London to Leeson House and RAF Hurn. They arrived at Worth on 14 May, and soon transferred the klystron equipment from the coach into R9490. Since arriving at Hurn on 12 April, this aircraft, like V9977 before it, had been undergoing ground flight check preparations prior to the installation of the H2S system. By mid-May it was practically ready to begin flight checks, but these were cut short, as were the tests carried out in V9977, by the orders to move from Leeson House and RAF Hurn, to Malvern and RAF Defford.

## Time to decide – then disaster

The disruption that the move to Malvern must have caused is hard to underestimate. The move took place from Sunday 17 May 1942, to the following Sunday, 24 May. Everything had to be stopped, packed in crates and moved by road across England to Worcestershire. The two Halifaxes had their H2S equipment removed by a combination of the EMI and TRE/TFU teams as well as the engineer fitters attached to TFU at RAF Hurn. V9977 and R9490 as well as the other aircraft attached to TFU were flown to RAF Defford where, upon arrival, they were subject to routine maintenance checks.

There had evidently been some problems with one of the engines on Halifax V9977 as it was decided at this time to replace it with another – on 25 May, the port inner engine (No. 2) which had originally been a Rolls-Royce Merlin Mk.20 was replaced with a Packard Mk.38. Though this was a lengthy procedure, it was not uncommon during the war, and the aircraft engineers at RAF Defford would have been able to carry out the operation in a matter of a couple of days. Between 25 May, and 1 June, all four engines of V9977 underwent a standard 30 hour inspection, and it was decided to adjust the tappets. It was also during this same period that the H2S units, both the magnetron version for V9977 and the klystron for R9490 were re-installed in the aircraft.

To add to the problems of the move itself, the rain and cold had been almost incessant during the entire time that everybody and everything had been on the road. There were major problems with accommodating everyone in Malvern and there was no provision for food other than to queue for ages in the rain outside the Winter Gardens Pavilion in Great Malvern while the Women's Volunteer Service cooked for the more than 1000 staff attached to TRE/TFU. There were builders everywhere trying to erect accommodation blocks, laboratories, workshops – it must have been mayhem. Moreover, amongst all this, there was the constant pressure to get the H2S system working properly. The first of the re-scheduled test flights was to be carried out in R9490 on 2 June; Lovell and Hensby were on board with Pilot Officer Vincent flying the aircraft and Wing Commander Saye along to assess the klystron H2S unit's results. Unfortunately the flight was not very successful as the klystron once again proved that it was not powerful enough resulting in poor range, this time just four miles. Saye later complained after that flight, in a letter to the Air Ministry, that all he could make out was a 'snowstorm' on the CRT, that he could not determine this as a town or anything else for that matter. The klystron system was not working out. It was time to try the magnetron again.

The magnetron version of the H2S system had not been immediately installed back into V9977, instead it had been taken to the workbenches that TRE were using on the top floor of the Preston Science School at Malvern College. Hensby and O'Kane had been working on the unit for some time since the flight of 17 April, which had been so disappointing. Now they thought they knew why. It was decided by Blumlein that another series of test flights in V9977 with the magnetron H2S system should take place on the weekend of 6–7 June, during which time his EMI team could assess the progress made by Hensby and O'Kane. Blumlein, Blythen, Browne and Trott checked into a hotel in Tewkesbury a few miles from the aerodrome at Defford and spent most of Saturday 6 June in discussion with Lovell, Hensby, O'Kane about the nature of the tests and how they were to be conducted.

On the evening of 6 June, Hensby and Lovell went up in V9977 as a pre-cursor to the following day's flight which had been designated for the EMI team to assess the magnetron system. Lovell recorded that evening that the latest modifications to the radar produced good results. Gloucester, Cheltenham and several other towns were all clearly displayed at greater images than Lovell had ever seen before. All looked well for the tests the next day. By now, the weather had turned fine again. Sunday 7 June was a clear, sunny day when Halifax V9977 took off from Defford to conduct the tests with the EMI team of Blumlein, Blythen and Browne on board. Hensby and Vincent were also aboard as were the six flight crew.

It can only be assumed that the results of the tests would have been favourable, as they had been the night before, but we will never know. During the flight, Halifax V9977 developed a fire in the No. 4 starboard outer engine, and in the course of trying to find a suitable location for an emergency landing, the aircraft crashed killing everyone on board.

# Chapter 10
# The loss of Halifax V9977

## Secrecy and speculation

The circumstances surrounding the tragic loss of Halifax V9977 and the 11 crew onboard have never been fully documented in an open publication. Needless to say, in time of war (and bearing in mind the sensitive nature of the work being carried out during the flight), much of the detail of the crash was hidden for years under a blanket of total secrecy. Some forty years after the crash, an investigation was carried out by engineers at the Radar Research Establishment at Malvern, but the document was intended as an internal publication and had been produced ostensibly to correct details which had originated from the crash investigation immediately after the flight, and which had subsequently been taken to be factual. The actual course of events and the reasons that led to the loss of the aircraft were therefore brought to the attention of only a few Ministry of Defence personnel, the civilian Air Historical Branch (RAF) archivists, and a select few who had personal or professional ties to the work that had been carried out by the team on that fateful flight.

This inadvertently led to much speculation as to what actually occurred on that Sunday in June 1942, and how it was, that on a perfectly clear day, a nearly new aircraft with a total flying time since delivery of just 64 hours and 45 minutes, could fall from the sky killing all aboard and, in the process, rob the world of untold scientific genius.

## Origins

The origins of the aircraft which became the Handley Page Halifax began in 1936 when the Air Ministry issued a specification P.13/36 inviting tenders for a twin-engine heavy bomber to be powered by engines each of 2000 hp giving a speed of 270 mph at 15,000 feet with an altitude ceiling of 28,000 feet. The new bomber would have to fly satisfactorily on one engine at 10,000 feet, have a 500 yard take off with 1000 lb of bombs, increased to 700 yards with 3000 lb of bombs as well as carry a maximum bomb load of 8000 lb. It should have a range of 2000 miles. The crux of the matter was that this tender was based around an engine which at the time simply did not exist. Aero engines, due to the complexity of their technology, took about five years from drawing board to service.

Ernest Hives (later Lord Hives) at Rolls-Royce circumvented the situation by combining two well proven in-service Rolls-Royce Kestrel V12 block, 21-litre engines onto a modified crankcase to produce a 42-litre X block engine producing 2000 hp. This was miraculously available by 1939 and became known as the Vulture.

Design proposals for the new aircraft were received from Handley Page (HP.56), and A.V. Roe (which would eventually lead to the Avro Manchester), though almost immediately afterwards Handley Page set about redesigning the HP.56 specification from a two-engined aircraft to a four-engined aircraft for four of the Rolls-Royce Merlin engines instead.

The new submission to the Air Ministry from Handley Page, HP.57, was for a substantially larger and heavier aeroplane than that which had been specified, but even at this early stage in its development the need for four-engined bombers had been foreseen by the ministry officials. Awarded a contract to build two prototypes on 3 September 1937, the first Halifax, L7244, was taken by road from Cricklewood where it had been built to Bicester for its maiden flight on 25 October 1939. That flight, made by Handley Page's Chief Test Pilot, Major J.L. Cordes, was considered to have surpassed all expectations for the aircraft; L7244 performed very well and the all-up weight achieved had reached 23 tons rather than the 18 it had been designed for. The second prototype, L7245, first flew on 17 August 1940, from Handley Page's own Radlett airfield in Hertfordshire, and was then passed to A&AEE for flight acceptance tests during October.

Designated Halifax Mk.I, with Rolls-Royce Merlin Mk.X engines, the aircraft had by now achieved an all-up weight of 24½ tons. The first aircraft of the production models (L9485) flew for the first time just weeks later on 11 October 1940. Somewhat alarmingly, the staff at A&AEE had discovered during this time that the twin rudders on the Halifax were rather inadequate at keeping the aircraft straight on take off. There seemed to be a tendency for them to overbalance at low airspeeds and especially when the aircraft made tight turns or when power was asymmetric (when a propeller was feathered for example). They also remarked that the rudders were unresponsive at speeds below 120 mph, a serious shortcoming in an aircraft of which so much had been expected and that had just become operational.

Halifax V9977 was built by English Electric under license from Handley Page at Preston, in Lancashire, in the autumn of 1941. The aircraft was part of a contract (No. B982938/39) for 200 aircraft that would be built by English Electric (who would eventually construct 2145 Halifax aircraft by the end of the war, some 35% of the total built) with other Halifaxes being constructed by its designers Handley Page Limited at their own factories (Handley Page only constructed 22% of the total number of Halifaxes built, some 1367 aircraft. The remainder were constructed by London Transport, Roots Securities and the Fairy Aviation Company.) V9977 was the second aircraft of a batch of 19 aircraft (Registration Nos. V9976–V9994) completed and delivered between September 1941 and December 1941. These were the first aircraft of a new batch of the bombers designated the Halifax Mk.2B, which had extra range, more powerful engines (the Mk.XX as opposed to the Mk.X which Halifax Mk.Is had used), a Boulton Paul dorsal turret providing better defensive armament and an all-up weight of 27 tons.

The Halifax aircraft had a wingspan of 98 feet 10 inches and its maximum all-up weight of 60,000 lb included a 14,500 lb bomb load. Intended to be flown with a crew of seven, the full compliment would be pilot, navigator, wireless operator, second pilot/engineer, and three gunners. With a maximum speed of 270 mph and a cruising speed of 150 mph, the Halifax Mk.2 had a range of 3000 miles depending on load and weather conditions.

The design problems with the Halifax that had first been noticed by A&AEE were known of long before the loss of V9977. In fact two aircraft had been assigned to A&AEE Boscombe Down for specific bomber performance test flights to evaluate the Halifax. By this time, several aircraft had been lost in very odd circumstances that had been narrowed down to the problems with the rudders. It would seem that when the rudder was used to keep the heavy aircraft straight, it could swing sharply over to the right or left and stall. If this occurred during take off the aircraft would swing out of control and off the runway. If it occurred once airborne, the aircraft would sideslip into the ground or enter an inverted, uncontrollable spin with the same result.

In order to correct this stalling the pilot would need to rapidly re-trim the entire aircraft and try to regain control before the stall developed further. This could be achieved if the aircraft was being flown by a pilot who was experienced on type, which meant he would have intimate knowledge of his aeroplane and its handling characteristics. The Halifax however, was a new in-service aircraft at the beginning of 1942, and precious few pilots had more than a few hours' flying experience on the type. In an emergency situation, this could prove to be an important factor in the recovery and control of the aircraft.

Many trials were carried out on the two aircraft allotted to A&AEE at Boscombe Down (W7922 and W7917), hoping to find a reason for the problem and correct it. During one trial on 4 February 1943, W7917 crashed killing the flight crew and a civilian scientist aboard who was conducting rudder tests at altitude. Ground observers saw the Halifax dive, go into a flat spin and then plummet into the ground. When the wreckage was examined, it was discovered that the rudders had overbalanced so violently that one of them had in fact broken away. This stalling of the tailplane facilitated an entire re-design on later Halifax variants starting with the Mk.3. This aircraft was a redesign of the Mk.I and Mk.2 series, re-engined with Bristol Hercules XVI radial engines, the mainplane spar was also increased to 104 feet. Gone were the triangular design of the earlier tailplane versions, to be replaced with a much larger rectangular design which gave the aircraft more stability at low speeds. The new rudder had an increased area of around 40%; changing the profile seemed to have been an acceptable solution to the problem.

Unfortunately, V9977 was of the Mk.2 variants and had been built with the triangu-

Figure 10.1
Halifax Mk.2 'triangular' tail fins which proved too small and unstable at low airspeeds.

Figure 10.2
Halifax Mk.3 'rectangle' tail fins which
were much larger and solved the
instability problems associated with
the earlier marks of the aircraft.

lar, unstable fin design. With these design defects only rectified after the loss of
V9977 and several other Halifaxes, the question of whether this specific design
defect played any part in the loss of the flight remains a mystery, though it would
seem that the pilot did maintain directional control right up to the last moment
before the crash. As serious as the rudder problem was on the Halifax, it was in fact
only one of a series of problems that continued to dog the aircraft through its early
operational years. Additional demands were placed upon the aircraft through
increasing bomb loads, which naturally affected its performance. Most seriously, the
take-off speed of the aircraft increased with the additional weight making the take
off run protracted to dangerous levels. When the maximum bomb load of two
4000lb bombs was carried, the take-off was further complicated by the fact that the
bomb-bay doors did not close completely, thereby increasing drag. The aircraft,
once airborne climbed very slowly and its operational ceiling dropped. The
problem grew to the extent that some squadrons started stripping down their
Halifaxes, losing any piece of unnecessary equipment carried in order to help the
efficiency of the aircraft.

There were however, other significant design defects with the Halifax that were not
rectified in later variations. Primarily amongst these was the fact that it had a very
poor cockpit layout with vital control panels and instruments spread over a large
part of the aircraft. In particular the main fuel management system was remote from
the cockpit, being situated well aft of the pilot's position. The main fuel management cocks that were to be shut off in the event of an emergency were located in
the centre fuselage of the aircraft between the wings.

The Halifax was, even by contemporary standards, outdated as an aircraft, when one
considers that it was designed in parallel to the Lancaster airframe. Although initiated under the same specification, the Halifax lacked the foresight of the many
safety and military operational features built into the Lancaster, which had an excellent cockpit layout with easily accessible system controls. In the Halifax this was considered fundamentally inhibitive to flight safety even at the time the aircraft was
being built, but in time of war and with the pressing need for the bombers, little
attention was paid to this.

Power for the Halifax Mk.2 bombers came from the four Rolls-Royce Mk.XX Merlin
engines, which were more powerful than their predecessor, the Mk.X. By mid-1941,
the Merlin had already proved its worth as reliable power plant for all aircraft types
– it was considered to be among the most reliable of all aircraft engines of the time.
The Mk.XX Merlin, was built by Rolls-Royce at their factory in Derby, and differed
from the Mk.X in having a two-speed supercharger capable of producing 1175 hp

at 21,000 feet. They were fitted to most of the heavy bombers of the time, Lancasters and Halifax aircraft, indeed the Mk.XX Merlin is still to be found in the last flying Lancaster of the Battle of Britain memorial flight.

At the time of its last flight however, Halifax V9977 was not fitted with four Rolls-Royce Mk.XX Merlin engines, but three. The fourth Merlin had come from the American Packard Company who were building them under licence in the United States. They designated the engine the Mk.38 Merlin, though it was the same as the Rolls-Royce unit.

V9977 was fitted with the following four units:

| Engine position | Mk. | AM No. | RR No. | Manufactured |
|---|---|---|---|---|
| 1. Port outer | 20 | 198338 | 37667 | June 1941 |
| 2. Port inner | 38 | 245179 | 10183 | Packard |
| 3. Starboard inner | 20 | 198334 | 37659 | June 1941 |
| 4. Starboard outer | 20 | 198341 | 37671 | June 1941 |

The three British made engines were built between 14 and 21 June 1941, with the fourth Merlin engine part of a shipment that arrived from America at around the same time.

## Flight trials and delivery

Halifax V9977 was the second aircraft to be built by the English Electric Company and its initial flight test on 24 August 1941 lasted the unusually long period of three hours. The flight trial took place at the local Samlesbury airfield and had been completed by 25 August 1941. The aircraft's Air Ministry Form 78 Log Sheet shows it was then transferred to No. 45 M.U. for armaments and signal mods (i.e. the fitting of guns, bomb racks, radio equipment etc.). On 26 November, it flew to RAF/TFU Hurn for the Design Office in connection with the GEE system, and on 24 December 1941 (presumably as a follow up to the visit of Dee, O'Kane and Hensby to A&AEE), it flew into Handley Page at Northolt for installation of the pending H2S scanner mountings.

The aircraft was allocated to telecommunications research along with a second Halifax Mk.2, identification number R9490, which was being built by Handley Page at their factory. Both aircraft were to be modified so that the scanner could be suspended under the aircraft, covered by a Perspex cupola. These instructions were relayed to the manufacturers following the visit by Lovell, King and Whitaker to the Handley Page works on 4 January 1942. Both aircraft had been authorized on the very highest authority and were to be delivered to TRE/TFU at RAF Hurn as soon as ready, which gives some indication of the urgency that Churchill placed on the radar development programme, bearing in mind how short the supply of aircraft to Bomber Command was at the time.

At Hurn the development of the new H2S system continued with flight trials carried out in Blenheim Mk.4 medium bombers. These were considered unsuitable for the work however because of the aircraft's limited maximum altitude and size, and therefore their ability to carry the scientific team and all the equipment. Halifax V9977 was delivered on 27 March 1942, and Halifax R9490 delivered on 12 April

Figure 10.3 Halifax V9977. (*Courtesy of Sir Bernard Lovell*)

1942. For the next ten weeks until the loss of V9977, a series of comparative tests took place both on the ground and in the air to assess the klystron and magnetron radar systems.

No complete records have survived of the day to day events at RAF Hurn for V9977 and R9490, though it has been possible to piece together some of the history of the two aircraft from the records of the maintenance crews at the airfield. Most of these are from the log of Ronald L. Hayman, one of a small team of civilians employed in the inspection department of TRE. Based in the workshop at Hurn and later at Defford, Hayman was stationed at Hurn from 29 January 1942 onwards and kept a rough work notebook in which he logged aircraft, including a few movements of V9977. His task, along with the rest of the ground maintenance crew, was to over-haul and fit out all TFU aircraft and prepare them for flight.

From arrival on 27 March 1942, until 15 April 1942, V9977 was in the workshop at RAF Hurn having the magnetron H2S system installed and ground tested. As this Halifax had been delivered first it was designated the number one demonstration air-craft. Therefore, the more promising magnetron system was installed in it, as there were problems surrounding the klystron oscillator, which was to be fitted in R9490. When Halifax R9490 arrived at Hurn on 12 April 1942, it remained in the hangar having the klystron H2S system installed and ground tested. However, during this period, the orders came through to transfer the entire TRE/TFU and RAF SIU (Special Installation Unit), over to RAF Defford and this disrupted the work.

On the evening of Thursday, 16 April 1942, V9977 made its first test flight with the newly installed magnetron H2S radar. However, the aircraft was forced to cut short any tests and return to Hurn because of trouble with the 80-volt alternator on the starboard outer (No. 4) engine. This alternator was important, as it was the primary power supply to the test equipment. After spending the night and the rest of the fol-lowing day in the hangar with maintenance crews working on the aircraft, V9977 was ready to make a second attempt at the flight in the evening of 17 April. This flight was trouble free and the observers reported the H2S radar giving good ground echoes at a height of 8000 feet.

Hayman's rough notes show that V9977 was pre-flight checked on the evenings of 23 April, 27 April and 28 April 1942. This would suggest that the test flights continued to take place during this time, but any actual record of this from the tower at Defford has long since been lost or destroyed.

From Friday 1 May until Saturday 16 May, the hangar records show that neither aircraft was pre-flight checked. This was probably because the orders to transfer to RAF Defford had come through on 5 May, and the TRE installation fitters were busy removing the radar equipment in preparation for the move. During this time V9977 and R9490 would have been given a total maintenance overhaul in preparation for the move.

On Thursday, 14 May 1942, the team from EMI, consisting of Eric White, with Felix Trott (and a junior engineer, Ivanhoe J.P. James, who, incidentally would go on to become very influential in the development of colour television at EMI), arrived back at RAF Hurn, to continue the preparations for the transfer of the H2S development equipment to RAF Defford. The actual transfer of TRE/TFU from Hurn to Defford was taken as a precautionary measure against possible parachute attack from enemy raiders; it was felt the Worcestershire location was more remote and therefore offered less chance of detection. The area around Defford also offered the scientific teams the kind of terrain that was ideal for testing the two radar systems. Unfortunately the orders to move and the need to complete the transfer before the end of May were very disruptive to the programme, and possibly even fatal to V9977.

Figure 10.4 Tappets being adjusted on the starboard outer engine of a Halifax Mk.2 in the summer of 1942.

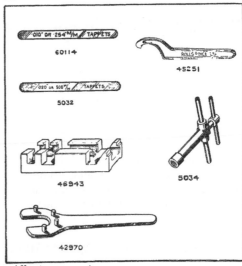

SPECIAL TOOLS (ENGINE)
(Headings and their accompanying tools are listed in alphabetical order.)
Additional tools for fitting inserts, generally, are listed and illustrated on page 105.

**Figure 10.5**
**Merlin XX tappet adjustment**
**tools.** (*Courtesy of William*
*Sleigh*)

| A.M. Ref. No., Sect. 36DD | Part No. | Description |
|---|---|---|
| | | **Boost control** |
| 46943 | HB.26179 | Gauge, for setting aneroid valve |
| | | **Camshaft and rocker mechanism** |
| 60114 | D.24663 | Gauge, feeler, for checking tappet clearance, 0·010 in. (inlet) |
| 5032 | F.45983 | Gauge, feeler, for checking tappet clearance, 0·020 in. (exhaust) |
| 5034 | E.32755 | Spanner, box, for adjusting valve tappet screw |
| 42970 | D.22789 | Spanner, pin, camshaft turning |
| | | **Camshaft drive** |
| 45251 | D.13126 | Spanner, "C", for nut securing lower bush of upper gearwheel |

The two aircraft were naturally flown from Hurn to Defford somewhere between 20 May, and 26 May, but this meant the klystron radar equipment being removed again from R9490 for the flight. This meant that the installation was not completed at Defford until 1 June. During those last few days at Hurn, V9977 underwent pre-flight checks on 17, 18 and 19 May. The crew of the aircraft (who, like all other crews, were almost totally new to this type) would continue with their training and evaluation of the Halifax. The move by road of the staff, airfield engineering equipment and the scientific teams took place from Sunday 17 May 1942, through to the following Sunday 24. White, Trott and Harker had found accommodation at the Swan Hotel in Tewkesbury, just five or six miles up the road from the airfield, and it was here that they would stay for the next few months.

Once at Defford V9977 and R9490 had their respective radar equipment re-installed by the TRE radar installation fitters. If any flights were carried out during these first few days they are not recorded, but after each flight it was normal maintenance practice to carry out post-flight checks on the entire aircraft. Between 25 May and 1 June, the four engines of V9977 underwent a mandatory 30-hour inspection which would include adjusting the tappet settings. On 25 May, the port inner engine (No. 2) which had originally been a Rolls-Royce Merlin Mk.XX was replaced with a Packard Mk.38. This engine only had a total running time of 20 hours. Between this inspection and the crash on 7 June, the aircraft only flew a total of 6 hours 40 minutes.

On Tuesday, 2 June 1942, R9490 now with its installed klystron radar equipment, made an initial test flight, though from a scientific point of view it was not very successful, with problems occurring in the display of the echoes on the cathode ray tube. It is likely that the problems experienced with the klystron system on 2 June were directly responsible for the EMI team, under the direction of Alan Blumlein, undertaking a comparison flight with the magnetron system on Sunday 7 June.

The records show that V9977 was pre-flight checked on Wednesday 3 June and again on Friday 5 June. These were possibly flight serviceability checks to support the continuation of the equipment test flights, though ground tests had proved that the magnetron H2S system was now showing signs of great improvement. During the evening of Saturday 6 June 1942, V9977 did make one final check flight before the designated full test the next morning. On board were Geoffrey Hensby and Bernard Lovell and both noticed the magnetron system responding to echoes of towns and features at a far greater distance than had previously been reported. The flight on 7 June was intended to demonstrate as much as anything to the personnel from EMI just what the system could achieve.

## Piecing events together – April to June 1942

As has been mentioned, several members of the EMI team were stationed on detachment to TRE, firstly at Hurn and later at Defford. Felix Trott and Maurice Harker had been stationed with TRE for approximately eight weeks before the accident took place, and had been responsible for the installation and flight testing of the EMI klystron-based H2S equipment during late April and into May, in Halifax R9490. As the results came in, these were usually sent by motorcycle courier back to Hayes, though on occasion Eric White would bring them back, to Blumlein, Browne and Blythen who were the department heads working on the project. During these weeks both test projects, that being run by EMI using the klystron valve as a transmitter and that by TRE using the magnetron valve, were given equal priority. H2S was so important that whichever valve, or even if both valves could be made to work, it did not matter. The key factor was to get it to work properly.

Following the move from Hurn to Defford, both Halifax aircraft required their inspection tests completed as well as all the experimental equipment to be reinstalled and checked. This took some time and was not without its hazards. Felix Trott recalled that on two occasions a klystron tube had been broken when being installed. The tube, which was fragile to say the least, had initially been installed in the H2S unit in the workshop at Hurn by Eric White who had inadvertently broken off one of the glass connectors. When a second klystron tube had been sent by armed motorcycle courier from London, this tube was fitted successfully by Trott. Each installation and test sequence of the H2S equipment would take the best part of a day to complete, by either team, EMI or TRE, and while these two groups were working in essence on the same project together, they rarely had much interaction as both projects were identified as 'secret' classification.

Halifax R9490 had been flight-checked and readied for a test on 2 June 1942, and most of the test flights had been carried out by Felix Trott. The results, as we now know, were not very satisfactory and it was reported back to Hayes that the klystron system was not producing anything like the desired results. The next day however, when the results of the flight of V9977 on 3 June were reported by TRE to EMI, Blumlein decided that this marked improvement warranted a visit from him to Defford to see the results personally. It was decided that Blumlein, Browne and Blythen would therefore drive from Hayes to Defford in an EMI company car (EMI petrol rations had been allocated specifically for such purposes), to join White, Trott and Maurice Harker who were already there, on the Friday morning, 5 June (Harker was also stationed at Defford as he was completing his work on AI Mk.VI at the time).

## The Miller integrater

That Friday, 5 June 1942, also happened to be a significant day for Blumlein in quite another manner. His latest patent, No. 580,527 was applied for on that day. The patent was for a circuit which over the years has become known as the 'Miller integrator', though many of Blumlein's colleagues of the time felt that it really should be known as the 'Blumlein integrator'. The story of the circuit and how it came to be improved by Blumlein is a complex one, but bears explaining because the circuit would eventually play an important role in the development of electronic equipment in the years that followed.

In the very early years of radio it was discovered that oscillations could occur if there was an inductive load in the anode circuit of the triode valves being used as RF amplifiers. Edwin Armstrong, the American engineer who would later go on to invent frequency modulation (FM), discovered in 1915 that these oscillations were caused by feedback from the anode circuit to the grid circuit via the inter-electrode capacitance. When a mathematical analysis of the triode valve as an RF amplifier was carried out by another American, John Miller of the US Bureau of Standards, in 1919, it was demonstrated that at frequencies below resonance, where the anode lead was inductive, this feedback from the anode circuit to the grid circuit could give rise to a component of negative input resistance. This negative resistance was in parallel with the grid tuned circuit and, under conditions of low damping a sustained oscillation could occur.

This problem was partly solved (for tuned RF amplifiers) by neutralization where an equal, but anti-phase signal, was coupled back to the grid. In later variants (which used screened grid valves or RF pentodes) the anode-to-grid capacitance could be reduced to a fraction of a picofarad. One of the most important features of this feedback became known as the 'Miller effect'. This was a phenomenon in triodes which caused an alteration of the input admittance in consequence of some change in anode-circuit impedance. The effect follows from the fact that the anode/grid capacitance of such valves represents a path through which a fraction of the alternating component of the anode current can flow.

In sophisticated radar systems such as those being developed during the early years of World War Two, a much more linear time base than those already available was

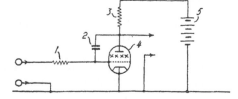

Figure 10.6
Detail from Patent No. 580,527 (1942),
the 'Miller Integrator'.

required with precise and adjustable time delays in order to search for and lock on to a target object. Blumlein took the basic elements of the Miller integrator and deliberately added an external capacitor to the stray capacitance which initiated a whole series of circuits using diodes as auxiliaries, including the celebrated 'phantastron' and 'sanatron'. He also extended the principle to the integration of more than one voltage at a time, as used later in analogue computers. Typical of the man, Blumlein decided that as the principle had been discovered by John Miller, the invention which he (Blumlein) had now patented should be called the 'Miller integrator', though as mentioned there are many who would say that it really should be known as the 'Blumlein integrator'.

There is a postscript to this story. During the war such patents were not published and it turned out that the basic 'Miller integrator' had been applied for a few months before the Blumlein patent by A.C. Cossor Ltd and J.W. Whiteley (Patent No. 575,250 dated 17 February 1942). Eric White however points out that 'Although the date of this Patent is 1942, the idea had been used by Blumlein some years earlier, and in fact appears in a somewhat elaborate form in his television frame scan circuit, Patent No. 479,113 of 1936.'

## The last weekend – Friday 5 June 1942 to Sunday 7 June 1942

Eric White, Felix Trott and Maurice Harker all recalled the events of that weekend vividly when asked about it over fifty years later. On the Friday morning, Blumlein had called White to tell him that he, Browne and Blythen would be arriving that afternoon from Hayes to prepare for the weekend flight tests of the magnetron system in V9977, in order to compare the results that Lovell and Hensby had been seeing with the rather poor results the EMI team had been getting from the klystron valve system. White was instructed to inform the RAF personnel of the EMI team's arrival, and arrangements were made for authority to clear the flight itself. Preparations were made for Blumlein, Browne and Blythen to stay at the Swan Hotel in Tewkesbury, which was where White, Trott and Harker were all staying.

It is not known which of the three men actually drove the car up from London on the Friday morning, though it is quite probable that Blumlein did in either his own Ford V8 or an EMI staff car for which petrol rations would have been allocated. Following their arrival, and after a short introduction to the work in progress, Blumlein was asked by Maurice Harker if it would be all right if he might return to Pinner, near Watford, the following morning as his parents' house had just suffered a burglary and he wished to try and help them sort it out. Blumlein told Harker that he thought it would be all right for him to return as there were plenty of extra EMI personnel at Defford with Browne, Blythen and himself there; and besides, Harker was finishing work on the AI Mk.VI at this time, and had not yet been brought into the H2S project. Blumlein told him to take the weekend, and to return on the Monday afternoon.

That Friday evening everybody ate dinner at the Swan Hotel and then all went for a walk as Maurice Harker recalled:

> 'I was walking with C.O. Browne and F.R. Trott, with Blumlein and White a way off ahead of us. Well, after a while of this, it was a lovely summer's evening you

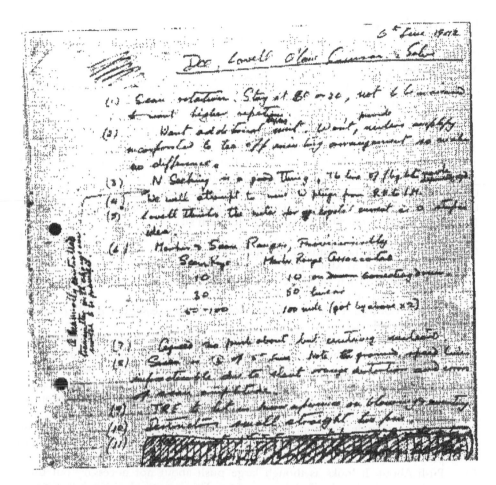

Figure 10.7 Detail from Blumlein's hand-written notes on H2S made at Defford dated 6 June 1942, the day before the crash of Halifax V9977.

see and not a bit dark yet, C.O. Browne says "I've had enough of this", and he'd been walking with some stick or something, so, he proceeded to tap this stick on the ground to get White and Blumlein's attention or I think they would have walked another five miles, they were that engrossed in conversation. Anyhow, we returned to the Swan, and turned in for the night, and in the morning I left to return to Pinner and my parents' house, while they were setting off to the airfield at Defford, and of course that's the last time I saw Blumlein.'

That Saturday was spent in preparation for the flight trial the next day, with everyone returning quite late to the hotel in Tewkesbury. Felix Trott recalled that despite the day's work; there were still several pre-flight preparatory checks to be completed the following day before the aircraft could take off, but they did not expect these to take long at that time. It is also worth mentioning that in the hotel that evening,

Trott recalled the presence of a quite high ranking naval officer in the bar, having 'at least three stripes on his arm'. Whether this man had anything to do with Defford or was there purely by coincidence, he did not know, but the man did stand out in his uniform that evening surrounded by civilians.

During Friday, after he had arrived at the airfield and been shown around V9977, Blumlein began to make some sketchy notes (which were probably meant as mere reminders to himself for when he had returned to Hayes), and continued to do so for the rest of the weekend. These hand-written notes (See Figure 10.7) concerning the tests to be carried out still survive in the vast EMI archives, and are, perhaps, the most poignant by him. Written in pencil on four pages which have been torn out of a standard memo pad, and dated 5 June, 6 June and 7 June 1942, they give us a quite unique insight into the thoughts of Alan Blumlein as the EMI team prepared for the tests that lay ahead that weekend. The four pages have been transcribed as follows (the text is often cryptic and the spelling Blumlein's):

(Page One)

O'Kane & Selves, Sq/Ldr Samson. 5th June 1942.

1) Motor Control. a) Motor comes up automatically when H.T. button pressed. b) Two pole switch on power unit to stop scanner. c) Lead to Motor 6B. Scramet. Puis ou rockets // 3 by 3.

2) Marker Control. To be switch not key. Marker is essential if RD confirms control fitted.

3) Control Box Shape. As shallow as possible – e.g. to fit 5" tray. Box for control on R.H. and W. plugs on L.H. Spikes & catches top and bottom. Presets (Height Zero, Range Zero, Brightness) on RH Side above (or below) boss if this can be arranged.

4) Scan & Range Marker Controls. Delete one switch and gaug. Decision on range to follow flights.

5) Indicating Angle Marking. Make as planned with thumb moved drift plate with degrees red against a vumb line at bottom. Plate to have 3 parallel lines plus 10° (920°) lines at either side forward see text opposite. Agreed.

6) Push About. It looks as though large push about not necessary but three ranges required and adjustable centring. This is not a decision but depends on flight trials.

(Page Two)

7) Subject to confirming test at Hayes, we fit transformer instead of lead amplifier.

8) Make up all units on basis of flight trials set as now modified at Hayes.

9) Decide use 2A plugs W196 for destructor.

10) Will decide out of hand at Hayes whether 18 or 12 for A.S.V.

11) Decided tests should be made with test equipment to allow orders to be placed

(Page Three)

6th June 1942. Dee, Lovell, O'Kane, Samson & Selves.

1) Scan Rotation. Stayed at 60 or 20, not to be increased to want higher repetition.

2) Want additional display unit. Won't provide unless amplify incorporated to tee off existing arrangement so makes no difference.

3)  N Seeking is a good thing, the line of flight maybe definitely rgd.
4)  We will attempt to move plugs from R.H. to L.H.
5)  Lovell thinks the meter for crystal circuit is a stupid idea.
6)  Marker & Scan rages, Provisionally

| Scan | Range | Marker Range associated |
|---|---|---|
| 10 | 10 on correcting & renn. | |
| 30 | 50 linear | |
| 50–100 | 100 rule (got by above × 2) | |

7)  Agreed no push about but centring useless.
8)  Screen as 5) of 5th June. Note ground speed lines unpredictable due to slant range distortion and errors of scan amplitude.
9)  TRE to let us know experience on blowing & mounting
10) Destructors small straight two pin
11) (Originally written: "TRE want 1st I.F. must be in T2R not on because Mk. VIII has got it & designs have been based to get Gram Co...", there Blumlein must have changed his mind as the entire passage is heavily crossed out and he re-wrote the following) T.R.E. will discuss advisability of having 1st I.F. in T2R.
12) Discussed use of new power Klystron, decided to carry on with existing modulator.

(Page Four)

Note. Warn Oura jack may be required on T2R.
13) Interrogator. Conference to be arranged to re-hash.
14) Motor on 6 pin shoe & stop switch
15) 12 & 18 way decision out of hand at Hayes
16) Go ahead to make prototypes.
17) EMI to provide mock ups of all units mid T2R & M

7th June 1942.
We Form Monitor. Mk. VIII type or Test set 31. (TRE to produce Mk. VIII type for discussion)
I.F. Signal Generator. (To be produced by us for 1st Tests only)
Wavemeter
Signal Generator R.F.
Field Strength Measurer.

There the notes end. Blumlein would have made these brief jottings at EMI Defford or Tewkesbury. The notes for the 7 June were most probably made during the morning at Defford.

## The day of the flight

Sunday 7 June 1942 was a bright clear summer's day, with patchy wisps of cloud ⁹⁄₁₀ths and excellent visibility up to 18 miles with cloud at 3000 feet. There was a light surface wind of 10 mph from approximately 300°, north, north west. It was an ideal day for flying, the kind of weather pilots enjoy the most. The test flight to evaluate and compare the magnetron H2S system was scheduled for the morning, but because of several preparatory delays they didn't actually take off from Defford until 14.50 h that afternoon. The object of the flight was to produce photographic images taken directly from the CRT of the ground echoes from the radar as the aircraft flew over the Severn Estuary, the coastline and Cardiff and Swansea.

The flight plan was chosen to highlight the prominent features of the area: large flat expanses on the Severn estuary, built up areas around Cardiff, Swansea and Newport, coastal plains and the sea, as well as several off-shore islands. This was part of the reason that RAF Defford had been so useful in the first place as a base for TFE/TFU as it offered close proximity to terrain of this sort. However, airspace was somewhat restricted by the 35 active airfields in the vicinity, most of which were engaged in the training programmes for aircrew. This was why a Sunday was such an ideal opportunity for this kind of flight. There would be little chance of running into the various training missions taking place at the weekend, and the high number of airfields around offered the possibility of many emergency landing sites should anything go wrong.

Bernard Lovell had intended to fly on V9977, as had Sam Curran, the head of the EMI/TRE liaison team. Curran recalled many years later the events of that afternoon:

'Clearly Geoff Hensby was required to go because of his in-built experience, and Brown and Blythen wanted to go, partly to see for themselves the conditions of operation as well as performance. With these three and myself, i.e. four civilians, plus seven RAF personnel, the total allowed, 11 in all, was reached. Then, out of the blue, Blumlein said to me he would 'love to go along' and I felt I had to volunteer to stand aside in his favour. Further, I said I would give him my flying suit as time had virtually run out, and I added that I had seen a system not unlike the one to be tried. So off went all 11 of them.'

As Blumlein climbed aboard the aircraft he turned to Lovell and said, 'Don't worry, we'll make it short eh?' Bernard Lovell had also given up his place on the aircraft to another of the civilian staff from EMI, either Cecil Browne or Frank Blythen (it is not certain which), with equally tragic consequences. Yet another member of the EMI team, Felix Trott, should have been on the flight, but being fed up with all the delays that had persisted before take-off he decided to go and get himself a cup of tea. This would have required walking to the far side of the airfield, so Trott borrowed a bicycle and headed off in search of his refreshment, evidently under the impression that the delays might continue for some time yet. When he got back from his refreshment, V9977 had left without him – surely the luckiest tea break in history.

Some sixty years later Felix Trott would recall the day vividly:

'It was a lovely hot, fine, day; and I don't know why there was a vast amount of delay .... and things dithered and dithered and dithered, and it wasn't known who was going to fly and so forth. A couple more people from TRE turned up and they wanted to go, and there wasn't really any particular point in either myself or the other people who were stationed down there going on this trip because we could do it anytime. So, off they went. I got on the bicycle which we had down there and cycled round the perimeter track to the Officer's Mess and had myself a nice, cooling cup of boiling hot tea.
    And when I came back they had gone, and one didn't expect to hear any-

thing because there wasn't normally any radio communication anyway for a good hour at least. And I suppose, one began to get a bit anxious after two and half, three hours. And, how long in fact the station had known, I don't know; but eventually someone came down and told us that there had been a crash, which turned out to be miles away in the Wye Valley. And whether they had done the standard tests and said "yes, this looks promising, now let's look at some more difficult terrain" – I don't know. And I doubt whether anybody knows. The unfortunate thing about this aircrash was that the aeroplane that crashed wasn't the one that had the EMI experimental equipment on board, but was one that had the TRE equipment aboard, and Blumlein, Browne and Blythen, who were the three EMI engineers who were killed, had gone up by invitation to see how this equipment was working. But, it crashed, and that was the end of them.'

When V9977 finally got clearance to leave Defford the aircrew, scientific team and observers all climbed aboard – in all eleven personnel. The complete details of those on board the flight are as follows:

### 1) Pilot – Pilot Officer D.J.D. Berrington (115095)
Pilot Officer Berrington was a very experienced pilot indeed for that period of the war, with over 3300 flying hours. He had been a civilian pilot before the outbreak of the war, accumulating over 2000 hours before joining up on 11 September 1939, as a Sergeant. Throughout his RAF career he was consistently assessed as above average and on 2 October 1940, with 444 flying hours behind him he was posted to SDF (Special Duty Flight) at RAF Christchurch, which later became the base for TFU. He had been pilot for many of the experimental radar flights in the previous year and a half, and was well known and liked by the TRE and EMI personnel.

By the time of the crash of V9977, he had completed a further 1050 hours in an additional 27 aircraft types including Wellingtons, Whitleys, Herefords and Boeing 247Ds. The Halifax however was a very new type of aircraft and Berrington had accumulated a total of 12 hours 30 minutes as first pilot, with an additional 2 hours 5 minutes as second pilot. This was perhaps, not quite as unusual as it might seem. Most crews in mid-1942, were becoming adjusted to the new aircraft being developed at the time, and very few other pilots would have had more time than this in the Halifax. Crews were pressed to take as many aircraft training courses as they could, but usually mission priorities meant that precious little time could be spared for acclimatizing. Though Pilot Officer Berrington would have been conversant and examined on Halifax emergency drills (fundamental to his clearance on type), he might not have been as familiar with the emergency drill, the location of vital control systems and instruments as, for example, he may have been on other aircraft in which he had accumulated more flying hours.

This must be taken into consideration along with the poor cockpit design of the Halifax, and the remote location of many of its vital systems. While there is no evidence to suggest that this compounded the situation on 7 June 1942, it may have.

### 2) Navigator – Flight Sergeant G. Millar (751019)
Flight Sergeant Millar would have been responsible for plotting the route chosen for the demonstration of the H2S system on 7 June. The flight plan is not known in

complete detail, but it is probable that the aircraft flew out from Defford in a south easterly direction to offer the radar equipment the most diverse terrain around the Severn estuary and the built up areas around Cardiff and Newport. It must be remembered that this sortie took place over well known local territory precluding the need for Millar to actually plan or chart the proposed route as such. Quite why the aircraft deviated from the flight path that had been flown during previous tests, is not known. It is possible that the results from the H2S unit over the coast were satisfactory and Blumlein wanted to see more difficult terrain displayed.

Millar may have been asked by Berrington to plot a course away from the coast, and back roughly in the direction of Defford but this is not known. Certainly this route, if actually flown, would explain why V9977 was allegedly seen by observers on the ground in so many different places that afternoon.

*3) Wireless Operator/Air Gunner – Aircraftman 2nd Class B. C. F. Bicknell (1271272)*
It would seem that Aircraftman 2nd Class Bicknell had not attained full aircrew status at the time of the flight of V9977 on 7 June 1942. However, Bicknell appears to have been a trainee of some potential and would therefore have been included, no doubt with Berrington, Phillips and Millar's complete approval, to increase his experience and knowledge of the Halifax type. He was an airman in the radio trade and his presence on a flight, the nature of which precluded the use for such equipment and being local only, did not affect the safe operation of the Halifax. Berrington therefore would most probably have been only too happy to give this radio technician the air experience.

*4) Flight Engineer – Leading Airman B.D.G. Dear (571852)*
Leading Airman Brian Dear is assumed to have taken the position of Flight Engineer despite also not being fully qualified for the position. He was an L.A.C. grade fitter Class II (E) (engines) and as such had attained the highest technical appointment in aero-engine craftsmanship in the Royal Air Force. Though his personal documents have not survived, it is thought that he was one of Halifax V9977's ground engineers and as such would have flown on post-maintenance tests.

He had been earmarked for remuster to Flight Engineer grade and while at the date of the fatal flight had not passed through the official flight engineers' course on paper, it can be accepted that his intimate knowledge of the Halifax engines, fuel systems and controls surpassed that requirement to Berrington's complete satisfaction. Dear would also have been familiar with the emergency drills associated with the Halifax aircraft.

*5) Second Pilot – Flying Officer A. M. Phillips (44185)*
Second pilot, Flying Officer Phillips, as the senior commissioned officer to 'Captain' Berrington would probably have been responsible for monitoring the Flight Engineer's position, especially as it was manned by an inexperienced crew member. As the role of second pilot was actually redundant on the Halifax, it is also possible that Flying Officer Phillips, like many aircrew at the time, was using the flight of V9977 as an opportunity to gain more type experience. At the time of the crash he had accumulated 700 hours of flying time on 16 types of aircraft, but significantly none of these were in Halifaxes.

*6) Squadron Leader R.J. Sansom (33372)*
Squadron Leader Sansom was aboard flight V9977 as he was Bomber Command's scientific liaison officer, and was currently on loan to the TRE H2S project from them. It would have been his responsibility to assess and compare the results of the magnetron flight with that of the klystron flight several days earlier and report this back to Bomber Command.

*7) Pilot Officer C. E. Vincent – TRE (110285)*
Pilot Officer Vincent was also a member of the TRE H2S project group. He had been placed directly responsible to Bernard Lovell and had carried out, as a flight observer, many of the first experimental flights in V9977 with Lovell, Hensby and O'Kane prior to the EMI team arriving.

*8) Mr Geoffrey S. Hensby – TRE*
A civilian scientist, Geoffrey Hensby, 24, had been working on the H2S project since its initial tests in 1940, under the directorship of Dr Bernard O'Kane at RAE and later at TRE. As radar systems operator it had been Hensby who had taken part in the initial demonstration flight over Southampton in November 1941, which had gone a long way towards convincing Bomber Command of the merits of the centi-metric radar programme. During the flight of V9977 on 7 June 1942, Geoffrey Hensby would have been the scientist in charge of the experiment and given Captain Berrington his route to follow for the day.

*9) Mr Alan Dower Blumlein – EMI*
Head of the EMI team contracted to develop and manufacture the production version of the H2S radar system. It was on Alan Blumlein's instruction that the flight on 7 June was authorized.

*10) Mr Cecil Oswald Browne – EMI*
Part of the EMI team intending to develop and manufacture the production version of the H2S radar. Cecil Browne had worked with Alan Blumlein at EMI since his earliest days there, in particular on the development of television. He shared six patents with him, including one with Frank Blythen, ranging from amplifiers for television and cathode ray oscilloscopes, to television transmission apparatus.

*11) Mr Frank Blythen – EMI*
Part of the EMI team intending to develop and manufacture the production version of the H2S radar. Like Cecil Browne, Frank Blythen had worked with Alan Blumlein at EMI since he had arrived there. He also shared three patents with Blumlein including the one with Browne. They worked on amplifiers, cathode ray oscillo-scopes and lastly on magnetic focusing arrangements for CRTs.

## From take-off to cruising altitude

Halifax V9977 started its engines and the crew went through their usual pre-flight safety checks. The ground crew pulled away the wooden chocks that held the wheels against sudden engine surges, and the aircraft slowly at first, building speed later, taxied onto the perimeter runways at Defford heading for the end of the main runway 22 (220° heading) ready for take-off. During this time the scientific team would be busy checking that their equipment was working satisfactorily, probably for the umpteenth time.

At the end of the runway the aircraft halted and Berrington would have normally prepared to apply full power to the four Merlin engines to achieve take-off speed, but Berrington was not happy, there had been a power failure in the radar system which was driven from a generator housed in the S/O No. 4 engine.

Pilot Officer Berrington radioed the tower and explained that there was a problem with the power supply to the radar equipment, and could somebody come out to the aircraft and check it to see if it could be fixed or if they could find the problem. Obviously with such perfect flying conditions, coupled with the fact that the scientific team had already had several delays during the morning, Berrington wanted to avoid aborting the flight if at all possible. A call was made from the tower to Derek H. Moseley, the workshop manager of TRE Defford, explaining the problem. Moseley and Ron Hayman (whose rough notes have been mentioned earlier) were requested to go out to V9977, which was still standing at the end of the runway ready for take-off with all four engines running.

Hayman and Moseley went out to the aircraft and after an inspection, they very shortly discovered that the problem with the power generator for the radar system was being caused by a faulty cable socket connection on the front of the voltage control panel (VCP), which was the electrical output to the radar. It would seem that the socket had not been properly screwed fully home on the front panel and this being quickly corrected, the radar functioned once again. Berrington was informed that all was well again and first Haywood and then Moseley climbed out of V9977 to allow it to continue on its way, thus Moseley became the last person in Halifax V9977 before the crash other than those killed.

With the power problem to the radar successfully overcome, Berrington obtained clearance for take-off from the tower and applied full power to the four Merlin engines, which thundered the aircraft down the runway. At around 14.50 h Halifax V9977 cleared the runway at RAF Defford and maintained its heading ahead in the climb with the engines throttled back to 2850 rpm at +9 pounds boost pressure, the standard climb setting of the four Merlin engines. At this steady climb setting giving 150 mph ascending at about 2000 feet per minute the aircraft would have reached 10,000 feet to the west of Gloucester and thence by visual reference commenced its run over to the south west coast of Wales. Several previous routes for demonstration flights had taken the aircraft over an area between Gloucester and Kidderminster, but possibly because of the nature of the test and the need to compare the results to the klystron H2S system, the route was not to be taken on this day.

By the time V9977 took off on its last flight most of the members of the EMI team were convinced that the magnetron H2S system was the one that should go into production, and were obviously keen to impress upon Bomber Command the results of the equipment. For this reason the 'target' for the day would need to be distinctive and so a series of passes would be made over Cardiff and Newport to get good ground echoes of built-up areas and the Severn flats to demonstrate identification of coastal shoreline and then eventually back to Defford to evaluate the results.

Very little is known for certain of the exact route chosen for the flight as radio would have been part of normal wartime procedure. The navigator's log and flight plans (assuming any were kept as Berrington may well have been flying 'visual') were all lost in the fire after the crash. However, V9977 was seen and recalled later by several witnesses on the ground (most of whom it must be stressed, were not called to appear before the Board of Enquiry as their accounts were considered to be non-verifiable). It is from these eyewitness accounts (accurate and inaccurate observations of aircraft on the day, many of which were probably not V9977), as well as deducing from the flying time and fuel remaining, that the following reconstruction of the events leading up to the moment of impact has been pieced together.

Having left Defford the Halifax would have required about 15 minutes to climb and reach its cruising height. However it would appear that V9977 flew at various altitudes on this day with ground reports (non-verified) placing it at between 1500 feet and 15,000 feet for parts of the flight. The matter of the actual altitude flown has never been definitively established. It is unlikely that it ever will be. However, it should be pointed out that at altitudes above 10,000 feet the use of oxygen would be required. Berrington would probably not therefore have taken the aircraft to anything like 15,000 feet, and besides the purpose of the flight was to test equipment which was designed to look at the ground. An altitude of 15,000 feet would have been in excess of its design limits.

## Sightings

It must be pointed out that the following altitudes reported on or about 16.00 h that afternoon were from unidentified members of the public who presumably attributed their sightings to those of the crashed aircraft once they had heard of the accident. These matters were not followed by the Board of Enquiry as the sightings were non-verified and aviation specialists of relevant experience have clearly identified them as non-contributory to the events which actually took place. There were in fact several aircraft movements in the area at the time of the crash and any reference to low flying aircraft which were not trailing smoke/fire from the right outer engine were ignored. That aside, several so-called 'sightings' of V9977 have been used to position the aircraft in the moments up to and immediately following the engine fire commencement.

At about 16.00 h an unchecked sighting of an aircraft was seen from the ground circling a point some 4 miles north west of Usk in Monmouthshire at a height of approximately 1500 feet. Some minutes later the starboard outer propeller was seen to stop, this would have been engine No. 4. This was probably not V9977, but, more likely, a service aircraft undergoing a post-maintenance air test which necessitates the feathering of a propeller. At around 16.10 h an aircraft, having left this position, was sighted heading in a south westerly direction, and some minutes beyond that it was seen again by ground observers near Coleford in Gloucestershire. Significantly, this time smoke was reported coming from the starboard outer engine. At this point the height of the aircraft was reported as approximately 500 feet. This was almost certainly V9977.

When next seen the aircraft was heading in a north, north easterly direction just 200 feet above the treetops passing over the roofs of the small village of English Bicknor, the observers on the ground all reported seeing large amounts of smoke and a fierce

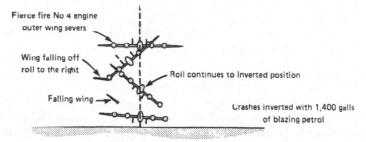

Figure 10.8 The assumed attitude during decent of Halifax V9977. (*Courtesy of William Sleigh*)

fire billowing from the No. 4 engine. Halifax V9977 finally crashed into a field on the southerly side of Coppet Hill, just south of the village of Goodrich. Its final resting-place was some 130 yards north of the River Wye, the structure having finally failed after being burnt through by the fire in the wing and the starboard outer engine.

## Eye witness to the crash

The direction the aircraft was heading at the time of the impact was roughly on course for a return to base, but it is far more likely that Pilot Officer Berrington was trying to attempt a crash landing on the flat ground just beyond the village of Goodrich, the other side of Coppet Hill, and approximately 3 miles from where V9977 crashed.

What had happened? Incredibly, considering the remote location of the crash site, the final moments of Halifax V9977 were actually witnessed – indeed, the falling air-craft only missed the eyewitness concerned by about 250 yards! Onslow Kirby was a farm worker who had rented and worked Green Farm on Coppet Hill from its owners. On 7 June 1942, he had been standing in the pasture field, just to the south of the farmhouse itself, some 150 yards from the edge of Coldwell Wood. The fol-lowing is a reconstruction of the final moments of Halifax V9977 based on Kirby's account which was given to the Air Ministry Crash Investigation authorities a day after the event.

From his position in the field, some 250 yards from where the main fuselage finally fell, Kirby first heard and then saw the stricken aircraft approaching him from almost due south. He recalled the time as approximately 16.20 h. The aircraft was flying very low and coming directly towards him from the direction of the village of English Bicknor which was hidden from view by the high ground of Court Wood and Raven Cliff, south, south east of Coppet Hill. At first glance he noticed that there was a fierce fire burning on the right wing (star-board), but he could not recall details such as whether or not the propeller was

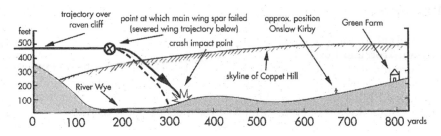

Figure 10.9 The trajectory of Halifax V9977 to the crash point. (*Courtesy of William Sleigh*)

actually turning on either No. 3 or No. 4 engine. Nor could he determine which of the two engines was the source of the fire because of the amount of flames and smoke.

At the point that he first saw the aircraft, Kirby estimated its height above the ground as approximately 500 feet, but by the time it was above Court Wood and Raven Cliff he was quite specific that V9977 only cleared the treetops by 20 feet or so. As the aircraft cleared the trees and passed over the south bank of the River Wye its altitude had dropped to just 350 feet above the ground. The height of the ground at Raven Cliff is approximately 320 feet above the level of the ground where Kirby was standing, so his estimates can be considered as quite accurate.

As the aircraft passed over the mid-point of the River Wye the structural integrity of the starboard outer main spar failed. This is the part of the aircraft that holds the wing to the main inner spar of the fuselage. The outboard mainplane disintegrated making the aircraft unstable and unable to maintain level flight. At this point the Halifax rolled over through 180°, ending up practically on its back, at the same time it dropped almost vertically towards the ground falling in an arc down towards the field in which Mr Kirby stood witnessing the entire event. Halifax V9977 struck the ground on the south side of Coppet Hill where the field rises quite steeply from the north bank of the River Wye. The main fuselage section hit the ground first 130 yards from the river just below Kirby's direct line of vision, which was slightly obscured by a low ridge. The severed piece of wing followed the crashing aeroplane, falling short of the main fuselage by 30 yards.

Kirby was standing between 250 and 300 yards away from all this and described the moment as an immense bang followed by an enormous explosion and a subsequent fire. Running over to the crash site, Kirby's initial reaction showed little concern for his own safety but, upon seeing the complete destruction of the fuselage coupled with the huge fire that now engulfed the remains of V9977 it became obvious that he would need to go for help. When it was self evident that there was little he could do on his own he ran back to Green Farm to telephone the Fire Service.

In the fierce inferno that followed the crash much of what was left of the aircraft was consumed. There was absolutely no chance of survival for any of the eleven members of the flight and it is assumed that they were either killed instantly upon impact or died very shortly afterwards in the fire.

## First news

At RAF Defford the aircraft was due to have returned at about 17.00 h. O'Kane, Dee, Curran, Trott and Lovell were waiting, along with the rest of the TRE staff and the TFU ground staff at the aerodrome for the flight to return. Lovell and O'Kane especially were beginning to get concerned when they had heard nothing from Berrington following his one solitary radio contact. O'Kane kept a diary at the time, and the following list of events are taken from his personal records:

17.00 Lovell rang control. Not landed.
17.30 Rang control. No news and no contact. No interest.
17.45 Rang control. No news. Rang a flight officers mess, control: Nobody about.
18.00 Rang G/C King [the CO of Defford]
18.10 W/C Horner [an officer on King's staff] rang up and wanted to know what was up. Told him.
19.00 Nothing done.
19.15 F/Lt. Reynolds [another on King's staff] had just left control having informed Group.
First news of crash.

So by mid evening on 7 June the first reports of the crash had started to come in and the various members of the TRE/TFU were organized into a small convey of vehicles, among whom were recovery teams, medical teams, security personnel and of course the remains of the scientific team who had to see what could be recovered of the H2S equipment. All now left for the crash site to see what could be done. From O'Kane's diary again:

21.00 Left for Welsh Bicknor.
Finding the less important cargo [that is the magnetron].

That evening, the small procession of vehicles meandered their way through the twisting narrow lanes of the remote countryside around the River Wye. As darkness fell at the end of that summer's day in June, and with the light in the west barely illuminating the scene, the teams from RAF Defford came across what remained of Halifax V9977 now spread across a significant part of the pasture field in which it fell.

Recalling the events of that night in 1977, for the BBC TV series *The Secret War*, Lovell explained:

'The reports of a crash in south Wales began to come in, and the rest of that night was just a nightmare. I was driven by the Commanding Officer of the aerodrome, a man called King, and winding through these lanes near Ross on Wye, searching for this wreckage and then the field with the burnt-out Halifax. And of course it was wartime. There was no time for emotions, our first duties were to search for the precious highly secret equipment, and collect the bits and pieces of it.'

Very little had been spared by the force of the initial impact, and the huge fire that had followed. With hindsight it seems rather macabre that the bodies of those that died were ignored and took second priority to the retrieval of the radar equipment that the aircraft had been carrying but, such was the sensitivity of the work being carried out, that this was the instruction that the search and rescue team was told to work under.

Bernard Lovell, who of course would have been on the flight, and presumably would also have been killed, had he not given his place to an eager member of the EMI team, led the men searching through the wreckage. The Halifax had disintegrated on impact and the pieces of the aircraft, those that were left, were spread over quite an area – nearly an acre in fact. The importance of finding the magnetron and the rest of the H2S radar system was not lost on the team, and eventually they did recover all the pieces of the equipment.

As day gave way to night, a stillness returned to the field where the bomber had fallen. Armed guards were posted, the remains of the eleven who had perished were covered with sheets, and the teams went back to Defford with the shattered remnants of the precious radar equipment – King and Lovell arriving there around half past midnight. The rest, including the recovery of the bodies of the aircrew and the TRE/EMI team would have to be left to the next day.

Lovell would write in his diary the next day of the enormous tragedy he must have felt losing so many close friends and colleagues:

> Monday, June 8 [1942]: Yesterday was terrible. Our Halifax V9977 crashed and killed Hensby, Vincent, Blumlein, Browne, Blythen, S/L Sansom, and a crew of 5 including the Pilot Berrington. All day Saturday (sic Sunday) we were having a meeting [at Defford] with EMI. I arrived at 2.30 to continue with Ryle etc.: they were just about to go off. They said they'd make it short. By 5 I began to worry. It took until 7.15 to stir up any interest in it. At 7.55 we heard it had crashed, at 8.35, 11 bodies were taken out, at 9, Group Captain King [the commanding officer at Defford] was driving O'Kane and myself down to salvage the apparatus. It was about 6 miles SW of Ross in the Wye valley. It was a mass of charred wreckage, quite unbelievable. We salvaged some bits but there wasn't much except the magnetron recognisable. Arrived back 12.30 at Defford and finally Malvern at 2.

The Air Ministry were informed by signal the next morning at 09.00 h from RAF Madley that the flight had been lost with all on board, and it was organized for a team of crash investigators to visit the site the same day, Monday 8 June. The EMI team also went out to the crash site on that Monday morning though with all the security they had some difficulty at first getting anywhere near the field where the Halifax lay as Felix Trott recalls:

> 'There was a guard at the entrance to the field where the aircraft had come down, and this fellow, who had probably been there throughout the night, was armed with a "Tommy gun" and wouldn't let us pass despite our protestations that we were civilian scientists. Anyway, I suppose this chap was only doing his job, but you see he would not respect anybody that didn't have a uniform, and we'd driven over from Tewkesbury where we had stayed the night, without anybody from Defford to accompany us.

Then, I suddenly remembered my military pass which I had with me, and decided it was worth a try showing him this. He looked at it and decided we were probably OK, at any rate, he wasn't going to argue with authority as he saw it, and we entered the security area; but without that pass I don't suppose we would have been allowed to enter. The aeroplane had hit the ground inverted and had struck a small hump in the middle of this quite large field, exploding on impact. I could see that the undercarriage had been put down by the pilot, though now it was sticking up into the air from the remains of the engines. There were bits of aircraft everywhere, twisted and burnt out and none very large at all. On the far side of the crash area, down towards the river, the army or whoever it was, had erected a small white tent and in there, the bodies had been laid out. I didn't go in, I didn't want to see them, but I did catch a glimpse as the tent flaps were pulled aside and I saw far more than I wanted to.'

Eric White also recalled the horrific scene:

'By this time there were people walking around all over the place. Some were military others were not. I expect that they were something to do with the crash investigation team. We walked over to this little white tent that had been erected and there were some people from TRE already there. Inside were these blackened, burnt bodies which were beyond identification and I left very quickly.'

Among the people walking around that morning was an official photographer,

Figure 10.10 Felix Trott's military pass which was used to gain access to the crash site on Monday, 8 June 1942. (*Courtesy of Felix Trott*)

though whether he was from Defford, attached to TRE, or part of the Accident Investigation Branch is not known for certain. What is known is that this man took at least six photographs of the crash site, which were then taken back to Defford for processing. They were then entered into an RAF log book of negatives as plate numbers 'G/59' to 'G/64' under the title 'Halifax Crash'. The plates were exposed on Wednesday, 10 June 1942, plate G/59 showing a 'general view', G/60 the 'port engine outer', G/61 the 'starboard outer engine', G/62 a 'field corner view', G/63 the 'fuselage' and G/64 the 'four engines'.

These photographs were logged as 'required by Wing Commander Horner', who was attached to TFU at Defford, and despite the log clearly stating that they were booked out and then returned, the plates have mysteriously disappeared. They are no longer to be found in the archives at Malvern. Indeed, it is quite possible that for purposes of secrecy, they were removed from the archive files and any prints made from the plates, even the plates themselves may have been destroyed. Other documents relating to the crash mention more photographs taken that day, in one instance 13 photographs were supposed to exist. Whether these are a completely separate set, or include the six mentioned above, is not known. What is certain is that the very highest authority in the land ensured that the secrecy surrounding the loss of Halifax V9977 was total. None of these photographs of the crash site has yet been found in any archive, and may not do so for many years to come, assuming that is, that they still exist at all.

Back at EMI that Monday morning, Maurice Harker arrived for work and happened to get into the elevator with a very subdued-looking G.E. Condliffe. Harker said 'Good morning' to Condliffe who turned to him and replied, 'I don't suppose you've heard yet then?' 'Heard? Heard what?', and it was then that Maurice Harker was told that Blumlein, Browne and Blythen had all been killed the previous afternoon. 'I simply could not believe it', he recalled,

> 'I walked into my office and just sat there staring out of the window. It was such complete and utter disbelief. I said to Condliffe "What do I do? Shall I just carry on as before?" And he said "Yes, carry on as before". And of course, I had to collect what things I needed and drive back to Defford that afternoon which I did. But I didn't go out to the crash site once there. I don't suppose there was any need, and I certainly didn't want to go.'

Three days after the crash, on 10 June 1942, Lovell would write to his wife Joyce in Swanage where she was then still living with their two young children, 'The accident has left a great aching sore in our minds which is always there. We try to forget it by working hard and thinking that we just have to make a tremendous job of it to make up for that enormous sacrifice.' And in his notes to himself he wrote, 'The loss of Blumlein is a national disaster. God knows how much this will put back H2S?' Even two weeks later Lovell's diary thoughts were still immersed in the crash, 'Last week we tried to forget about the crash and get on with the work. This week we have perhaps succeeded better.'

## The investigation

Following the crash of Halifax V9977 and the incalculable loss of the scientific team that was aboard, Churchill personally ordered that the findings of the investigation to determine what had happened be shown to him. He had been handed a single page preliminary report of the crash (as events were understood at the time), on 13 June, by the Chief of Air Staff. In this letter the Prime Minister was promised that a further investigation would be carried out as soon as possible. Churchill wanted to have it explained to him why Halifaxes were crashing, and especially why this one had crashed, and what could be done to prevent any more unnecessary losses. The detailed crash investigation was rapidly set in motion by a Court of Inquiry, and once the wreckage of V9977 was removed from the crash site it was sent away to be minutely inspected to determine what had gone wrong.

On 1 July 1942, the crash investigation report was issued as a secret document and only distributed to the heads of various involved departments. The report, Service Accident Report No. W-1251, remained secret and hidden from the general public for decades after the incident (bearing testimony to the importance of the H2S work by the Ministry of Defence); to my knowledge, until now, it has never been published in full. The following conclusions therefore are drawn directly from Service Accident Report No. W-1251 and several independent investigation reports drawn up at the time to support its claims:

## 1 The fire

The fire in the starboard outer engine No. 4 was obviously the cause of the crash, but that engine, which had been built less than a year before the date of the accident, only had a total running time of approximately 70 hours. The fire in the engine was recognized as the worst possible type of fire that could have happened being known as a 'blow-back' fire. This is where the fire itself consumes the casing and housing on which the engine is mounted to the wing and crosses over the fire wall in the wing igniting the fuel tanks inside.

To prevent fire the Merlin Mk.XX engine has a fire extinguisher bottle located at the rear of the engine casing just above the supercharger. This should operate as soon as any fire is detected either by the automatic systems installed as safety measures or manually from the cockpit. In the case of a blow-back fire the flame and heat is so intense in such a short period of time that the extinguisher bottle can be considered almost useless. It can become punctured or prematurely exhausted by the fire with no reserve facility available. With the extinguisher bottle empty or malfunctioning so that it did not discharge correctly, the fire would quickly spread through the engine housing and cross the fire wall separating the fuel tanks in the wing. It was proved later, after the investigation, that this shielding in the Halifax was quite inadequate and rather than prevent the spread of fire actually aided it.

The fuel held in each wing was located in six tanks of varying size and capacity. These were located between the outer and inner engines, between the inner engine and the main fuselage and in one smaller tank situated just behind the leading edge slats of the wing. In all each wing held 943 gallons of high-octane fuel, giving a total of 1886 gallons when the aircraft left Defford. During the flight of 1 hour and 30 minutes, around 420 gallons of fuel would have been used, leaving 1466 gallons still in the aircraft. Once the fire had ruptured the fuel tanks the structural integrity of the aircraft was very limited, perhaps only a few minutes. Berrington would have known this and that is why he was heading for flat ground as soon as he could in order to take the risk of a crash landing.

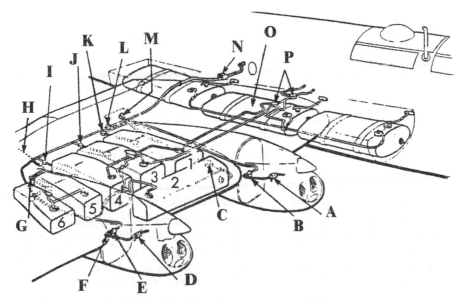

A Engine Supply Cock, B Filter, C Fuel Cock No. 2, D Engine Supply Cock, E Filter,
F Main Fuel Pipe passes through the Fire Wall, G Fuel Cock No. 5, H Wing Joint,
I Fuel Cock No. 4, J Fuel Cock No. 3, K Fuel Cock No. 1, L Wing Balance Cock,
M Drain Cock, N Main Balance Cock (situated on rear spar), O Front, Centre & Rear
Long-Range Tanks (these were not fitted on V9977), P Distribution Cocks

Figure 10.11 Halifax Mk.II fuel tank and fuel supply system detail. (*Courtesy of William Sleigh*)

As the fire in No. 4 engine passed the fire wall and ruptured the first of the fuel tanks the heat was so intense that the supercharger, carburettor and casing mounts, which were all made of aluminium exploded. When the crash investigation teams arrived on the scene of the wreckage of V9977, they discovered a 200-yard long trail of molten metal shed in flight, leading away from the main fuselage down towards the River Wye. As the fuel tanks erupted, the structure of the outer wing spar, which holds the end of the wing to the main fuselage of the aircraft, finally snapped through. At that point, the outer wing spar broke loose at a position roughly where the starboard outer engine is mounted. Unable to maintain level flight, the aircraft then fell to the ground.

## 2 Rolls-Royce engine strip report No. 73

The question for the crash investigators having now pieced together what had happened was why had it happened in the first place? When the wreckage was removed from the crash site, the engine that had caught fire and probably caused the subsequent crash was sent to Rolls-Royce, who had manufactured it, for a strip down examination. Rolls-Royce produced from this inspection a report (Rolls-Royce strip down report No. 73) which identified the reason the engine had failed.

It seems that during the last 30-hour inspection carried out at Defford between

25 May and 1 June, a fitter had not sufficiently tightened one of the locking nuts that held the rear inlet valve in place. With this locking nut loose, the vibration of the aircraft in flight eventually caused it to unscrew allowing the inlet valve to travel far more than normal. Under enormous pressure, the stem of the valve finally broke causing the charge of incoming fuel to have free access to the outside of the rocker cover of the engine. As the fuel was pumped into the valve space and ignited by the spark plugs it would be able to freely spread past the engine cover and out into the space around the engine itself. All of this would have happened in a matter of seconds after the initial failure of the inlet valve stem.

Once out in the open, the fire would quickly spread being continually fuelled by more incoming petrol to the inlet valve space. As the fire grew it passed backwards along the engine casing igniting any combustible material, oil, grease and the like, until it grew to such proportions that it was completely out of control.

At this point the remaining time left to Halifax V9977 was determined by how long it would take the growing fire to burn through the fire wall that protected the fuel tanks in the wings from the engines and ignite them. Once this had happened a landing of any kind was required in minutes or, as happened in the event, the igniting fuel would eventually compromise the structure of the aircraft.

Rolls-Royce determined then that a single nut left untightened had caused the crash and recommended that in future a senior Non Commissioned Officer should be responsible for checking that all tappet locking nuts were tightened after the fitter had worked on them. The report also, sadly, highlighted the fact that the loss of an aircraft due to a single tappet nut being left unscrewed was known to them, but the contingency plans that were designed to alleviate this problem had not yet been put into operation and Rolls-Royce had not seen fit to warn the fitters.

The question of prevention was then looked into; could the pilot have done anything to prevent the crash happening?

More importantly perhaps, could the personnel aboard V9977 have parachuted out to safety as soon as it was realized the aircraft was doomed? Reconstructing the last few minutes of Halifax V9977 from eyewitness accounts is one thing, trying to guess at what actually was happening inside the aircraft itself is quite another. The following is what is most likely to have been heard and seen from inside the aircraft in the moments leading up to the crash.

Of the eleven people on board V9977, eight were non-involved crew and scientific personnel. Only the pilot, flight engineer and possibly the second pilot would be directly involved with the emergency situation. Pilot Officer Berrington, as has been seen, had a total of 14 hours and 35 minutes in Halifax aircraft prior to the flight of 7 June 1942. His second pilot, Flying Officer Phillips had no flying time in Halifax aircraft at all.

This set of circumstances would dramatically affect the reaction time of the pilot and crew to the emergency situation as they became aware of it. Their speed and drill experience, had it been any other aircraft type, may well have prevented the loss of the flight. Unfortunately, coupled with the poor cockpit layout in the Halifax, and the remote location of the fuel shut-off cocks, it is fair to assume that a certain amount of wasted time and energy would have ensued when the fire was first noticed.

Initially unknown to Berrington (the very nature of that class of devastating detonation which initiated an immediate serious engine fire would most certainly not have been pre-indicated to the crew on the engine's speed, oil pressure and temperature as well as coolant temperature instruments), he was also faced with the worst possible type of engine failure, and even if his experience and reactions were better and quicker it is still possible that V9977 was doomed from the moment the blow-back fire began.

The first immediate indication that something was wrong would have been a sudden yaw of the aircraft due to the propeller's output changing from thrust to drag as the power ratio and efficiency of the Merlin engine dropped off. At this point the first indications of smoke may have been noticed from a visual inspection to his right out of the window. His probable immediate reaction would have been to re-trim the flying controls of the aircraft, feather the propellers and shut-down the engine concerned (assuming this was determinable).

There is no way of knowing if Berrington was aware or not if the fuel cock had isolated the fuel feed to the engine as it should have done. In any case the remote location of the fuel shut-off cocks and the growing fire in the engine may well have led him to falsely believe that it had done so automatically and that therefore the fire bottle had prematurely discharged. This however, is immaterial as the fire extinguisher was never designed to cope with this kind of fire.

The Flight Engineer would normally have played an equally important role in the emergency as the pilot, but of course Leading Airman Dear was an engine fitter and had not yet passed an official flight engineers course. It is possible that he would have waited for instructions from his senior officer, Pilot Officer Berrington or Second Pilot Phillips. If he was given any instructions to check on the fuel shut-off cocks (because of his familiarity with the aircraft fuel systems it is safe to assume he knew where these were), he would have to travel back through the fuselage of the aircraft to a position roughly between the wings where the No. 3 and No. 4 engine fuel tank shut-off valves were located.

This was a design flaw in the Halifax and would have seriously contravened fire prevention and safety laws today, endangering as it did the crew and the aircraft. It should also be remembered that Halifax V9977 was packed full of scientific equipment at the time and, as anybody who has spent any time inside the fuselage of Second World War bombers of a size comparable to a Halifax will be aware, space inside the aircraft was very limited indeed.

With the blow-back fire developing in engine No.4 it was vital that it be isolated within a time span of no more than 30 seconds. In the event, the location of the instruments and controls, coupled with the lack of overall type experience of the crew meant that the isolation process took far too long. The fire had time to blow back and burn its way through the protective fire wall and into the wing cavity itself where the six fuel tanks held roughly 750 gallons of unused fuel.

As the fire crossed the bulkhead and came into contact with the thin walls of the No. 5 and No. 6 fuel tanks it was only a matter of time before these would ignite. The design layout of the fuel tanks in each wing of the Halifax was such that the No. 2 tank, located in the leading edge of the wing, was considered vital to the structural integrity of the aircraft. If the No. 2 tank was ruptured by the fire, the starboard main spar holding the wing to the fuselage would fail. With this positioned right next to the window, Berrington would have been only too aware of the growing urgency of the situation.

By the time a few minutes had passed from the initial flash point and the start of the fire, the crew, probably now fully aware of the failure of the emergency procedures they had already attempted, would realize the immediate need to find somewhere to land. It is at this point that the one and only brief message which was supposed to have been sent from V9977 was received at the tower at Defford. It should also be stated however, that this message is not recorded anywhere and there is serious doubt if indeed it was in fact sent or received. The message, if we assume that one was actually sent, would almost certainly have come from Second Officer Philips. It supposedly reported that an emergency was in progress and that the fire in the engine was out of control. A rough fix was given and information that an attempt at a crash landing would be tried. Unfortunately, and incredibly the detail of the message was not recorded, but it would seem that the report gave the aircraft's height as between 2000 and 3000 feet when the fire began.

With time running out, and the runway at Defford over 20 minutes away – far beyond the structural survival time of the aircraft – Berrington turned onto a northerly heading of 340° and rapidly began to lose height trying to find a flat area of ground on which to crash land the stricken Halifax. From the estimated position of V9977 at this point he probably had two options – the flats of the Severn estuary or the flat ground of the Wye valley just north of the Forest of Dean – he chose the latter.

As the aircraft lost height and continued on its path across the Forest of Dean, Berrington would have been able to see the flat fields about 2 miles north of the village of Goodrich. These would have made an ideal place for the aircraft to attempt a wheels-up landing, a manoeuvre which was extremely dangerous, but with time running out he had little option. At this point the crew would almost certainly have been made aware of Berrington's choice and would probably have informed their civilian passengers, all of whom would have been taking the necessary precautions in readiness for such a crash landing.

However, the question of the danger in attempting such a manoeuvre would not be Berrington's to decide. With just under 4 miles to go to a safe landing point, at a position almost directly above the River Wye just north of the village of English Bicknor, the structural integrity of the starboard main plane gave way causing a large section of the wing to separate from the aircraft. Though only at a height above the ground of just 350 feet, with all ability to maintain level flight gone there would have been nothing Pilot Officer Berrington nor anybody else could have done. The aircraft inverted and crashed into the ground at a speed estimated at over 150 mph and was disintegrated by the impact. With over 1600 gallons of fuel left in the wings the ensuing fire consumed almost everything including the 11 occupants of Halifax V9977.

Many have speculated without full appreciation of the technicalities of the situation whether the incident was avoidable. What is certain is that Halifax V9977, whether under the control of Pilot Officer Berrington, or (say) the Chief Test Pilot of Handley Page, would never have survived that event. This was a situation Berrington must have quickly appreciated, as would his flight engineer who was the most technically qualified member of the crew.

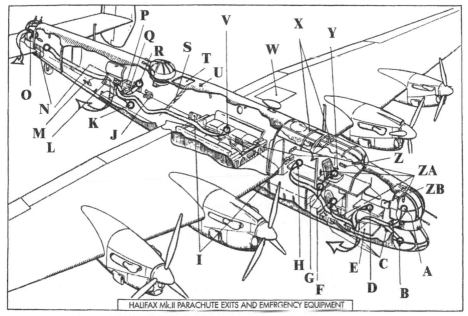

A Front Gun Turret, B Bomb Aimers Station, C Parachute Stowage, D Navigators Station,
E Parachute Exit (open), F Wireless Operators Station, G Pilots Cockpit, H Engineers Station,
I Parachute Stowage, J Marine Distress Signals, K Flare Launchers Station, L Parachute Exit
(open), M Crash Axe, N Parachute Stowage, O Rear Gun Turret, P First Aid Stowage,
Q Emergency Rations Stowage, R Mid Gunners Station, S Escape Ladder, T Escape Hatch,
U Dinghy Release, V Rest Station, W Dinghy Stowage, X Window Escape, Y Flare Release
Handle, Z Push Button for the Fire Extinguishers, ZA Fire Extinguishers, ZB First Aid Stowage

Figure 10.12 Halifax Mk.II escape route and normal parachute stowage details.
(*Courtesy of William Sleigh*)

### 3 The question of exiting the aircraft and parachutes

One of the most perplexing questions that arose from the accident was why had the
crew not attempted to bail out when it was realized that the situation was beyond
control? In the event of an emergency in a Halifax there were three possible exits.
The first was the forward escape hatch located towards the nose of the aircraft by
the navigator's station. The second exit was located in the mid-section of the fuselage, a profiled door on the floor of the fuselage on the starboard side of the aircraft, just aft of the top turret. The third exit was the rear gunner's turret which
could be rotated and ejected from the aircraft allowing escape. As the flight of
V9977 did not include the Halifax's normal compliment of three gunners, it is fairly
safe to assume that the rear turret was not in use and can be discounted as a primary
route of escape.

A fourth exit on the top of the fuselage just forward of the top gun turret was
designed to be used in the event of a wheels-up landing. It is reached by means of
a ladder inside the aircraft, but would not normally be considered as a potential
escape route, in the event of a parachute jump being required, because of the
danger of the escapee hitting the superstructure of the aircraft whilst still in flight.

Therefore, on 7 June, any escape of the 11 personnel on board V9977 was restricted to the first two exits. As it was probable that by the time Berrington realized the gravity of the situation the fire in engine No. 4 was extreme, and may well have spread to the wing, any escape via the lower profiled escape hatch would have been very risky indeed. The fire may have spread along the wing to the fuselage and potentially burnt any person attempting to drop out of the hatch. This left just the front escape hatch which also had the same risk attached, but to a lesser extent.

There was also the question of parachutes. The official crash investigation report states that the remains of eleven parachutes were found in the wreckage of V9977, though it does not detail whether any of these were actually found on the remains of the crew or the civilians aboard, or whether they were assumed to have been stowed. What it also does not outline is whether any of the scientific team on board were aware of their stowage positions and deployment, or if they had any previous emergency drill experience. Bearing in mind the larger than normal number of people in the aircraft, and the large amount of equipment being used, it is likely that the parachutes would have been stowed in the aft of the aircraft away from the confined crew compartment.

The circumstances of the parachute remains in the wreckage of V9977 are also open to speculation. After the crash the Government naturally kept much of the detail of the incident from the public including the immediate family members of those who had perished. Not withstanding the need for secrecy in time of war, this did unfortunately lead to considerable suspicion focused on just how many parachutes actually were in the aircraft when it went down. Almost immediately after the crash of Halifax V9977, a directive from Churchill himself ordered that any future flights with civilians or other scientific personnel on board, must carry enough parachutes for each person involved.

While it cannot be used as conclusive proof that the accident investigation report was 'doctored' to show the remains of eleven parachutes, it does seem very odd that a directive from the very highest in the land should focus on the issue of numbers of parachutes, if the flight in question had indeed been carrying the correct complement.

### Conclusion

When all the elements of the investigation into the crash of Halifax V9977 are assessed it would seem that the following factors were to blame:

1 The initiation of the worst type of engine fire caused by prematurely introduced mechanical failure. The following four points refer:
   - The cause of the crash was initiated by the apparent insufficient tightening of one of the 48 tappet adjustment stud lock nuts of an inlet valve of the starboard (right) cylinder block's valve mechanism of the No. 4 starboard outer engine. This relatively new engine's date of manufacture and low flying hours were irrelevant.
   - The tappet lock nut, through insufficient tightening, had freed itself and rotated anti-clockwise under the action of engine resonance. Its movement was equally followed by the tappet adjustment stud, which operates the inlet valve,

thus the widening of the critical clearance gap seriously reduced the valve's travel. That in turn reduced the fuel charge flow's cooling influence with inherent overheating of the valve steel by the combustion gases. The continuing 'hammering' of the heat weakened steel valve on its seat, still under spring tension, caused fatigue with subsequent failure.

The failure of the inlet valve permitted the ignited combustion gases the freedom to gain access to the inlet manifold flame trap. Under those circumstances this would disintegrate thereby permitting the instant igniting of the whole volume of highly combustible fuel/air charge mixture downstream of the carburettor and supercharger. This mixture, identical to the detonation of fuel in the engine's cylinder, could not be contained in the cast aluminium ducting, which exploded with devastating consequences.

- The fire, free to engulf confines of the power plant bay, would be enhanced by the continuing gush of petrol from the still active wing tank booster pumps and fanned to torch proportion by the ram air feed to the engine's heat exchangers. This inferno would then proceed to bypass the wing mounted fire bulkhead supporting the power plant structure. Time scale for this devastating chain of events would be in the order of three seconds from the valve failure.

No fire suppression system and its associated extinguisher bottle(s) is designed to counter a fire of such intensity and, as such, comment on the system's performance is irrelevant.

- The ferocity of such a major fire, carried by the airspeed slipstream past the engine bay's fire bulkhead and over the mainplane area accommodating to the three outboard fuels tanks of capacity of 400 gallons. These tanks clearly ignited contributing to the inferno, and caused thermal degradation of the strength of the spar attachment which distorted under flight loading leading to the severance of the outer mainplane immediately before impact.

2 Questionable cockpit design in the Halifax aircraft relative to the rapid extinction of an engine fire and fuel shut-down systems.

3 A new in-service aircraft, the nature of which crews had very little experience irrespective of previous aircraft types flown.

4 A compliment of type inexperienced crew members.

5 The cause of the crash had no connection whatsoever with the engineering design standards of Rolls-Royce Merlin aero-engines, or perhaps that of the Halifax bomber (in spite of the latter's general design shortcomings that did little to retard the chain of events). The 1942 Accident Board of Enquiry did not call for any major modifications to either engines of airframes which most certainly would have been the case if their findings had been otherwise.

## Service Accident Report No. W-1251 (see Author's note on p. 344)

The following is the complete service accident report on the loss of V9977 produced by the Accident Investigation Branch immediately after the crash itself. This document was kept secret for nearly fifty years after the actual events took place, which bears testimony to the degree of importance placed upon the development of the H2S radar system by the British Government.

It should be considered however, that the area of confusion and conjecture over the cause of the loss of Halifax V9977 was clearly within the scientific fraternity itself where, in recent years, they were trying to understand the reason for that event. Unfortunately, those in that fraternity had (understandably) given credence to everything and anything written, including clearly erroneous matter held in the Ministry of Defence, Air Historical Branch Archives, the basis of which is contained in the following report:

**Secret**

*Service Accident Report No. W-1251*

**Accidents Investigation Branch.**

| Aircraft: | Halifax II, No. V9977 | Engines: | Merlin XX. |
|---|---|---|---|
| | | Port Outer | No. 37667/198338 |
| | | Port Inner | No. 54917/245179 |
| | | Stbd Outer | No. 37673/198341 |
| | | Stbd Inner | No. 37654/198334 |

Unit:   Offensive Section, Telecommunications Flying Unit, Defford

| Pilots: | First Pilot: Pilot Officer D.J.D. Berrington | Killed |
|---|---|---|
| | Second Pilot: Flying Officer A.M. Phillips | " |
| Crew: | Flight Sergeant Millar, G., Observer | " |
| | Leading Aircraftman B.D.C. Dear, Flight Engineer | " |
| | Aircraftman II, Bicknell, B.C.F. | " |

| Passengers: | Squadron Leader R.J. Sansom (Attached T.R.E.) | " |
|---|---|---|
| | Pilot Officer C.E. Vincent (Attached T.R.E.) | " |
| | Mr. G.S. Hensby, Civilian T.R.E. | " |
| | Mr. A.D. Blumlein, Civilian E.M.I. | " |
| | Mr. C.O. Browne, Civilian E.M.I. | " |
| | Mr. F. Blythen, Civilian E.M.I. | " |

Place of Accident:   The Green, Welsh Bicknor, Herefordshire.

Date and Time:   7th June, 1942 at 1620 hrs.

*1. Notification.*
By signal from R.A.F. Station, Madley at 0900 hrs on 8.6.42. The scene of the accident was visited the same day.

*2. Circumstances of the Accident.*
The aircraft took off from Defford in good flying weather at 1450 hrs on a flight authorised for secret experimental work.

The normal route for these flights was between Gloucester and Kidderminster but the pilot acted under the instructions of the Scientist in charge, in this case Mr. Hensby of T.R.E., and deviations from this route could be requested by him during the flight as requirements necessitated.

At about 1600 hrs the aircraft was seen to circle a point four miles North West of Usk, Mon. at a height of 1,500 ft. A few minutes later, the outer starboard propeller was seen to stop. At 1610 hrs the aircraft left this point in a South-easterly direction and at 1620 hrs it was observed near Coleford, Glos., with smoke issuing from the starboard outer engine. The height was then stated to be 500 ft. and the aircraft was flying in a N.N.E. direction which was its track for base. Fire was seen to break out some seconds later and within two minutes of this the aircraft crashed.

None of the occupants survived.

### 3. Further Details.

(a) The aircraft was passed out from Handley Page's on 28.9.41 and delivered to O.S.T.F.U. Hurn on 27.3.42. Its total flying time was 64 hrs 45 mins.

Three of the engines were installed on 1.9.41 and their running time was approximately 70 hours each. The port inner engine was installed on 8.5.42, it's running time being 20 hours.

A 30-hour inspection was carried out on the engines between 25.5.42 and 1.6.42 and the tappets were adjusted. Between this inspection and the date of the crash, the aircraft flew 6 hrs 40 mins.

(b) Pilot Officer Berrington, it is understood, had flown over 2,000 hours as a civilian pilot before joining No.5 S.T.F.S on 11.9.39 as a Sergeant. No flying log books for this period were available.

Throughout his RAF career he was assessed as 'Above average' and on 2.10.40 with a total of 444 hours, he was posted to S.D.F. (Special Duty Flight) Hurn, which later became T.F.U. Up to the date of the crash he had a total of 1,050 hours flying on 27 types of aircraft, including Wellington's, Whitley's, Herefords and Boeing 247 D.
His time in Halifax aircraft was 12 hrs 30 mins as first pilot and 2 hours 5 mins as second pilot.

The second pilot had flown a total of approximately 700 hours on 16 types but had no experience of Halifax aircraft.

(c) The weather conditions observed at Pershore at 1600 hrs on 7.6.42 were as follows: -
Low cloud ⅗ths at 3,000 ft. Visibility over 18 miles. Surface wind N.W.W. at 10mph.
These conditions were general over the area concerned.

(d) On inspection at the scene of the accident, it was found that the aircraft had crashed on the steep slope of a grass field overlooking the West bank of the River Wye at a point 4 ½ miles S.S.W. of Ross. The aircraft struck the ground at an inverted attitude and was shattered on impact. The wreckage, distributed over an area of approximately an acre was almost totally destroyed in the ensuing fire.

A narrow trail of melted metal, shed in flight, was discovered leading from the River to the wreckage, a distance of 200 yards. Expert examination of this melted metal revealed that the fire in the air had been intense and that it was not confined to the starboard engine but had spread along the leading edge to the nose tank.

Inspection of the wreckage did not disclose any defect in the airframe which could not be attributed to impact or fire. The settings of the undercarriage and flapjacks were noted but detailed examination of the flying controls was not possible.

The remnants of eleven parachutes were found in the wreckage.

(e)   Evidence as to whether the Graviner fire extinguishing apparatus has worked was inconclusive.

The engine nacelle extinguisher bottles were located amongst the general wreckage. Two of these had burst open by high internal pressure generated, presumably, by the ground fire, indicating that they were in the charged state. The third and fourth bottles were too badly damaged to supply any evidence. As the bottles are similar, it was not possible to say which came from which nacelle.

The extinguisher spray pipes on the engines did not show any traces of ejection of extinguisher fluid but it is understood that Methyl Bromide does not leave any deposit.

(f)   All four engines had been torn from the airframe and damaged by impact and fire.
Examination of the starboard outer engine showed that there had been a fierce pre-crash fire in the super-charger induction casing.

A further inspection of the engine showed that there had been no loss of coolant and that the bearings of all the connecting rods were in good condition and adequately lubricated.

It was not possible to carry out a more detailed examination at the scene of the accident and the engine was dispatched to the makers for a strip examination.

(g)   The following is an extract from the Rolls-Royce strip report No.73: -
"On removal of the starboard side rocker cover, it was found that No.2a rear inlet valve tappet locking nut has unscrewed and was subsequently found lodged in the oil well adjacent to the cam shaft drive at the rear end of the block. The tappet screw had screwed right back and the excessive tappet clearance so set up had fractured the inlet valve stem at the reduced collet section. The retaining safety circlip prevented the valve from falling right into the cylinder.

On examining the flame traps all the elements were found to be burnt along their entire length on the cylinder and induction pipe side. Several of the sparking plugs, in particular those removed from A2 cylinder, were heavily coated with zinc oxide, a constituent of the engine flame trap material. The remainder of the engine was mechanically sound and beyond suspicion, but the supercharger steel rotating guide vane rotor has been passed to our laboratory for examination, although on initial inspection it is thought that the broken vane had failed in tension due to overheating.

It is very evident that the primary cause of failure was the unscrewing of the tappet nut, giving rise to excessive clearance between the tip of the valve and the tappet screw. The high opening and setting velocity of the valve due to running with increased tappet clearance resulted in the valve stem fracturing. Following this, the valve would be allowed to drop so far into the cylinder and the hot exhaust gases would impinge on the flame trap element causing the latter to burn, finally igniting the incoming charge in the induction manifold.

The cause of the tappet nut unscrewing was in all probability due to it not having been tightened up correctly. It is understood that the tappets were adjusted by the Service some three hours before failure occurred. Moreover, the thread form of the tappet screw and tappet arm have been checked and found to be satisfactory."

### 4. Opinion.
The cause of the accident must be attributed to failure of the starboard outer engine, with subsequent intense fire, when flying over country which precluded the possibility of a successful forced landing.

The engine failure and resulting fire was due to the fracture in fatigue of an inlet valve stem as detailed in the makers report (para. (g) above).

### 5. Observations.
The following is a further extract from the makers report: -

"It has been realised by the makers for some time that complete engine failure can occur due to a tappet nut being left loose. Development is proceeding to overcome this contingency on various schemes which will eliminate the possibility of failure due to one tappet being inadvertently left loose.

Meanwhile, a modification has been approved introducing full nuts for locking the tappet screw in place of standard half-nut. This will eliminate the possibility of failure due to overtightening the nuts with consequent stripping of the threads in the nuts, but will not safeguard against the nuts being left loose."

### 6. Recommendation.
It was learned at the makers that in spite of every care being taken in its adjustment, the lock nut could become loosened of its own accord. It is nevertheless recommended that the Service be instructed that when adjusting tappet clearance, a senior N.C.O. should be responsible for checking tappet nuts for tightness after the fitter has completed his adjustments.


Vernon Brown (signed)
Chief Inspector.
C.I. (Accidents), Eastern Avenue,
Air Ministry, Gloucester.
1st July 1942

**Distribution List.**
Professor L. Bairstow, C.B.E., F.R.S.
Aeronautical Research Committee.
D.D.M.O.
S. of S.
U.S. of S.(C).
P.U.S.
S.4 Stats.
A.M.T.
D.T.F.
D. of S.
D.S.R.
Sir Henry Tizard, K.C.B., A.F.C., F.R.S.
D.G. of E.
D.A.I.
D.T.D.
O.7
T.G.2
Inspector General
D.O.R.
D.T.O.
M.A.4
R.M.4
D.S.M.
C.R.D.
D.E.D (written by hand)
C.A.S. (written by hand)
V.C.A.S. (written by hand)
A.A.E.E. (written by hand)

## Aftermath

Naturally the loss of the scientific personnel aboard V9977 was a crushing blow to British radar development, but in time of war there is precious little space for sentimentality. The development of H2S was continued even before the official crash investigation report was printed. Halifax V9977 was replaced with Halifax W7711 on Friday, 26 June 1942, and along with R9490 the two aircraft completed the trails of the magnetron and klystron systems.

As predicted by the EMI team that had died, the magnetron system proved far too good not to use as a production system, and when, on Wednesday, 10 February 1943, W7711 returned to active duty with the RAF it became the first aircraft that had been fitted with the H2S radar system for operational use.

When the war was over the Government did allow a brief tribute to the men who had lost their lives throughout the conflict in the Telecommunications Flying Unit. But a full and fitting memorial was not erected until the dedication of the memorial window at Goodrich Castle on 7 June 1992, the fiftieth anniversary of the crash of V9977.

In the aftermath of the loss of the aircraft and crew, and the subsequent investigation that followed, no blame was attached to any person or single factor as prime cause of the crash. True it was a degree of negligence that had resulted in the tappet lock nut not being fully tightened; the Rolls-Royce report clearly indicates they were aware of the implications of such an oversight. Having said that, no other Merlin engine suffered from this particular problem and no other documented case of this type of failure is known. It would seem that the loss of V9977 was due to a freak accident that was literally one in a million.

In the years following the crash of V9977 and long after the end of the war, speculation among the family and friends of the aircrew and the scientific team aboard the flight did not abate. With little if any factual evidence being made available to them beyond the obvious, that their loved ones had died in the service of their country, it was not surprising that many, including Doreen Blumlein, believed, wrongly, that the crash was the result of sabotage by the Germans.

In the early 1980s, an engineer from the historic successor to TRE, the Royal Signals and Radar Establishment at Malvern, attempted to piece together the facts about the loss of V9977. The document, 'Aircraft for Airborne Radar Development', was published in mid-1987, and a limited number of copies were produced. Copies are known to be held by the Defence Evaluation and Research Agency (DERA), Malvern, which RSRE later became, the RAF Historical Branch, the library of the RAF Museum, Hendon and the library of the Imperial War Museum, London.

The author is William H. Sleigh, MOD/RSRE former Chief Engineer, RSRE Airstation Pershore. Sleigh had been a chartered engineer at Rolls-Royce Aero Engines, and led an interesting and colourful life as a pilot and test engineer on many radar development flights from the early 1940s onwards. Though he had originally worked for Rolls-Royce as an apprentice engineer extensively on the Lancaster and Lincoln bombers, he was soon promoted and transferred to TRE and TFU at Defford. From 1945, he took part in, or witnessed, many of the advances in radar development. Later, he became attached to the historical division of RSRE, and it was from here that his work began on the documentation of airborne radar. As the loss of the Halifax at Welsh Bicknor represented an essential part in this development, it would also be necessary to investigate all of the events that led to the loss of Halifax V9977. Sleigh had read in various scientific magazines after the war, readers' letters exclaiming their astonishment at how Blumlein could have been allowed to die in this aircrash. The Government must surely have known of his value, and that the Germans might target him for a sabotage mission. These letters had been replied to by none other than Bernard Lovell, yet he made no mention of the fact that it was mechanical failure and not sabotage, that had brought the aircraft down. The document Sleigh eventually wrote would take many months to compile, and although it was only intended to publish it to a limited audience it did catalogue for the first time, drawing from still secret (at that time) military and government files, the final conclusive proof of the events which led to the loss of the flight.

Within the main work of the document, there is a detachable section entitled 'Annex 4, The Loss of Halifax V9977'. It is within this section that Sleigh explains, 'The object of the Annex is to fill the existing vacuum in information, as well as correct the brief surviving records held by the Ministry of Defence, Air Historical Branch 5 [RAF] and the Royal Air Force Museum, Hendon, which suggest the cause of the accident could have been other than what is now known to have occurred.'

Annex 4 was produced in the absence of the official Air Accident Investigation Report. Therefore, most of the direct contributions that Sleigh used to substantiate the work came from four main areas. First, he met with and interviewed at length Onslow Kirby, the Herefordshire farmer who had of course actually witnessed the accident as it happened. Mr Kirby was considered by the Accidents Investigation Board to be a primary contemporary witness. Second, Alec Harvey-Bailey, a former colleague of William Sleigh and the senior Rolls-Royce engineer who had established the cause of the engine failure in 1942, was found and re-interviewed. He too was considered by the Accidents Investigation Branch to be a primary contemporary witness.

Third, Rolls-Royce engineers who were involved in the manufacture and test-flying of Merlin engines of the era were found and spoken to and, finally, the former Chief Engineer of the Radar Establishment's Aircraft Department was spoken to as he would be conversant with engineering operations on the unit which earlier operated Halifax V9977.

The annex was written in the hope that it would prove beyond all doubt that Halifax V9977 was lost due to the unwitting human error of an Air Force engine fitter. It was this fitter who had inadvertently overlooked the security of one of 192 ⅚-inch BSF lock nuts when undertaking a valve clearance safety inspection that was directed at preventing the very failure which initiated the tragic events that were to follow. In the conclusions at the end of the Annex it is stated that this was the 'probable' course of events: that during the routine 80 hour inspection that was carried out at Defford in mid to late May 1942, a fitter, working under extreme conditions imposed by wartime activities, had insufficiently tightened a lock nut on one of the inlet valve stems on the No. 4 engine, the starboard outer, of Halifax V9977, and this had instigated the events which led to the loss of the aircraft.

The final Annex, which is, as stated above, only part of a much larger document ('Aircraft for Airborne Radar Development'), is a fine piece of investigative work carried out with the direct help of many of the contemporary witnesses to the accident itself as well as the subsequent accident investigation process. It is worth pointing out that many of the people mentioned in Annex 4 have since passed away or are retired and might prove extremely difficult to trace. In 1984, however, many of these people were to be found in similar positions to those they had occupied in 1942. A decade or more after and the work was published and this was no longer the case; in fact had William Sleigh not undertaken the writing of 'Aircraft for Airborne Radar Development' when he did, much of the final conclusive evidence for the loss of Halifax V9977 may never have come to light at all.

Despite all his fine investigative work however, 'Aircraft for Airborne Radar Development' was never envisaged from its inception for general publication, so to this day it remains an internal publication. It was produced by the engineers from RSRE for their records and for the Air Historical Branch 5 (RAF) to be used as future reference material. By the time I first visited the library at DRA, Malvern, the shadow of secrecy surrounding the documents produced during World War Two had begun to fade. Many thousands of secret documents that pertained to that period were slowly but surely becoming more available to researchers, and for the first time it was possible to shed some light on events such as the loss of Halifax V9977 that had hitherto been impossible.

There were in fact people still on the site in 1996, who, albeit semi-retired, had worked with Blumlein and the others and who remembered that June of 1942, and spoke to me of it as if it were yesterday. They each had a look of some distant loss as they recalled the faces of friends now dead for over fifty years.

In the DERA Malvern Information Centre, the staff have kept, and added to a small file of documents and references, newspaper clippings and vague hand-written notes on Alan Blumlein as well as the loss of Halifax V9977. (It was pointed out to me by William Sleigh that these cuttings by journalists and non-technical people give most spurious reasons for the crash, hence the need for Annex 4. They are interesting, but of little use with regard to the crash and in many cases have propagated the incorrect information regarding the circumstances that surrounded the events of Sunday, 7 June 1942.) Any reference to material published in the years since the crash has somehow found its way by the hands of who-can-say how many former information centre managers, to this file.

Luckily, it also contained many references to documents kept in the archive at DERA from 1942, and even before. Much of the supporting material in this chapter of my book has come from these documents. They have enabled the gaps between the account found in Annex 4, and the previously reported events of 7 June 1942, to be filled. These documents would of course have been readily available to William Sleigh, and were undoubtedly used when putting 'Aircraft for Airborne Radar Development' together.

Having spoken to William Sleigh at this time, he was of the opinion that his specific work, in the form it eventually took, should remain as part of the historical collection at DERA. It is primarily an internal document written about airborne radar development in very broad sense, and does not set out to right the wrongs of historical inaccuracies surrounding the loss of one aircraft. Having said that, he was naturally very keen to see that the information he uncovered, coupled with any supplementary supporting information, should be brought to as wide a public as possible. This should be done in order to correct the assumptions and speculations of the past, brought about through lack of available and accurate information.

## The crash site today

I also wanted to visit the site of the crash myself, if for no better reason than to try to understand the layout of the terrain and the harrowing experience that Onslow Kirby must have gone through that day in June 1942. The land on which V9977 fell is part of the Courtfield Estate, owned by Patrick Vaughan Esq., who kindly gave me permission to visit the crash site in the company of Douglas Kirby, Onslow Kirby's son. I would have liked, of course, to have visited the site with Onslow Kirby, but sadly, he had passed away some months before my visit, before I had had the chance to talk to him personally.

The roads around the villages of English Bicknor and Welsh Bicknor are windy and narrow, and it would be easy to get lost trying to follow a map to the small private road that leads down to Green Farm and Coppet Hill. It was in the high summer of

1996, when I made my first visit to the crash site. On a Sunday, when the weather was much as it must have been on 7 June 1942, I stood in the exact spot where Onslow Kirby had witnessed the stricken aircraft fall. It was almost fifty-four years to the day since the crash of Halifax V9977. There is no monument or marker to indicate the location in the field itself, which is still farmed and regularly ploughed. Had it not been for the fact that I was looking for a place where an aircraft had crashed, the actual site could be said to blend perfectly into the rest of the countryside unnoticed. If you look more closely however, and if you have a guide who is aware of the significance of the place, then you can find the point that I had come to see.

There is a clearly visible indentation in the earth at the exact spot where the fuselage hit the ground which, despite the years that have passed, acts a poignant reminder of the events that took place there. Onslow Kirby kept a small plastic bag in which tiny fragments of V9977 were wrapped, brought up in the years that followed the crash during the ploughing of the field. There were twisted cartridge shell cases, a small fragment of Perspex and a twisted and burnt piece of metal, probably part of a strut. After his death, Kirby's plastic bag and its contents disappeared, and with it the last tangible memories of the crashed aircraft.

The path that V9977 took as it headed across the treetops towards Kirby would have taken only few seconds of time to pass. His view of the River Wye as it meanders through the shallow valley would have been obscured by the small ridge just away in front of him on the route to the riverbank. Beyond this, on the far bank, are the trees themselves which were cleared by just 20 feet. The treeline hides any view of the road or the village of English Bicknor, which lie just behind the ridge. At this point, the ridge forms part of a large, steep embankment on the south side of the river, and rises up to Court Wood and Raven Cliff. A narrow public footpath, part of the Wye Valley Walk, follows the southern bank of the River below the ridge, as it passes directly opposite the point where the Halifax fell. It is a path surprisingly little walked today, even in this picturesque part of Monmouthshire so often used by ramblers.

I found myself standing there in that field for quite some time, silently wondering what must have gone through the mind of Onslow Kirby faced with witnessing so cataclysmic a scene. Other than the knowledge that I possessed of the tragic events that had taken place on that day in June of 1942, it is fair to say that there is nothing of great distinction about this particular spot. Only that it is a quiet, beautifully serene, picture of English country life; timeless and peaceful, probably unchanged from that day to this – a day when the peace was so cruelly and suddenly shattered.

**Author's note**

Accident report W-1251 contains material which, due to the many hundreds of accidents at the time, cannot be considered scientifically correct. In addition, spurious aircraft sightings such as that over Usk at 16.00 h can cause confusion for future scholars. Therefore, accident report W-1251 should only be considered in the light of the misleading evidence which, due to the constraints of the time, were placed before the board.

# Chapter 11
# Legacy

## Aftermath

After Halifax V9977 was lost at Welsh Bicknor, just south of Ross on Wye, on Sunday 7 June 1942, there was little time for grieving. The loss of the aircraft with the one and only prototype magnetron-based H2S unit was bad enough, but the loss of nearly the entire H2S development team was incalculable. Lovell questioned at the time whether H2S could even be continued – he considered the loss of Blumlein especially a 'national disaster'; but this was war, and H2S was far too promising a development to be discarded or even put on hold. Doreen Blumlein had been at home on the Monday morning when the news of Alan's death was brought to her:

'I was alone in London with the children. Actually, I was putting clean sheets on the bed, because he was coming home that day. The afternoon before, I'd been to see his mother who was staying in Ruislip, and driven over with the children, and of course that was when he was killed, but they didn't tell me.

It was ten in the morning, and I saw two cars draw up. One was Shoenberg and Lady Shoenberg, and the other was Condliffe and his wife. I thought I knew what they had come for. And of course, Shoenberg's face would've told anybody anything. They didn't know a lot then. They only knew there had been a crash you see. So, he said to me, "There's just a chance that one of them may have bailed out", so I said, "It wouldn't have been Alan would it?" and he said, "No, it wouldn't have been Alan". But apparently, Alan was in the tail of the plane, and I think someone recognized him. But the others were ... I mean we had three coffins, but there were three masses of bones I think ... They didn't know.'

Alan Blumlein's death certificate, which is dated 25 June 1942, simply states that cause of death was 'Multiple injuries received while an authorised passenger in an aeroplane which crashed as the result of an accident. Accidental Death'. It was received from the deputy coroner for South Herefordshire following the inquest into the cause of death on 23 June 1942.

The secrecy that surrounded those days is perhaps hard for us to understand now, more than half a century later. Death seemed so much more an everyday occurrence then, that many of the harsh realities of life appear incomprehensible now.

'"I remember he said to me one day, in the war ..." Doreen Blumlein later reflected, "... Now if anything happens to me, you accept the facts and face up to it and carry on", he said, "And marry again within six months". I said, "Oh no, I couldn't", he said, "Oh yes. You'll have to live on, but I shall have gone like the snuff of a candle". He said, "You see, I believe my after life comes from any good or bad that I have done in the world, and in my children". Now that was his belief. I don't think he was right, but then that was his right. That's what he believed. So, he thought that there was nothing after. It was what you'd carried on, what you've done, which is quite a logical thing I suppose for those who believe in that sort of thing. But that was his view of life and religion.'

There was to be no official announcement that any of the men had died. No obituaries. No mention of any kind, in case it leaked to the enemy who were well aware of what it would mean to lose so important a part of the team as Blumlein, Browne and Blythen. Blumlein especially was well known to German intelligence; his work before the war had singled him out for special attention, and leaked knowledge of his death needed to be avoided at all costs.

' "He did tell me that his name was on a list that he would be taken to Germany, if the Germans came" Doreen recalled, "And he told me exactly what I was to say, so that the boys and I would be all right. I can't remember now exactly what I was told, but I was told that I would be all right, but that he would be taken to Germany, he knew".'

It was well known that Hitler had been complaining that British radar was the ruin of his U-boat campaign and the government was desperate that the news of the loss of such an important radar development team did not bring Hitler any solace.

New teams were rapidly deployed by both TRE and EMI, and arrangements made for another Halifax bomber to replace the one that had crashed. Halifax W7711 replaced V9977 on Friday 26 June 1942. This aircraft, which continued to serve with TRE/TFU until its return to the RAF on Wednesday 10 February 1943, completed the H2S research with Halifax R9490. When W7711 returned to active duty with the RAF it was the first aircraft that had been fitted with the H2S radar system for operational use.

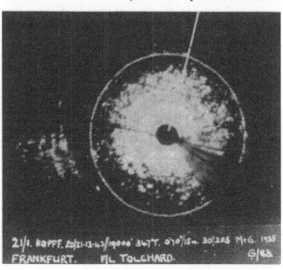

Figure 11.1
An H2S radar screen plot of a raid over Frankfurt, December 1943. (*Courtesy of EMI*)

Figure 11.2
An RAF radar operator
with the H2S indicator unit
in a Lancaster, 1943.

Somewhat ironically, on 17 June 1942, just ten days after the accident another of Blumlein's patents was applied for: No. 574,133 was a series of improvements that he had been working on to the transversal filter. It was applied for in the name of Doreen Blumlein who now acted as proxy to her late husband.

At Hayes, in the weeks immediately following the loss of Blumlein, Browne and Blythen, a discernible hush descended upon the laboratories. There was an utter disbelief that such a thing could have happened; the war, which had always seemed so far away, had now come home to the people of EMI. It was also just beginning to dawn on everybody what an incredible loss these three men had been, and how, how on earth would they, could they be replaced? The H2S project needed to be completed and EMI appointed Eric White to take over the heavy electrical engineering from Blumlein. Felix Trott returned to Hayes and took over the mechanical engineering from Cecil Browne and Maurice Harker stayed at Defford to complete the design and installation work which had been carried out by Frank Blythen.

On Wednesday 10 June 1942, Sir Archibald Sinclair wrote to Isaac Shoenberg from the Air Ministry:

> It was with deep regret that I learned of last Sunday's tragic accident involving the death of three members of your staff, Mr. Blumlein, Mr. Browne and Mr. Blythen. It would be impossible to over-rate the importance of the work on which they were engaged. We, of the Royal Air Force, are deeply indebted to them for all that they have done and were doing to provide us with vital articles of equipment. Their loss must have come as a sad shock to you and it will be deeply felt in the development work that they were carrying out.

Nine days later, on Friday 19 June 1942 at Hayes Parish Church, a memorial service was held for Alan Dower Blumlein, Cecil Oswald Browne and Frank Blythen. The service was of course held with the greatest of attention to security, none the less, it was well attended with far more people wishing to be present than the little church could hold. The service, given by the parish priest, began with the following words:

> 'We meet here together to pay a last tribute of affection to those whom we wish to honour; to express our sympathy with their homes and families in their sorrow; and to declare and strengthen our faith in our Lord Jesus Christ, who is the Resurrection and the Life. To those whose faith rests in the Resurrection

to eternal Life, death is in very truth the gate through which we enter into our glorious heritage, which is the complete answer to all that is hard to understand in man's pilgrimage here on earth.

Today we are commemorating those whose lives were not ended in the ordinary course of nature, but by the cruel stroke of war. We take comfort and hope from our Lord's words, "Greater love hath no man than this, that a man lay down his life for his friends". In the Name of the Father, and of the Son, and of the Holy Ghost. Amen. Let us remember before the throne of grace our brothers Alan Dower Blumlein, Cecil Oswald Browne and Frank Blythen, who have been called to their eternal rest.'

The service continued with several readings, Psalm 23, Psalm 27:1–6, the Lesson from Wisdom 3:1–9, Hymn No. E. H. 428 *'Let Saints on earth in concert sing'*, and Hymn E. H. 426 *'Lead us, heavenly Father, lead us'*. This was followed by two minutes of silence for remembrance. The service was ended with what must have been a very solemn rendition of *'Jerusalem'*. A small, two-page booklet of the memorial service had been hastily printed by EMI and handed out to the mourners at the church. Because of the secrecy surrounding the events, however, these booklets had to be collected in after the service and so they survive, some of them, to this day in the archives at EMI as fresh as the day they were used.

The memorial ceremony of 19 June 1942 would be the only memorial for Alan Blumlein for the next thirty-five years.

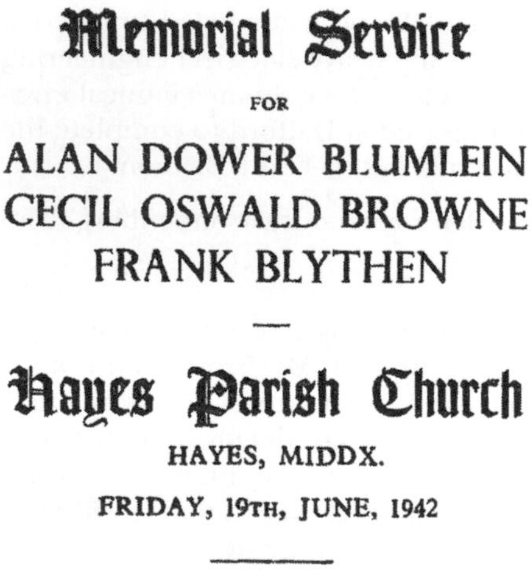

# 𝕸emorial 𝕾ervice

FOR

## ALAN DOWER BLUMLEIN
## CECIL OSWALD BROWNE
## FRANK BLYTHEN

—

# 𝕳ayes 𝕻arish 𝕮hurch

HAYES, MIDDX.

FRIDAY, 19TH, JUNE, 1942

Figure 11.3
The Memorial Ceremony Programme,
19 June 1942.

## Still a war to win

In the weeks following the accident the relevant legal proceedings that surround the deaths of employees of a company such as EMI took place. Doreen Blumlein had taken Simon and David to Cornwall to be near her family some weeks before the accident. She now engaged Messrs Vivian Thomas & Son, Solicitors, of Alverton Street, Penzance, to take care of the Blumlein family's legal needs. On 4 September 1942 Vivian Thomas & Sons received a letter from EMI informing them that the directors of the company had agreed that an investment trust of £1500 for Simon and David Blumlein be set up, payable to them on their 25th birthday, or at any time after the age 21 at the discretion of the trustees. As to the drafting of the deed, this could be done either by a solicitor acting for Mrs Blumlein or a solicitor acting for the company.

In early December 1942 Vivian Thomas & Sons sent a letter to the secretary of The Columbia Graphophone Pension Fund at EMI, to which Alan Blumlein had contributed, in which they thanked them for the cheque of £230 payable to Mrs D. Blumlein. They also mention:

> The Corrective Account referred to in our letter to you of the 15th September last has not yet been filed as we are waiting for details concerning further possible correction in another matter, but this will be done shortly. In the meantime may we part with the cheque to Mrs Blumlein, or do you wish us to keep it here until all the duties are paid?

The Secretary of the Pension Fund wrote back on 10 December 1942:

> In reply to your letter dated 8th December, regarding the cheque for £230, we have discharged our liability by paying Mrs Blumlein through you. It is for you and Mrs Blumlein to decide whether the cheque should be held or otherwise, but we suggest that it be held pending payment of all duties.

Following a final short note dated 16 December 1942 in reply to the letter from the pension fund, the official correspondence, or at least that which was kept in the files at EMI, stopped.

There was of course, still the day-to-day matter of winning the war. EMI had enormous commitments and besides, Alan Blumlein, Cecil Browne and Frank Blythen would have wanted, expected their colleagues to carry on without them. The work was the most important thing here, was it not? Ironically on the day of the crash Prime Minister Churchill had written to Sir Archibald Sinclair, delighted to hear of the initial trials of H2S:

> I have learnt with pleasure that the preliminary trials of H2S have been extremely satisfactory. But I am deeply disturbed at the very slow rate of progress promised for its production. Three sets in August and 12 in November is not even beginning to touch the problem. We must insist on getting, at any rate, a sufficient number to light up the target by the autumn, even if we cannot get them into all the bombers, and nothing should be allowed to stand in the way of this.

Winston Churchill summoned Bernard Lovell to the Cabinet Rooms for a meeting to discuss the matter on Wednesday 17 June 1942. However, during the evening of Tuesday 16 June Lovell was informed that the Prime Minister had, at the very last moment, postponed the meeting despite its urgency, as he had made arrangements to fly to Washington for a meeting with President Roosevelt. Churchill left Stranraer on the evening of 17 June, arriving in America early the next morning.

The meeting between Churchill and Roosevelt had been arranged in order for the

President to explain further developments in what was known as 'the Manhattan Project', America's programme for the development of the atomic bomb. While Churchill was in the United States, in Africa, Tobruk fell on 20 June. This was probably the darkest day of the war for Britain. A fully equipped army of 33,000 men had lain down their arms to Rommel's Afrika Korps of less than half that number. Churchill returned to England on 25 June, following one of the truly historic meetings of the Second World War. He had been briefed by Roosevelt that the atomic bomb was very likely over a year, possibly two, from completion, and that plans needed to be drawn up for the possible invasion of Europe in the spring of 1943, which was considered the very earliest possible date. The war was going to last a lot longer.

The meeting with Lovell to discuss the need for H2S took on a new urgency and Churchill rescheduled it for Friday 3 July 1942. Despite the fact that just the previous day Churchill had survived the end of a furious, three-day debate in the House of Commons as Egypt looked about to fall to the Germans, he turned his attention back to problem of H2S. Lovell's recollections of that day are extraordinary:

'Today if one has business in Downing Street one has to produce passes and the policeman at the Whitehall barrier first finds out by telephone if one is expected. On the morning of 3 July 1942, there were no such safeguards. A few minutes before 11 a.m. Dee and I walked into Downing Street and no one asked us who we were. We then walked through the door of Number 10 and there was no one to ask us why we were doing so.'

The meeting, held at the Cabinet Room at Number 10 Downing Street, was attended by Lovell, P.I. Dee, J. Llewellin (the Minister for Aircraft Production), Sir Archibald Sinclair (the Secretary of State for Air), Arthur Harris (Commander-in-Chief, Bomber Command), Professor Lindemann (Lord Cherwell), Robert Watson-Watt, Isaac Shoenberg from EMI and several others involved in the proposed production or deployment of the bomber force.
Lovell continued:

'Churchill appeared wearing a boiler suit of RAF blue zipped up the front to the neck. "Good morning gentlemen. Come in." He seated himself, back to the fireplace, isolated on one side of the table in the Cabinet Room. He pointed out to us in turn that he must have 200 H2S sets by 15 October. It was explained that H2S did not exist – the only working system had crashed weeks earlier and Shoenberg said it was impossible for him to make 200 sets by mid-October, even if a prototype was available.
      The first of many unsmoked but chewed cigars was ejected over his shoulder to impact on the surrounds of the fireplace behind him. "We don't have objections in this room. I must have 200 sets by October." Four times, he pressed the buzzer by his right and four times the same high ranking army officer entered. "Where are the notes I was reading before I came into this room?" "They are on the table before you Prime Minister." The fourth time he believed that they were in front of him. Another buzz: "Where is the secretary to take minutes?" No one came and so the Minister of Aircraft Production commenced to fulfil this function until a Major General entered. "Prime Minister the Cabinet is waiting", "Tell them to wait and come here to take minutes."
      No one was offering the Prime Minister much encouragement with regard

to producing 200 H2S sets by October. Cherwell was the nearest person to him on his left. He turned, "What does the Professor think?" "They can be built on breadboards", announced Cherwell to our complete astonishment. The PM diverted his attention to Harris. "What does the Air Marshall think?" We were well-acquainted with Harris' fierce objections to the presence of unreliable gadgets in his aircraft and his meek response, "I must have them in Stirlings", finally turned the proceedings into a level of fantasy that bore little relation to the real world outside the Cabinet Room.

Now it was nearly 1 p.m. The Prime Minister said he must see the Cabinet and we must decide how to produce 200 H2S sets by mid-October, "Our only means of inflicting damage on the enemy". Supported by Cherwell's assertion that it could be done on "breadboards" he dispatched us to an adjoining room and said we were not to emerge until we had agreed how to equip two squadrons with H2S by October.'

And so a feverish period of work began again at Hayes and at TRE. Bernard Lovell was brought back down to earth by Churchill who was determined that the work must continue at all costs. Lovell, having rallied what was left of the development team together, then set about picking up the pieces where Blumlein, Browne and Blythen had left off. Flight testing began again in Halifax R9490 on Wednesday 7 July 1942, exactly one month after the crash. Eric White had travelled to Defford the day before to begin the assessment process for the two valves and carried with him a new 1500 volt transformer which needed to be fitted and tested. On the Wednesday two flights were scheduled, one in the morning with White, Curran, Pilot Officer Ramsay and Pilot Officer Davis. While Eric White's notes report problems adjusting the reflector to fill the gap between 1½–4 miles ground distance, they were able to follow strong echoes. In the afternoon Harker and O'Kane made a further flight which reported even better results.

Over the next few days various meetings were held at Defford to determine a course of action. These were held with Lovell, Dee, Lewis, O'Kane, Curran, Cherwell, Bennett, King, Air Commodore Tait and White. It was decided that a new system of reporting the results obtained at Defford back to EMI should be instigated, with Curran in charge. On Friday 9 July 1942 it had been urged that a 5kW magnetron be used in W7711 and Eric White made a note to himself to check on the progress of the 50kW magnetron that EMI were then developing. That weekend White reported that work carried out during the week and bench-tested on Saturday had resulted in a 10–15 times signal strength being achieved. A flight was arranged on the Sunday afternoon with Group Captain Bennett, as pilot, O'Kane, Houchin and White, during which a bridge was seen at a distance of over 30 miles. Eric White wrote in his notes: 'Range about 30 miles (on a bridge, not a town!)'. H2S was beginning to work again with a magnetron as a transmitter.

The following day, Monday 12 July 1942, while Group Captain Bennett went for an early morning flight in the TRE Halifax W7711 (to see the previous day's results for himself), it was decided that Halifax R9490 should have the klystron system removed. In its place a magnetron was to be used as a transmitter, with a spark gap modulator, and the remaining bits and pieces left over from the original EMI prototype units. It constituted a rudimentary H2S system and the first test flights were carried out later that day. Now with both the Halifaxes airborne again, with the magnetron installed and EMI engineering units once again, Lovell and TRE had a system they believed in. The only problem now was how to make it work quickly and accurately enough, and then how to produce enough of the units for installation in the aircraft of Bomber Command before the end of the year.

There were still so many development problems that needed to be resolved during the months following the crash. The magnetron system had several disadvantages (other than the security aspect surrounding it) that precluded its immediate use. For example, it was especially difficult for a navigator using H2S to identify a specific town just from its radar image. This image on the PPI (plan position indicator) might not necessarily bear any physical relationship to the shape of that town on a regular map. Therefore, H2S interpretation was a highly skilled business, often requiring the confirmation of approximate location from other sources before H2S could be used at all. Large features such as coastlines, lakes and on occasions even railway lines were relatively easily identifiable, as was the proximity of the aircraft to a built-up area. However, being specific, as Bomber Command had to be, that was another thing altogether. If the aircraft flew over a very large built-up area the radar returns from the ground were so strong that the screen simply filled up with an intense blur of light, making any identification almost impossible. This was to prove especially true later during operational use in Berlin and its suburbs.

Despite this there was no doubt that H2S would significantly improve results from the likes of Gee and even Oboe on occasions, but it was the cost of these potential improvements that worried the Air Ministry. The work at Defford, EMI in Hayes and GEC in Wembley went on quite literally night and day. That any units were available before the end of September 1942 is something of a minor miracle. Yet ready they were. By 13 September 1942 the first Stirling bomber, N3724, had been fitted with an experimental H2S system and on 21 September Halifax N7808 was assigned to be fitted with the first of the EMI production apparatus. This work was completed on 30 September and from the results a somewhat improved picture on the PPI was dispatched to the Bomber Development Unit (BDU). The BDU navigators found that the serviceability was hopelessly bad, but thought that H2S would be 'valuable to a high extent both as a navigational aid and as an aid to locating targets'. Though Churchill's demands for two squadrons of H2S bombers by October had never been remotely possible, by the last day of 1942 there were 24 H2S-equipped aircraft in 8 Group – 12 were Halifaxes at 35 Squadron, Graveley, and 12 were Stirlings at 7 Squadron, Oakington. H2S, despite having one of the most unfortunate and costly, in terms of lives, development programmes of any airborne device through-out the war was finally ready to take to the air and see action. It was time to put the theory into practice.

With the introduction of H2S radar Bomber Command losses dropped considerably for a while, though eventually German counter-measures were found to jam the H2S frequency. Bomber Command crews however, who had been keeping their H2S sets on all the way to Germany, soon learned their lesson, at great cost at first, but eventually radar operators would only switch their H2S sets on for a few seconds at a time, just enough to get a fix, every six minutes or so. In this way the German night-fighter defences found it very difficult to obtain a fix on the main bomber stream and so direct the German night-fighters on intercept courses.

The first of a series of devastating raids was carried out over Germany in the early part of 1943, in particular the first operational sortie of H2S to Hamburg on 30 January 1943. The weather that night was atrocious and 40 Lancasters of the main force had to turn back because of it. Of the aircraft carrying the precious H2S sets, four of the six Graveley Halifaxes and three of the Oakington Stirlings turned back. This meant that of the aircraft that made it to Hamburg, just six were actually equipped with H2S.

The raid however, from a bombing point of view, was an unqualified success. When the first H2S aircraft returned the crews were enthusiastic. They reported that they had no difficulty in identifying Heligoland, Zwolle, Bremen, Zuider Zee, Den Helder, East and North Frisians, Cuxhaven and Hamburg itself. The average range at which the towns were identified was about 23 miles, the maximum being 33 miles and the minimum being 11½ miles. Coastlines, estuaries and rivers were described as appearing on the cathode ray screen 'like a well-defined picture of a map' and the six navigators who reached the target claimed positive identification of the docks, stating that they appeared as 'fingers of bright light sticking out into the darkness of the Elbe'.

Two nights later, on 2–3 February 1943, 10 H2S 'Pathfinder' aircraft marked Cologne for the main force, while 11 marked Hamburg again on 3–4 February, and again on 4–5 February 8 marked Turin. All seemed to be going well and a telegram congratulating Lovell arrived on his desk from the Operations Commander Pathfinder Force, RAF Wyton: 'Heartiest congratulations from myself and the users to you and you collaborators on the development of the outstanding contribution to the war effort which has just been brought into action'. By normal RAF standards of the time, the Hamburg raid suffered minimal aircraft losses, just eight aircraft failed to return. This was almost certainly due to the accuracy of the new radar and the minimized flying time over the target.

Inevitably, an aircraft carrying the magnetron system was lost and then recovered by German Intelligence. Luckily this didn't happen until March 1943, over Rotterdam. The wreckage of the aircraft had been inspected thoroughly and the recovered equipment taken to the Telefunken Laboratories to be re-assembled. However, by total coincidence, just as this was being completed, the facility was raided by Bomber Command and the captured H2S equipment badly damaged again. Once again, the engineers at Telefunken Laboratories, headed by Professor Brandt, rebuilt the battered magnetron oscillator, this time in one of Berlin's many air raid bunkers. It was here that Brandt and his team, having discovered how the magnetron worked, linked it to a CRT and were able to watch, in amazement, one night in March 1943 as an actual air raid took place, exactly as the aircrews of Bomber Command flying above would have seen it. Using time-lapse photography, they filmed the raid from the CRT. That night 33 aircraft were lost during the raid.

The significance of what he had seen was not lost on Brandt and he wasted no time in informing the German Intelligence authorities, who realized the magnetron oscillator being used against them could go a long way towards explaining why Bomber Command were suddenly so accurate with their bombing missions, and why Coastal Command were as accurate when attacking their U-boats at sea. Over the next year the allies shipping losses steadily fell and U-boat losses to anti-submarine warfare increased dramatically. Wellington aircraft of Coastal Command, fitted with 10cm ASV, could detect submarines long before the aircraft itself was seen or heard and as a consequence U-boat commanders adopted a new strategy of remaining submersed, only surfacing in the dead of night or at a safe port on the French Atlantic coast.

Once the allies re-occupied France after 6 June 1944, the U-boats lost the use of these ports too, and with practically every bomber coming off the production line being fitted with H2S by the autumn of 1944, the U-boat offensive ground to an almost complete halt. Although H2S had originally been designed as a 'blind-bombing' device, its use throughout the war as a navigational aid, a detection system for U-boats as well as a targeting device for the bombing of German cities, had proved its versatility. When the war finally ended in 1945 H2S was considered by Churchill as the single most important scientific development of the entire conflict. As Rowe had exclaimed when holding the still wet photographs of those very first cathode ray tube images from 1941: 'This is the turning point of the war'. It was.

During the last years of the war two further Blumlein patents were filed posthumously. The first of these, Patent No. 582,503, is dated 15 October 1943 and resulted directly from the work that was carried out with Frederick Calland Williams, Eric White and Maurice Harker on the AI Mk VI system. The patent is co-written with Williams and Eric White, with Doreen Blumlein as legal representative of the deceased Alan Blumlein. The work is actually Blumlein's second largest patent (second only to No. 517,516 of 1938) and comprises some 24 pages, with 14 diagrams and 37 claims. The invention comprises the basis of the entire radar system which was derived from the 'Spot display and walking strobe' system. It was capable of searching for and locking on to a target and then tracking that target as it varied in range. The system was designed so that it could be carried in an aircraft and thus became the most advanced form of AI that Blumlein worked on. It was capable of producing strobing pulses of varying lengths that would scan a full range of frequencies until an object was detected. Once an echo was received this would be locked onto and tracked. The radar could determine if the object that had been tracked was friend or foe, whether it was an aircraft or a ground station transmission and, as it moved in range or height, could give the observer constant indications on cathode ray tube.

The second of Blumlein's patents applied for after his death, is his last, and 128th patent, No. 595,509. It is somewhat ironically dated 12 May 1945, just days after the end of the war in Europe. Co-written with EMI, Eric White and once again with Doreen Blumlein given as legal representative of Alan Blumlein, the specification describes a modification to an earlier patent, No. 535,778 (not Blumlein's). By comparison to some of Blumlein's achievements this patent is somewhat plain, describing as it does a simple circuit for the stable production of pulses using a resistance and in another example a simple network. Blumlein, as he had throughout his career, had returned to his experiences with telegraphic and telephonic networks.

With the war over and won, there was hardly likely to be anybody asking questions such as was the enormous effort, the ultimate price, the sacrifice made by Blumlein, Browne, Blythen, Hensby and the crew of V9977, not to mention everyone else involved in the project, the months of tireless work they carried out getting H2S into Bomber Command's aircraft, was it all justified? In the end, H2S proved to be every bit the device that the team of scientists who had given their lives in its pursuance believed it to be. Not only did the unit become fully operational with the Royal Air Force, and many others throughout the Second World War, but variations of the magnetron-based H2S radar system continued to be developed, improved and used right up to the Falklands War of 1982.

At this distance in time, there is little point in asking if it was justified. It happened. I personally believe that Blumlein, his colleagues and every person involved in the project would not have changed a single occurrence, even those that were tragic, if it meant that the ultimate achievement of winning the war was, in any way, jeopardized.

## After the war – occasional mentions

At the end of World War Two life slowly began to get back to some semblance of normality for the people of Britain. The one formula which the government wanted very much to reinstate in order to generate some relief from the trials of rationing and rebuilding, was that of television. In June 1946, television returned with a fanfare and, of course, has continued to be broadcast ever since. That same year the Rank Organisation released a film entitled 'School For Secrets' which starred Ralph Richardson, Raymond Huntley, Richard Attenborough and John Laurie. The film also happened to be the directorial debut for one Peter Ustinov who co-wrote the screenplay which centres around a group of scientists, 'boffins' as they are called, who are responsible as a team for many of the radar breakthroughs during the Second World War. Though the film uses fictitious names in places for the boffins, many of the key personnel, such as Henry Tizard for example, are referred to. The movie has four main themes, the development of Chain Home, Airborne Interception, H2S and the raid on Bruneval to capture German radar technology.

Though Alan Blumlein is not mentioned at any point in the movie, his character, Edward Watlington (played by David Tomlinson) is greatly involved in the design and development of H2S. There is a Halifax flight which ends in disaster when the Halifax flies too low during the first test flight of the new H2S system, and inadvertently hits the guy-line on a barrage balloon, causing the aircraft to crash, killing everybody on board. Sadly, the film has rarely been seen, and it is not available on any video medium yet. Though its historical accuracy is questionable, it is, none the less, an interesting attempt just one year after the war ended to explain to a curious public how some of these marvellous inventions came about. While somewhat melodramatic in places (and curiously disjointed in terms of dialogue), as a directorial debut for Ustinov, 'School For Secrets' is an important piece of cinematographic history (in America it was released as 'Secret Flight' referring more directly to the loss of the Halifax perhaps?). As it is not readily available, but well worth investigating, the film can be seen for a modest fee by making an appointment with the British Film Institute in London.

At EMI in 1946, work restarted on consumer interests as well as the electronic components which had served the radar industry so well during the war. EMI turned their attention again to music as their major source of income and this would eventually lead to the company becoming one of the largest manufacturers of recorded entertainment in the world. Blumlein's recording/reproduction system was also finally phased out in 1946, being replaced by the RS-1 (Recording Studio Model One) disc cutting units, which had been developed from Blumlein's original by engineers Barry Waite and Harold Davidson. The RS-1 extended the high frequency response to 10kHz to match Decca's stunning new FFRR 78s which had appeared the year before. EMI, however, still had no place for binaural recordings, and in 1949, the rights to Patent No. 394,325, lapsed. This was extended a little due to a Government decree that compensated manufacturers for losses incurred during the war years. By December 1952, however, when the rights lapsed again, for the last time, EMI had still not found a use for Blumlein's masterpiece of invention. Perhaps EMI were only too aware of the prediction made by K. de Boer of the Philips Research Laboratory in 1941, 'When the time comes to make use of binaural reproduction...it will become necessary in the first place to find a process of making binaural records on a large scale.'

In the years that followed Alan Blumlein's death, his family, friends and colleagues slowly found ways to deal with their loss as time made it more tolerable and eventually bearable. They had their own lives to re-build after all. Careers, which had been put 'on-hold' so-to-speak, now needed to be picked up again and attention turned to the less dramatic interests of a post-war world. In 1948, Henry Clark and Philip Vanderlyn published an IEE paper titled 'Double-ratio AC bridge with inductively-coupled ratio arms', in which they gave the majority of the credit for the work rightly to Alan Blumlein. The paper was based upon the unpublished documents, which Blumlein had worked on in 1941, describing his bridge in its various forms, but which had lain dormant in some EMI folder since his death. Clark and Vanderlyn 'discovered' this work and produced the paper which was presented before the Radio Section of the Institution on 11 January 1949, proclaiming: 'The material in Part 1 of the paper is the original work of the late A.D. Blumlein, B.Sc. (Eng.), Associate Member, and the present authors have used verbatim a considerable portion of a hitherto unpublished memorandum written by him in January 1941.'

In 1952, Stanley Preston published an article titled 'The Birth Of A High Definition Television System', in the *Journal of the Television Society*, in which Blumlein's work was referred to throughout. At the end of the paper, Preston writes:

> You would, I am sure, like me to say something about Alan Blumlein, who was such an outstanding member of the EMI Research team. I first met him when he came to EMI from Columbia in 1931. He was then an experienced telephone engineer and an expert in acoustics. He had completed the development of an entirely new sound recording system for Columbia and this system was soon used for all EMI recordings. When he turned his mind to television at EMI, he produced so many inventions of importance that it is difficult to select from them. He invented resonant return scanning as early as 1932 and, together with E.L.C. White, he added the now well-known H.T. boost in 1936. He had a large share in the conception of the DC reinsertion circuit and the many pulse circuits, which had to work out for the successful application of electronic television. His inventions were not confined to the circuit field and his name appears on many of our patents relating to vacuum tubes and acoustics. Although Blumlein made many inventions, I think he always regarded himself as being an engineer rather than an inventor. His view was that a good circuit engineer should be able to design circuits to meet a given specification without any need for extensive experimental adjustments after they were constructed. He was consequently a great advocate of negative feedback to overcome variations on performance, which

were normally caused by valves and other components having variable characteristics.

Although he had the gift of lucid exposition, unfortunately, he did not find time to write many papers. The few he did write are models of clarity and I think his paper on the EMI waveform, which he read to the IEE in 1938, is a classic. He was a man of amazing energy and enthusiasm. He was not content simply to invent solutions to problems but always wanted to try out the solutions in a practical way and then to put them on a proper engineering basis if they showed any promise. This same energy and enthusiasm were equally apparent in all his other activities. He was most stimulating in his conversation on any subject. He was also a great favourite with young children because he could enter into their fun especially if they happened to have toys of a mechanical or electrical nature. When the war commenced in 1939, his talents were applied to matters of greater national importance than television. He was mainly concerned with the most difficult airborne radar problems and he met his death while flight testing experimental radar equipment in 1942. He was then only 38 years old. Two of his colleagues from EMI, C.O. Browne and F. Blythen also died with him. Had he lived he would have risen on the highest ranks of his profession. Even though his life was so short he will long be remembered as a man who made outstanding contributions to British television.

On the last page of Preston's article in the *Journal of the Television Society,* there appeared a photograph of Alan Blumlein giving his address before the IEE in 1938. Incredibly this is but one of only four known photographs taken of Blumlein that exist today. For some years after this, very little was heard about Blumlein, though of course his work continued to be applied in all manner of applications. It was not until the mid-1950s that the name of Alan Blumlein would come to the fore again when experiments with dual-channel audio were once more being carried out by a number of companies intent on developing a new high fidelity audio reproduction system.

Suddenly, with the introduction of reasonably priced consumer tape machines in the mid-1950s, all the companies who were involved in music production were looking for a method of cornering this new and lucrative market. EMI once again turned their attention to binaural recording with Philip Vanderlyn saying on 18 January 1954: '...it is desired to take the first steps towards building up a library of binaural tapes'. EMI, however, were not the only company who were working on two-channel audio at the time. At Decca, their long-time rivals, engineer Arthur Charles Haddy and a team of electrical and audio engineers, had spent five years producing a two-channel audio system, which they called 'stereophonic' sound. At almost the same time as EMI were resurrecting Blumlein's binaural, Decca were making the first stereo recordings (by pure coincidence the two companies commenced recording sessions one day apart; Decca in Geneva on 17 May 1954, EMI at Abbey Road on 18 May). Haddy and the Decca team were however, unaware that Blumlein and EMI had already been granted a patent for the work as long ago as 1931, and when Decca, who had spent quite a considerable amount of money by the time the system was ready, came to apply for a patent, they discovered that they were infringing upon Blumlein's work in every way.

Though Arthur Haddy was naturally disappointed that the work of the team at

Decca had not achieved the desired patent, he was none the less satisfied that the work carried out in experimental recordings by both companies from mid-1955, through to early 1958, led to the introduction of commercially available stereo recordings by 1960. Sadly, Blumlein's original name of 'binaural' was not considered appropriate for the new system, and 'stereophonic' sound adopted as a kind of compromise. In 1957, Henry Clark, Philip Vanderlyn and Geoffrey Dutton published a paper for the IEE titled 'The "Stereosonic" Recording And Reproduction System'. Again, rightful reference was made at length to the work of Alan Blumlein. One of the more ridiculous events that surrounded the commercial launch of stereo was the press originally calling it an American invention, something which the editor of *The Gramophone*, in the issue of April 1959, took some pleasure in a scathing attack on the press, pointing out that Blumlein and binaural were British through and through. In 1980, Arthur Haddy paid the ultimate tribute to Alan Blumlein, saying, 'Everything he did was twenty years ahead of his time.'

While there were still no references in any text books, much less tributes, to the technical achievements of Alan Blumlein in the years following the introduction of stereo, there were occasional articles in which his work was recalled, usually by a colleague who had known him and who felt that some form of written account should be forthcoming. One such article appeared in *Wireless World* in the September 1960 issue written by the highly acclaimed engineer M.G. Scroggie, who, though he had not known Blumlein personally, acquired much information about him from those that did.

Encouraged by others such as Stanley Preston, Henry Clark, J.B. Kaye and various other colleagues, Scroggie produced a six-page article entitled, 'The Genius of A.D. Blumlein', in which gave an all-too brief synopsis of the work Blumlein had carried out from 1927 to 1942, with an even shorter account of his early life, education and qualifications. Short though the article is, it represents the first serious attempt at documenting the life and works of Alan Blumlein. While it contained many inaccuracies (probably brought about by the passage of time and the lack of direct communication with the Blumlein family, although Scroggie did know, and was considered a friend by Doreen Blumlein), it has all the basic elements in place and included and explanation of several of his patents, many which had not previously been published.

Scroggie accompanied the article with several diagrams, mostly circuit diagrams, which had been taken directly from Blumlein's patents and which served to better illustrate the points which were being outlined. The article points out that Blumlein possessed a genius which had not yet been appreciated and Scroggie used Carlyle and Edison to highlight the point:

> Genius was defined by Carlyle as, first of all, transcendent capacity for taking trouble. It was analysed by Edison as 1% inspiration and 99% perspiration. These both corrected popular ideas on the subject by emphasising the part played by hard work. But that 1% is just as essential. If there have been people whose brilliant originality died with them, there were many more whose slogging failed to make up for their lack of imagination. The thinking of most of us is shaped by concepts we have received from others. When the genius comes along, with thoughts that break out into new concepts, his fellows often find him hard to understand. They may even oppose him, because he doesn't

conform to their ways of thought. Matters are made worse when, like Heavyside, he is unable or unwilling to make his ideas clear to the less intelligent.

If, as I am convinced, the genius of A.D. Blumlein is not yet widely enough appreciated, that is certainly not the reason. His exposition was exceptionally lucid. The trouble is that so very little of it was published. His contribution to technical literature amounts to little more than two IEE papers, the first shared with Prof. Mallett and the second with several colleagues. (Both of these papers incidentally, were awarded IEE premiums). He was too busy to write. So technical literature is the poorer and his name seldom seen by his successors. Then the last few years of his work were shrouded in wartime secrecy and his career was cut short at the early age of 38 in the service of his country. Even that fact was not published until more than three years later and then only briefly.

Besides this, he avoided rather than encouraged publicity to such an extent that photographs of him are almost non-existent, unless one includes a back seat view of him addressing the IEE in 1938. Some originators are commemorated in the name of a device, law or discovery – for example, the Hartley circuit, Ohm's law, and the Hall effect. Unfortunately, none of Blumlein's frequently mentioned inventions bear his name. Although it appears here and there in literature, probably very few even of the workers in the same field, and especially the younger ones, have any idea of how far ahead he was in so many important developments. How many present-day stereo fans, for example, realise that the system of recording brought on to the market in the last year or two was invented by Blumlein in 1931?

I have therefore attempted a review of the more important of his inventions, as a modest tribute to his memory. Not having had the privilege of knowing Blumlein personally, and lacking close acquaintance with some branches of work in which he excelled, it is a regrettably poor one, but I hope better than nothing. Lest my emphasis on technical achievements give the impression of a one-sided individual, it should be mentioned that, notwithstanding them all, Blumlein was thoroughly human and found time for such relaxations as flying, practical astronomy, enjoyment of music and theatre.

Scroggie's article would undoubtedly have been widely read by many of Blumlein's former colleagues at EMI and ST&C; *Wireless World* was, after all, the foremost publication of its kind at the time, and the tribute certainly rekindled an interest in the work of Alan Blumlein at a time when one of his most important inventions, binaural sound, was being enjoyed as 'stereo' sound for the first time by the many. Whether as a direct result of the article written by Scroggie, or as part of the general interest in stereo sound, or even perhaps from the persuasions of a fiend or colleague who had known Alan Blumlein, read the article in *Wireless World* and realized that here was a man deserving of recognition we cannot know. What is certain is that there now followed a period of activity which continued throughout the early 1960s which would lead to the first attempts being made at writing a serious biography of Blumlein's life and an appraisal of his work.

On 25 January 1963, Sir Isaac Shoenberg died at the age of 82. In a tribute given some eight years later at the inaugural Shoenberg Memorial Lecture, J.D. McGee said of him:

'He was an intellectual in the best sense of that much-abused word. He loved ideas, and to discuss ideas, in a wide range of subjects. His logic was rigorous, but his conclusions were tempered by a very human kindness. He did not mix easily with strangers and seldom received from the press the sort of notices that his achievements merited. Only once did I see him relax with the press – at the 25th anniversary of the opening of the BBC Alexandra Palace station. His great contributions to British science and technology were eventually recognised by the award by the IEE of the Faraday Medal in 1954. He was appointed to the board of directors of EMI in 1955. His great work for television was finally recognised by his adopted country when he was knighted by the Queen in 1962 – alas, within only a year of his death.'

## The 'Blumlein biographer' dilemma

At first, the question of who should write a biography of Alan Blumlein was seemingly easily solved. In 1966, just as interest in his work was surfacing again for the first time in over twenty years, it became apparent that one Basil J. Benzimra, an engineer and a Fellow of the Institute of Electrical Engineers, had decided that enough time had passed without a satisfactory account of Blumlein's work, and took it upon himself to write the book, he thus became the first 'official' Blumlein biographer. Benzimra suggested that if anyone had information pertaining to Alan Blumlein, that it be sent to his eldest son, Simon Blumlein, for safe keeping, during the time that he (Benzimra) prepared the material for a biography.

Benzimra's first published article on Blumlein appeared in *Electronics & Power* magazine in June 1967, under the title 'A.D. Blumlein – an electronics genius'. It is a fairly comprehensive account of every one of Blumlein's 128 patents, which are listed (incredibly, Benzimra actually gives three patents with the wrong numbers. This may however, have been a result of a typing error by *Electronics & Power* magazine), with a brief explanation of their content in this seven-page article. It was also the first published document to contain the now obligatory photograph of Alan Blumlein, wearing the tie of an old Cholmeleian, and looking very serious (it reminds most people of the kind of picture they have made for their passport photograph). This picture, believed to be the only such photograph of Blumlein, would appear on the pages of just about every article about the man in the years to come, and is, incidentally used as the cover photograph for this book.

The article has obviously been researched well with additional photographs of John Collard in Switzerland in 1927, several of Blumlein's inventions including the twin ribbon microphone, the binaural recording head and a copy of the letter he sent to Columbia in September 1931, thanking them for his, and his colleagues bonus. On the last page of the article is a reprint of the hand-written notes by Blumlein, clearly dated 6 June 1942, the day before he died. Benzimra ended his article by expressing his thanks to Doreen and Simon Blumlein, and requested that any additional information should be sent to Simon Blumlein at his London address.

The article received great acclaim, and everybody who had known Blumlein or waited for an official documentation of his life looked forward eagerly to the publication of Benzimra's forthcoming biography.

In 1968, Mr Rex N. Baldock, an audio consultant and member of both the IEE and the British Kinematography Sound and Television Society, had an idea for a joint, belated tribute in conjunction with the Royal Television Society, which would eventually be held on Wednesday, 1 May 1968.

The meeting was completely over-subscribed with more than 200 members and their guests turning up to hear speeches from a number of speakers, each of whom was given (as it turned out a totally inadequate) 10 minutes. Douglas Birkinshaw (then Assistant to the Director of Engineering at the BBC) was Chairman of the meeting; Professor James McGee (who was then Professor of Applied Physics at the Department of Physics, Imperial College of Science and Technology, London) spoke on Blumlein's contribution to television; Dr Eric White (then Chief Scientist at EMI Electronics Limited) talked about circuit design; Eric Nind (of the Power Klystron Division of EMI Electronics Limited) gave a speech about Blumlein's contribution to sound; Miss Joan Coulson (the Chief Recitalist at EMI Electronics Limited) spoke of early stereo recordings; Peter Clifford (the Head of the Standards Laboratory at Hawker-Siddeley Dynamics Limited) talked on measurements; L.H. Bedford (then Director of Engineering in the Guided Weapons Division of British Aerospace Corporation) gave an appreciation of a man he was privileged to call a friend; and Simon Blumlein, Alan Blumlein's eldest son, closed the meeting.

An account of the events of that evening were to appear in the July 1968 edition of the *British Kinematography Sound and Television Journal*, which detailed the proceedings and gave a complete listing of the speeches. Once again the frontispiece of the article had the standard photograph of Alan Blumlein, accompanied next to the text of their speech with a smaller photograph of the speaker in question, and in the case of E.L.C. White and P.M. Clifford, various circuit diagrams to illustrate the inventions they mentioned. In the case of E.A. Nind, several photographs were reproduced of the moving coil microphone that Blumlein had invented as well as elements of the Columbia Recorder/Reproducer.

Douglas Birkinshaw, who had of course been given the job of Chief Engineer at the BBC London Television Service at Alexandra Palace in 1935, introduced the other speakers that evening with a short account of Blumlein's life, his achievements and the tragic events of 7 June 1942:

'Many of us have subsequently speculated what would have happened in the science of television after the war, and of course in radar as well, had Blumlein lived to continue his inventions and contributions to science. This, however, was not to be. It is startling to discover that during his working life his vast range of activities resulted in no fewer than 128 patents. In my view, and in the view of the many colleagues with whom I have often discussed Blumlein, sufficient honour has yet to be paid to his life and work. It is with that object that we are here tonight.'

Following his speech, Birkinshaw introduced Professor James McGee, who had known Blumlein since joining EMI in 1932. McGee proceeded to praise the man he had, at first, not always seen eye to eye with on some occasions, but who he grew to know, like and respect greatly during the years they worked together developing the high definition television system at EMI:

'It is a great privilege but a sad task for me to speak about the late Alan

Blumlein. The twenty-six years since his untimely death have tempered the tragedy for us but they have not in any way dimmed the lustre of his life and his work. I feel very privileged to have known him as a friend and to have worked with him as a colleague for the last ten years of his life, and I will say a little about him in both of these relationships.

He was not a great Adonis, he had no great social graces, but he had an integrity of character and of mind which I think was the key to his personality. It was impossible to imagine him ever taking a mean advantage of anyone else. Over and over again I have found that he made his decisions only after the most careful and unbiased consideration of all the facts that he could discover about the subject, and his decision was taken quite irrespective of whether it was to his own immediate advantage or not. This characteristic was I believe reflected in his work. He strove for perfection, fundamentally sound work which should stand the test of time, and he abhorred the superficial, flashy, spectacular or anything in the nature of a stunt. You have already been told that he was educated at Imperial College as an electrical engineer – I thought that he was educated as a light engineer – and that he did some post graduate work in communications. When I joined Imperial College thirteen and a half years ago there were still some teachers active on the staff who vividly remembered him as a student; the kind of student that makes a teacher's life a happy one.

He and I first came into contact in 1932 when I joined EMI and in those days there was, as someone has put it, an iron curtain between the engineers who used electron valve devices and the physicists who were concerned with the 'innards' of these tubes. I think perhaps one should call this a glass curtain. In those days I think science and technology were more compartmentalised than they are today when radio engineers become astronomers, nuclear physicists become power engineers, and I even know one architect who became a world-famous electronic engineer, and so on. However, I had been trained as a nuclear physicist and it was only the exigencies of the great economic depression that threw me into industrial research on the physics of electron devices; this was a field at the time almost unknown to me but I always regard that as one of the lucky breaks of my life. So it was that while I and my many colleagues were concerned with the internal workings of the tubes required for television, for example cathode ray tubes, photo-cells, photo-multipliers, and later on, television camera tubes – this was mainly physics – Alan Blumlein and his colleagues were responsible for the integration of these into a complete practical engineered system of television. I have no doubt that he would today be called a systems engineer.

But Blumlein's questioning mind recognised no barriers, either iron or glass, and he very soon became very interested in our tube problems which were very pressing at the time, I can assure you. I vividly remember how he would push open the door of my office and say in a diffident tone of voice, "Do you mind if I make a silly suggestion?" Of course, his suggestions were anything but silly and this was usually the prelude to a very interesting animated discussion of some idea or problem. These discussions were of immense help to me and demonstrated quite clearly that even if he had not been trained as a physicist he knew more physics than most physicists. I am sure that my tube colleagues, such as the late William Tedham and the late Leonard Klatzow, and those who are happily still with us, like Dr Broadway, Dr Lubszynski, Dr Bull and others, would all agree that these discussions contributed much by way of stimulation and clarification of ideas. Anyone who has tried to do research will know how valuable such discussions can be, especially if they are with a person whom you can trust not to snatch ideas and it is in this respect that Blumlein's integrity was of paramount importance; it gave everybody that confidence.

He contributed many valuable ideas and was inventor or co-inventor of a number of important new features of television tubes of different types. One of these inventions, that of the cathode potential stabilisation of television camera tubes, has a rather interesting history, which I have not published previously but I think it is worth mentioning in the context of this evening's meeting. There had been some confusion in the company about the procedure for patenting ideas and as a result of this I had had rejected a couple of ideas that I thought might possibly be useful. One of these was the idea of cathode potential stabilisation. I was certainly not clear exactly how to do it at the time. Very soon, Mr Condliffe, the manager of the laboratories, came to see me with exactly the same proposal, which had been formally submitted by Blumlein. Now this might easily have led to friction. However, as soon as Blumlein was told of the situation he very generously and quite amicably agreed that it should be patented by himself and myself as co-inventors, an example of his generous attitude in any such situation.

Another example of his tremendous interest in anything experimental is illustrated by a story of our wartime activities. As you know, he was involved in the development of airborne radar, which resulted in his having to do a lot of flying by day as an observer. At this time, I was involved in other work, which involved me flying as an observer in night fighters. Of course, Blumlein was a pilot himself – he had taken his pilot's licence – and he was very interested in, and of course made a detailed study of, aeronautics as well. I remember telling him about my particular work one evening driving home from Hayes to Ealing and after a while he turned to me with a look of genuine friendly envy on his face and he said, "Oh Mac, I do envy you the chance of flying at night in a night fighter. It must be very interesting". He was so interested in anything of this sort that was going on around him.

He had a tireless enquiring mind. Every subject to which he turned his mind was analysed in great detail and he was not satisfied until he had it all quite clear. He was forever going back to fundamentals. One remembers how, in discussing circuit problems over the lunch-table, he would repeatedly – I remember it as clear as if it were yesterday – be going back to Thevenin's theorem. It is for these reasons I believe that so many of his ideas have been so important and in many cases have only borne fruit many years after his death. Only last week in his Royal Society lecture on technology, Dr F.E. Jones, Managing Director of Mullards, commented on how greatly Blumlein's ideas had contributed to the practical development of television since 1945. The development of his ideas on stereo recording and reproduction is of course another parallel story.

It has been said that Blumlein was weak in mathematics. This may have been so in a technical sense but it was also true of Rutherford, Bohr and Einstein. Einstein was not a good mathematician; he used to have to go to his mathematical friends to get help in solving his mathematical problems. However, Blumlein had what I believe is the essential attribute of mind for success in scientific invention and discovery, which is the power of vivid and accurate imagination, almost a sixth sense, which seems intuitively to understand how Nature works, and a mind at the same time disciplined by careful study of all that is already known. Now in this respect I believe it is not stretching credibility to compare Blumlein to Rutherford. I came from working under Rutherford to working with Blumlein, and so I was able to compare the two men at close quarters. Professor Denis Gabor suggested in his inaugural lecture at Imperial College that Rutherford might, *mutatis mutandis*, have been a great inventor – that is, a Blumlein. It is my opinion that Blumlein, given the right circumstances, could have been a Rutherford.

I think my conception of him and the view of him that I would like to leave with you is best summed up in a few words spoken about him by Sir Isaac Shoenberg a few days after his tragic death, which were as follows: "There was not a single subject to which he turned his mind that he did not enrich extensively." '

McGee was followed by Eric White who proceeded to mention the Benzimra article which had appeared in *Electronics & Power* in the June of the previous year:

'Benzimra's article ... lists the 128 patents with very useful brief statements of content. I should like to illustrate my theme by reference to a small but diverse selection of patents, taking first a few from television equipment, and then a few from radar equipment.'

White then went on to list and evaluate seven of Blumlein's patents, illustrated with circuit diagrams, which were: Patent No. 400,976 (1932), Energy Conserving Scanning Circuit; Patent No. 462,530 (1935), Constant Resistance Capacity Stand-Off Circuit; Patent No. 482,740 (1936), The Long-Tailed Pair; Patent No. 517,516 (1938), Transversal Filter; Patent No. 589,127 (1941), Delay-Line Modulator for Radar Transmitter; Patent No. 580,527 (1942), The Miller (or Blumlein) Integrator; and Patent No. 585,907 (1939), Improvement of Signal/Noise Ratio in Radar.

Eric White concluded his presentation with the following words:

'I hope that these few examples have given some idea of the breadth of Blumlein's interests in circuit design and his ability to produce radically new and aesthetically pleasing solutions to circuit problems. In conclusion, I should like to say that I shall always remember with gratitude and affection the nine years from 1933 to 1942, when I had the great privilege of working closely with Alan Blumlein, first on television circuits and later on radar.'

It was then the turn of Eric Nind to speak about his recollections of Alan Blumlein:

'I am sure that after listening to those very interesting lectures there can be no doubt left in our minds that Blumlein's work was a major factor in getting television off the ground in this country. But we must certainly not lose sight of his important contributions to audio engineering in general and to gramophone recording in particular.'

Eric Nind went on to describe, in detail, the work that had been conducted on the recording and reproduction system for Columbia, followed by the binaural work and the moving-coil microphone. The contribution to television sound was not overlooked either, with mention of the fact that Blumlein had developed television sound for Alexandra Palace based on his own system.

Following Eric Nind, Miss Joan Coulson, who had not known Blumlein personally, but who had joined EMI just in time to write the script to introduce the first stereophonic recordings being made there, played to the audience two of Blumlein's binaural recordings made in 1933, with Westlake and Jessup. These were followed by two tape recordings made in 1955, when EMI were 're-discovering' binaural, or stereo sound, as it had by then been named. Peter Clifford, who like Miss Coulson had not known Blumlein personally, then explained several of Blumlein's contributions to electrical measurements. Clifford outlined in simple terms the Wheatstone bridge and compared this with Blumlein's contributions. Several diagrams were reproduced to demonstrate this and other circuits.

Finally, a personal note was added to the evenings events from L.H. Bedford who, like J.B. Kaye, had known Blumlein since his days at Standard Telephones & Cables. There followed a discussion period held by the Chairman and finally an address from Simon Blumlein on behalf of the Blumlein family, thanking everybody for turning up and their contributions to what had been a most successful event.

Following the British Kinematography Sound and Television Society tribute in conjunction with the Royal Television Society, on Wednesday, 1 May 1968, Benzimra continued to acquire even more material for his biography which, he hoped, would be ready in 'a couple of years'.


## Another biographer?

Then in 1970, just as it seemed that a biography of Alan Blumlein might at last be on the way, Basil Benzimra became quite ill and announced that he was unable to continue work on the project. It looked as if the Blumlein biography would not happen after all. Out of the blue, a second writer came to the attention of those who sought a book about Alan Blumlein; his name was Francis Paul Thomson, and he was destined to become one of the most controversial characters in the entire Blumlein story.

Francis Paul Thomson was a junior member of the EMI team during the days when Blumlein was working on the high-definition television system, and only decided to take over the mantle of biographer following the 'prompting' and eventual 'persuasion' of others behind him. It would seem that Thomson, who had written a book in 1968, titled *Money in the Computer Age*, was contemplating writing another book on the early pioneers of radio and television, including Alan Blumlein. When he had heard that Benzimra had been forced to give up the project, Thomson decided he would take up the reins. He therefore became Alan Blumlein's second 'official' biographer before a word had, proverbially, been published. So began a nearly thirty-year saga of inactivity, and eventual controversy that has continued to beguile and frustrate academics, researchers, family and friends of Alan Blumlein, myself included, to this day.

Born in December 1914, Thomson worked as a laboratory junior at EMI from

16 May 1934, until 1942, when he joined special operations forces, remaining with them until wounded during 1944. After the war, Thomson began the work which would eventually lead to the setting up of the Post Office and Bank Giro system in Britain, an achievement for which he was awarded the OBE in 1975. Thomson took it upon himself to write the biography of Blumlein through persistent conversations with Doreen and Simon Blumlein.

Thomson, like so many others at EMI during this time later explained that he felt,

> '...privileged to work in the EMI Research Laboratories of which Alan Dower Blumlein was the Chief Engineer. Like many others, I was immediately impressed by his abilities and, indeed, by his entire personality. When secrets of the 1939 to 1945 war could be discussed in public, and when stereo, television and radar were helping to shape a new world order, I searched for a biography of Blumlein, but found little except brief obituaries. I was persuaded to write a biography and, during research for it, found there was a widespread feeling that he should be commemorated nationally.'

Francis Thomson eventually received approval from both Doreen and Simon Blumlein, and began his research in earnest in early 1972. Around Easter of that year, Thomson learned from a phone call he received from Russell Burns, at Trent Polytechnic, that Burns was writing a thesis on Alan Blumlein, and that if they could help each other in any way, he would be willing to offer what information he had gathered to Thomson since 1971, when he started work on the project. For a while Thomson contemplated giving up the idea of writing a biography, but when he was given an assurance by Burns that he had no intention of writing a biography of his own, and that Basil Benzimra had also given up the project, Thomson once again spoke with Simon Blumlein, and the biography idea was 'born' again.

There is no doubt that at this stage in his work Francis Thomson very probably had every hope of producing a concise work on the life and achievements of Alan Blumlein. He even had specially printed headed note paper drawn up to proclaim the fact that he was the official biographer – it read: 'Alan Dower Blumlein 1903–42, Engineer Extraordinary. A biography by F. P. Thomson, 39 Church Road, Watford, WD1 3PY, Herts, England. (telephone number given) In association with Simon J.L. Blumlein, 13 Heathfield Road, Petersfield, Hants, England.'

In June 1972, Thomson began his research with a series of appeals, one of which, in the form of a letter to the editor, appeared on the pages of *IEE News*, dated 23 June 1972. It read:

> Sir: During his comparatively short career, Alan Dower Blumlein (1903–42) contributed to electronic engineering progress a stupendous wealth of ideas, in addition to his many inventions. There is scarcely a radio, television, radar, or other electronically actuated instrument or system which does not utilise at least one of his ideas.
>
> Although known globally as a name on patent specifications, A.D. Blumlein is practically unknown as a man, except to a fast-dwindling small band of engineers, physicists and chemists, and Armed Serviced, Post Office and broadcasting personnel who had contact with him during his years as chief engineer of EMI research department at Hayes, Middx. A biography of Mr. Blumlein is

in preparation. All who had contact with him, particularly those who knew him from his schooldays onwards, or would like to make an assessment of his position in the history of technology, are invited to write to me. Communications received before the 30th November 1972, will be specially welcome!

In reply, Eric White sent Thomson a copy of his BKSTS address in May 1968, and Thomson followed this with a letter dated 9 September 1972, in which he outlined a draft of the material he was preparing for his dissertation, asking if White would like to contribute additional information. As this is the first known contact between Thomson and White, it is somewhat strange that Thomson addresses White with the impression that he, Thomson, knew Blumlein well, mentioning three times in two pages 'we who knew, or worked with (him)', and later, 'those of us who knew Blumlein'. Strange, since few of Blumlein's contemporaries can recall Thomson at EMI (he occupied an extremely junior position), or the work that he carried out while there. Eric White, who worked with Blumlein closely for many years, and who was renowned for his excellent memory recalled that, 'Thomson was so very junior, I am surprised that he ever spoke with Blumlein at all. For a while Thomson worked with Maurice Harker and myself, but he was the lab boy and one doesn't imagine a lab boy in conversation with the Chief Engineer'. Maurice Harker pointed out, 'I remember hearing that Thomson had undertaken to write Blumlein's biography, and being absolutely amazed that he would even consider such a task. He rarely met Blumlein to my knowledge, and as a very junior lab assistant, he would hardly have been in regular contact with the man.'

The remainder of Thomson's letter goes on to state that it would be: '... of great help to hear from you not later than November 30, 1972, even if you have not assembled all data.' What was the significance of the 30 November? Surely Thomson was not intending to give some form of publication deadline so soon after taking on (and announcing for that matter), a biography of this complexity? The letter then contains a series of 'Notes for the guidance of contributors', essentially 56 points, some of which are data Thomson had already collected, and which he evidently needed corroborated; the remainder formed the basis of a questionnaire.

Eric White telephoned a few days after receiving the letter, and a few days before leaving for Australia, to inform Thomson of what help he might be able to give. On 12 September 1972, Thomson wrote to White once more thanking him for the help he had given, and then promptly accounted over two pages a synopsis of how he, Thomson, was in some remote way related to Blumlein; how he too had attended Gothic House School and that Miss Chataway, the principal and supposedly Blumlein's 'godmother', was a '...good friend of some relatives' of Thomson's, keeping in touch until 1932. Quite why Eric White would want to be informed of this information is unclear, however, Thomson then goes on to give a potted history of his working career, ending with a further request for information from White regarding the administrative system at the EMI Research Department between 1930 and 1945.

Eric White's reply on 20 September 1972, included photocopies of some notes: '... almost certainly in Blumlein's writing, dated 5, 6 & 7 June, 1942' and points out that he had little to do with the administrative system at EMI in the period mentioned. Attached at the back of the reply were the answers to the first 18 (those relevant to White), of the 56 points Thomson had requested in his original communication.

Thomson's activities over the next few months included a visit to the hillside at Welsh Bicknor in the Wye valley in April 1973, to see for himself where Halifax V9977 had crashed, and further contact with EMI informing them of his progress. EMI even sent out a memorandum on 13 April 1973, informing staff that any help that could be given to Thomson would be appreciated. It had been decided early on between Thomson and Simon Blumlein that, for archiving purposes, all material associated with Alan Blumlein and his work discovered by, or donated to Thomson, should be sent directly to his address in Watford.

The year 1975 saw the holding of the 50th Audio Engineering Society Convention, in London, from 3 March to 7 March, at the Cunard International Hotel, Hammersmith. As part of the celebrations for the society, a special meeting of The Royal Society was proposed for Wednesday 5 March 1975 as a commemorative programme on the life and work of Alan Blumlein entitled, 'A.D. Blumlein: Inventor Extraordinary', with much of the organization once again done by Rex Baldock. Throughout the autumn and winter of 1974, Baldock had been contacting former colleagues of Blumlein to see if they would be willing, as many of them had done in 1968, to appear before the Audio Engineering Society for the purposes of giving a speech on Blumlein and his achievements. Naturally, all agreed. One of the people Baldock was very keen to have speak again was Eric White, and wrote to him on 17 November 1974, to ask if he could attend. Curiously, in his letter to White, Baldock mentions that: '...two other speakers will take part, viz. your colleague Mr P.B. Vanderlyn and Mr F.P. Thomson, who, as you know is writing the Blumlein biography (I gather he was your assistant).'

The meeting, which was held at The Royal Society, 6 Carlton House Terrace, London at 7 p.m. in the Wellcome Lecture Hall, was again well oversubscribed. The hall itself seated 274, but many stood in the aisles surrounding those lucky enough to have seats, when George Millington, Chairman for the evening began by introducing the planned events. The first speaker was Francis Thomson who gave a speech on the early life of Alan Blumlein; he was followed by Eric White who talked on Blumlein's leadership in television. Philip Vanderlyn was next, speaking of Blumlein's contributions to developments in audio, and this was followed by a short demonstration of some of the early sound recordings that Blumlein had made including some binaural work. There was a period of general discussion, which was followed by an appreciation of the meeting by Simon Blumlein and the proceedings closed by the Chairman. The meeting was extremely well attended, and well received. A week later, on 12 March, Simon Blumlein, using the same headed paper that Thomson used to proclaim the fact that he was the 'biographer' of Alan Blumlein, wrote to Eric White: 'My Mother joins in me in thanking you for your contribution to the meeting last Wednesday. Your comments on informal discussion among a development team certainly brought the hands together and I hope they will return to their work and put their affirmation into practice. I was also pleased to hear about the long-tailed pair as this was one of the main circuits used by Open University in the course I did last year.'

Following the AES Convention in London, it was generally thought that some form of official commemoration should mark the work that Alan Blumlein had achieved. To this end a petition was begun, probably Thomson's idea, but certainly with help from others, for a blue plaque to be placed on the wall of the house in Ealing that he had called home before he died. The petition however, would take the best part of a year before the Greater London Council, responsible for such things, would agree to the choice of person.

## Blue plaque

By the beginning of May 1976, the Greater London Council had been persuaded by some of Blumlein's former colleagues, Barry Fox, EMI and Francis Thomson to honour Blumlein's memory by placing a blue plaque on the wall of his former home at 37, The Ridings, Ealing. Although the process of approving the recipient of this, one of London's nicest forms of landmark, had been initiated, the wheels of bureaucracy would take yet another year before everything would be in readiness.

By early 1977, the process for the placing of the blue plaque at 37, The Ridings, Ealing, Alan Blumlein's last home, was well under way. Thomson was quite insistent that: '... every endeavour must be made to integrate up to, say, 12 student or trainee electronic engineers, of either sex and of any race or colour, nominated for their self-motivation, initiative, sense of humour, natural versatility, and outstanding academic record. The 12 will be chosen by an impartial panel.' EMI made an official announcement to that effect in early March and the nominations began to come in from the many companies who had candidates that they felt might suit the occasion. Thomson began sending out invitations to the event at the end of March 1977, though his earlier letter to W.E. Ingham, Director, EMI Research, of 16 March, gives evidence that the person chosen to unveil the plaque had, at that point, still not been made: 'The plaque will be unveiled by a person of great eminence on Wednesday, 1 June 1977, and I shall be sending you further details of time, etc. when these are known.' The original guest list ran to some 125 people, not all of whom were expected to attend; though it seems some provision was made for remote possibility that they all would. The Ridings, in Ealing, is a small road, and 125 people standing in the middle of it would bring the entire neighbourhood to a complete halt.

By 10 May 1977, the choice of who was to unveil the plaque had been made, and Dr Percy Allaway, Chairman, EMI, sent an internal letter to the directors informing them that Professor Sir Alan Hodgkin OM, FRS, Nobel Prize winner and former President of The Royal Society, had accepted the honour. The impartial panel had sat and chosen the 12 candidate engineers to attend; they were informed at the beginning of May: 'On June 1st, 1977, a GLC 'blue plaque' is to be unveiled on the house in Ealing where (Blumlein) last lived. It is the first such plaque ever to commemorate an 'electronic engineer and inventor'. The Organising Committee asked academic and industrial bodies to put forward the names of young people whose start in the field of electronics suggests that they, in their turn, may have the ability to make noteworthy contributions to the electronics industry, with a view to inviting then to be present at the ceremony and so to extend the 'living memory' record of the occasion. Your name has been put forward and accepted.'

One of the first publications to announce the forthcoming unveiling of the blue plaque for Alan Blumlein was *New Scientist* magazine in its 26 May 1977 issue with the following notification: 'The GLC is to unveil one of its blue plaques, at 37 The Ridings, Ealing next Wednesday. This was the last home of audio pioneer, Alan Dower Blumlein who, before dying in a wartime air crash, made pioneer inventions in the field of television and sound recording. Stereo records as we know them

Figure 11.4 The 'blue plaque' at 37, The Ridings. (*Courtesy of EMI*)

today were first conceived by Blumlein in the early 1930s, and probably the world's first orchestral stereo music was recorded by Sir Thomas Beecham at Abbey Road in January 1934.'

EMI had a press release drawn up in late May, which read:

> Tribute to a genius – Alan Dower Blumlein 1903–1942. Today, Wednesday, 1 June (1977), a GLC commemorative 'blue' plaque was unveiled at No. 37 The Ridings, Ealing, the home of the brilliant engineer Alan Blumlein, who was killed in 1942 in the 2nd World War at the age of 38. One of the most outstanding technologists of his generation, Blumlein has never received the recognition which he deserves for his revolutionary work in the fields of tele-

phone engineering, gramophone recording, stereophonic and quadraphonic sound, television and airborne radar, most of it carried out at EMI's Central Research Laboratories at Hayes, Middlesex.

It was while working on the vital development of airborne radar – which was instrumental in helping win World War II – that he was killed. Blumlein was one of a small party of engineers who died in an aircraft, which crashed in the West Country while flight testing airborne radar. Blumlein's death had to be kept quiet. British radar was far in advance of that of the axis powers. It was essential that Britain should keep its lead in this new technology and avoid any leak of a set-back reaching enemy intelligence. It is hoped that the unveiling of this plaque will belatedly bring to the public's notice Blumlein's historic achievements in the field of electronics.

The day of Wednesday 1 June 1977 was bright and sunny, quite unexpected for a British summer since remembered for being so wet. The plaque unveiling ceremony was begun by Dr Percy Allaway, CBE, Chairman, EMI Electronics Limited at 12.30 p.m. who then handed over to Sir Alan Hodgkin, OM, KBE, FRS, SC.D Cantab, Fellow of Trinity College, Cambridge and a Nobel Prize winner, to unveil the small blue curtains. Sir Alan Hodgkin then began to speak on behalf of Blumlein's critical role in the U-boat war:

'One of the pleasant things about British cities is the custom of erecting blue plaques on houses which once were the homes of the great. In London, the decision about which individuals should be commemorated rests with the Greater London Council. As you can imagine the selection committee's task is a difficult one and searching inquiries are made into the candidate's history and achievements. It is a source of great satisfaction to me that Alan Blumlein, whom I admired both as a person and as an extremely talented electrical engineer, should be honoured in this way. In my brief talk I shall try to give you some idea of Blumlein's achievements as well as of his qualities as a highly inventive scientist and engineer.

But first I should answer what may seem a somewhat irrelevant question. Some of you may be wondering why I, who am a biologist, should be speaking about an electrical engineer? I would guess that one reason is that I was caught up in centimetric radar for five years during the war and was much involved in applying it to various devices in fighters and bombers. Several people that I knew, including Alan Blumlein as well as a close friend of mine, Geoffrey Hensby, were killed when the Halifax bomber carrying an important radar prototype crashed in the Wye valley. The Halifax crash, which was probably due to a fire in one engine and caused the death of 11 people, was a particularly bad one. The plane contained the prototype of a centimetric radar set which was eventually installed in a large number of bombers and coastal command aircraft. In bombers the purpose of the set, which was called by the code name H2S, was to enable aircraft to find German towns at night. In this it was very successful. But the device was to have an even more important role against submarines when used in coastal command aircraft.

The first 10 cm sighting of an enemy submarine was made on 17 March 1943, and the first attack on 18 March. By June 1943, the effect of this and other measures was decisive and shipping losses which had risen to 400,000 tons in March 1943, were reduced to less than 50,000 three months later – only a year after Blumlein was killed. Soon afterwards U-boats were withdrawn from the North Atlantic and, as Churchill remarks, Hitler complained that this single invention (centimetric radar) had been the ruin of his U-boat campaign (this was an exaggeration).

Besides Blumlein, those killed in the crash included his EMI colleagues C.O. Brown and Frank Blythen, as well as a distinguished young physicist from TRE, Geoffrey Hensby. I would also wish to refer to the pilot, Squadron Leader Barrington (sic), who had been involved in flying centimetric radar since the beginning of 1941. I mention these reminisces of the war because they illustrate the importance of the work in which Blumlein was engaged at the time of his death. But now I must tell you something of Blumlein's life and of his great achievements in electrical engineering. For much of this information I am indebted to Mr Blumlein's colleagues at EMI and in particular to his biographer, Mr F.P. Thomson.

Alan Dower Blumlein was born in Hampstead in 1903. His father was a naturalised British subject and his mother the South African born daughter of a Scottish missionary. Alan Blumlein's father died in 1914 and he determined to help his mother by condensing his education. After three years at Highgate School, he won a bursary to the City and Guilds College where he obtained his B.Sc. as an external student. This must have been a great strain but he obtained first class Honours and was appointed a junior demonstrator. At 21 years of age, he joined the International Western Electric Corporation in London, the first which eventually became Standard Telephones & Cables. He was soon put in charge of a group of engineers and sent to France to resolve problems on the Paris to Madrid telephone circuit. This involved commissioning the first high-quality telephone lines, the so-called music lines. From Standard Telephones, Blumlein moved to the Columbia Graphophone Company and hence to EMI where he spent most of his working life.

Just over 50 years ago, Blumlein applied for a patent on the novel method of reducing cross-talk interference in telephone systems. Eighteen months later his second patented invention, the closely coupled inductor ratio arm bridge, laid the foundation for a family of important measuring devices. These two patents were the first of a series extending into a wide range of fields. They include such familiar electrical circuit devices as the long-tail pair, the Miller integrator in a time base and use of the cathode follower to reduce the input capacitance of a valve. There were also many important mechanical and acoustical inventions to whose application I shall refer shortly. By the time of his death, 128 patents had been granted – that is an average of about 1 every 6 weeks. Several of the later inventions had important application to the air defence of Britain, first in connection with sound-location and later with radar.

In a sense Blumlein's versatility makes it difficult to do justice to his genius. Today he was what might be called a systems engineer, that is a person who looks at the engineering and cost effectiveness of some development as a whole, and does not confine his attention only to the transmitter or receiver. In the 20s and 30s such people were rare and Blumlein was very much a pioneer – indeed one might refer to him as the first systems engineer. I can best illustrate his quality by referring to two major achievements. Less that a month ago the BBC inaugurated a series of quadraphonic broadcasts. This development with which we are all so familiar is a linear descendant of Blumlein's invention of stereo sound recording and reproduction. Blumlein called this technique binaural sound and it is only latterly that we have used the word stereo. But not only did his patent specification refer to the invention of stereo; it also included an outline of quadraphonic and ambisonic techniques. Stereo recording would have been useless without improvements in the methods of making gramophone records to which Blumlein made a series of very important and ingenious contributions.

The next field to which he turned his attention was the development of a high-definition television system with 405 lines as opposed to the 240 lines which a government committee had recommended. Forty years ago, the BBC chose the Marconi-EMI television system devised by Blumlein and his research team. On 12 May 1937, this system proved itself to the world by transmitting the coronation procession of George VI successfully in spite of a very dark overcast day. Much of Blumlein's system can still be found in our present system. We may equate his name not only with the first high definition television public service but also with the foundation of subsequent television systems.

Apart from being a brilliant engineer, Blumlein was a good friend to many people and a man of great charm, generosity and intelligence. I did not know him well but saw enough of him to realise that he was a wonderful and inspiring leader of the world-famous EMI team. When he talked about electronic circuits in his vigorous way, they seemed to come alive. Not only did this make working with Blumlein a lot of fun, it also created the climate in which new ideas are generated. If a patentable invention resulted from joint work, he would share the credit equally with all who had helped. He had absolutely no side and would work alongside a junior assistant throughout the night if there was a crisis. He hated humbug and laziness but would put himself out to help anyone he felt was genuinely trying. He enjoyed drinking beer, had a colourful range of expressions and a quick temper which was equally quick to subside. He clearly enjoyed life and communicated that feeling to anyone near him. Blumlein was devoted to his family and we are very happy that his widow, Mrs Walker and her husband Major Walker are with us today. We extend a warm welcome to them and to his two sons, Simon and David.

I have referred several times to Blumlein's width of interest both as an electrical engineer and as a scientist. He was a truly multi-disciplinary engineer of the kind badly needed in an age of specialisation. His death 35 years ago was a tragedy but his family and friends can derive some consolation from the great benefit and pleasure which Blumlein's inventions have given to countless people both in this country and abroad.'

Following the speech from Sir Alan Hodgkin, Francis Thomson gave a short speech also:

'Some of you who have been corresponding with me about this event have been expecting to receive an invitation from me, rather than from EMI, so I should like to give a slight history of the circumstances that led up to this ceremony. From 1935 to 1942, I was privileged to work in the EMI Research Laboratories of which Alan Dower Blumlein was the Chief Engineer. Like many others I was immensely impressed by his abilities and, indeed, by his entire personality. When secrets of the 1939 to 1945 war could be discussed in public, and when stereo, television and radar were helping to shape a new world order, I searched for a biography of Blumlein, but found little except brief obituaries. I was persuaded to write a biography and, during research for it, found that there was widespread feeling that he should be commemorated nationally.

A petition to the Greater London Council, to ask them to help commemorate Alan Blumlein in his native London, by erecting a blue plaque where he had lived, was successful, but I soon found so many people wished to attend the unveiling of the plaque, that it was beyond my means to shoulder the responsibilities. Hearing of my dilemma, the Chairman and Managing Director of EMI Electronics Limited, Dr Percy A. Allaway, who is this year's President of the Institution of Electronics and Radio Engineers, very kindly agreed to be your host for today. While reviewing the list of guests, it seemed that today's gathering might all be drawn from the older age-group, so we felt it was necessary that the torch lit by Alan Blumlein should be carried on to the coming generation. With the advice of EMI Electronics, a small committee was established to select 12 bright young engineers from nominations made by the Post Office and various companies.

These young people have been invited to join us today, and to regard themselves as the likely Blumleins-to-be of the latter quarter of this century. I wish to acknowledge with heartfelt thanks the immense encouragement and help given in the planning and preparation of this event by Mr Felix Trott, late of EMI Electronics Limited, by the Blumlein family and Simon J.L. Blumlein in particular, by Mr John Earl of the Greater London Council's Department of Architecture and Civic Design, and by the staff of Ealing Town Hall.'

The twelve young engineers chosen from various organizations in the private and public sectors of industry and from universities, whom Thomson referred to as the 'Blumleins of the future', were: T. Williams, Marconi Radar Systems Limited; R.G. Brockbank (24), Post Office; N.J. Playford (24), Imperial College of Science and Technology; P.R. Miles (23), Marconi Communication Systems Limited; J.M. Alexander, University of Strathclyde; Mrs. L. Waters (24), Standard Telephones & Cables Limited; P. Pitcher (25), Standard Telephones & Cables Limited; G.R. Knight (19), Post Office; D.E. Eastwood (24), EMI, R&E Division; C.J. Barratt (21), EMI Central Research Laboratories; B.J. Rook, EMI SE Laboratories; and G. Ash (23), EMI Electron Tube Division. Everybody then retired to the Carnarvon Hotel where a buffet lunch was held.

Before the unveiling had even taken place however, there was talk of a further commemoration of some kind for Alan Blumlein, and in a letter dated 26 May 1977, Thomson wrote to EMI suggesting that they discuss on 1 June, the possibility of a permanent memorial: 'The installation of a memorial plate in either Westminster Abbey or St Paul's Cathedral', or 'The raising of a fund that would provide sufficient interest from invested capital to provide a suitable large amount of money to encourage radar engineers to present papers annually'. On 3 June 1977, W.E. Ingram replied to Thomson apologizing for the fact that the matter had not been discussed during the unveiling two days earlier, and that he was not sure that he would able to contribute much to the proposed committee looking in to the possibilities of such a memorial.

Following the unveiling, there was a short period of intense public and journalistic interest in Alan Blumlein, much as Sir Alan Hodgkin had hoped there would be when he made his speech. A small article with a picture of the plaque appeared in

Figure 11.5 The Blumlein family at the unveiling of the blue plaque on 1 June 1977. Left to right: David Antony Paul Blumlein, Doreen Walker (formerly Blumlein), Simon John Lane Blumlein, his wife Anne M. Blumlein and their children Alan Blumlein, James A. Blumlein and Charles R. Blumlein. (*Courtesy of EMI*)

the July 1977 issue of *IEE News*, a letter to the editor appeared from Rex Baldock in the July/August edition of the *Journal of The Royal Television Society*. Baldock, who had been present at the unveiling, pointed out how gratifying it was to see the blue plaque finally as a memorial to Alan Blumlein, and that perhaps The Royal Television Society should hold a meeting devoted to Blumlein and his work in television since there were evidently many in this field who knew him still active and able to contribute.

The August and September editions of *Hi-Fi News & Record Review* contained a two-part article written by Adrian Hope which examined the work of Alan Blumlein and this was followed by a short article in the September 1977 edition of *The Cholmeleian*, the internal magazine of Highgate School, which Blumlein had attended of course, written by I.F. Davies, who had also attended the unveiling in June; the article is a brief account of the life of Alan Blumlein and his achievements. Much of the text, which ran to just two columns, was taken directly from Sir Alan Hodgkin's speech, the transcript of which had been made available to the press following the June event. One gets the impression that others knew a great deal more about their celebrity former pupil than the staff of Highgate School did, and only at the end does Davies add: 'Highgate can indeed be proud of a man whose work not only made one of the major contributions to the Allies' success during 1939–1945, but which today, and will for many years to come, have an influence on our lives in the broadest sense.'

That month, the *New Scientist* carried the following article in 'Ariadne', in its 22 September 1977 issue:

This year has been the year of Alan Blumlein, probably the least appreciated but most impressive electronics genius this country has ever produced. The GLC has now finally honoured his memory with a blue plaque and EMI, his old employer, has publicised his pioneering work on stereo sound, television and radar. In fact, it was Blumlein's involvement with radar that prevented earlier recognition of his talent. He was killed on a secret research flight in 1942, which was for many years hushed up. Even now, the full truth may not have emerged. The official story goes that Blumlein was testing H2S radar when the plane crashed. But why was the plane flying so low if it was testing this type of radar? One informed suggestion is that the plane was not in fact testing radar at all. It was, instead, testing what was probably Blumlein's only unsuccessful invention, an altimeter that worked on ground capacitance. Such an altimeter could only have been accurate at very low levels, and would inevitably have been upset by clouds and humidity. Did Blumlein's plane fly just that bit too low? Surely there must be some contemporary engineers still around today who know the answer.

Even the government got involved with written material on Blumlein. The October 1977 issue of *Spectrum*, the magazine produced by the Science Unit of the Central Office of Information, a publication which was sent to British Embassies all over the world, contained a piece by C.L. Boltz, entitled 'An unsung British genius'. Then, on 12 October 1977, Mrs Rachel Nelson, group public relations officer, received a letter from William Paul Lucas (who was always known as W. Paul Lucas), the administrations manager at EMI, Central Research Laboratories which raised the first serious questions about the work that Francis Thomson was supposedly carrying out on the 'Blumlein biography':

Mr F.P. Thomson some years ago was very active in collecting information for a biography on A.D. Blumlein. The story he was writing was to include a number of other personalities and cover the whole range of technical activities and personal impressions of his day. He contacted a lot of people and collected quite a mass of data I believe. Mr Thomson's interest in A.D. Blumlein has not waned and he was a prime mover in promoting the memorial plaque in Ealing you remember. It occurs to me to ask you if anything is known about the progress being made with the book. If perchance he has run into some difficulty or for some reason decided to abandon the project I trust the information collected could be retrieved because it must be of some historic interest to EM. If you could tactfully ask Mr Thomson how the biography is coming along we might learn about his intentions.

In the interim period, while EMI tried to find out exactly what it was Thomson was doing with all the material he had collected, and just how the biography was 'coming along', Lucas decided to organize a small exhibition on the second floor of the EMI building at Hayes. In a letter to him of 6 December 1977, it was reported that much of Blumlein's original equipment, the Columbia cutter and various microphones among it, had been delivered to Philip Vanderlyn for inspection as some was suspected of being damaged slightly. Vanderlyn contacted the AES, who were keen to honour Alan Blumlein in the forthcoming centenary edition of the *Journal of the Audio Engineering Society*. It had become obvious fairly quickly that the British section of the AES not only had little reference to Blumlein's audio work, but possessed inadequate knowledge of his achievements, a matter which Vanderlyn proposed should be rectified. By January 1978, work on the exhibition at EMI had been finished and the company were busy inviting students and anybody who had an interest in not only Blumlein, but the entire EMI

Figure 11.6
William Paul Lucas. (*Courtesy of EMI*)

team from that period, to come along and see for themselves what had been accomplished.

Sir Alan Hodgkin's speech which had been given at the unveiling the previous June was now reproduced in its entirety in the January/February 1978 edition of *National Electronics Review*, and then in September 1978, following an intense year of Blumlein activity, Philip Vanderlyn's article appeared in the *Journal of the Audio Engineering Society*, entitled 'In search of Blumlein: The Inventor Incognito'. Vanderlyn had written with Henry Clark and G.F. Dutton, in September 1957, an article for the IEE entitled 'The Stereosonic Recording And Reproduction System', which attributed the majority of its content to the work of Alan Blumlein and specifically his binaural patent, No. 394,325, of 1931. Now, in 1978, Vanderlyn was to write, rather enthusiastically:

> Of all the great engineers who have contributed significantly to the improvement of our living standards that we are pleased to call civilisation, few have suffered more from persistent anonymity than Alan Dower Blumlein.
>
> Engineers as a class do not rank high in the consciousness of their fellow men who are their beneficiaries: most people have probably heard of Brunel and Telford, Watt and Stephenson, the Wright Brothers and perhaps even a few more, but these are the exceptions. The fact is that engineers are specialists who have to do with 'things' rather than people, and the majority of people are too busy with other people over matters such as politics, finance or merchandising to concern themselves with people who deal with 'things'. The 'things' themselves – be they water supplies, civic works, oil tankers or airliners – are gratefully accepted, but the engineers who design them remain unknown or forgotten. How many people today could name the inventors of the transistor or the hovercraft? Engineers are fortunate in that they mostly derive great personal satisfaction from their work, nonetheless they do comprise an appreciable part of the Honors List. So who, you may ask, was Alan Dower Blumlein?

The article, which is six pages long then goes on to document Blumlein's early history and education before outlining his achievements under the headings 'Telephony', 'Recording', 'Stereophony', 'Television', and 'Radar'. Apart from the same photograph of Blumlein that Benzimra had used in his article in 1967, several archive photographs from EMI of elements of the Columbia Recorder/Reproducer are used to illustrate the work.

One might well ask why had Vanderlyn chosen to use the *Journal of the Audio Engineering Society* to publish his tribute? The answer is that of all the engineering fields in which Blumlein worked, audio engineering and audio engineers were among the very few who, by 1978, had continued to attribute his name directly to inventions and techniques that they used. To this day, many audio engineers use the term 'Blumlein microphone technique', to describe stereo or M-S microphone placing. It also likely that most audio engineers will be aware, perhaps above others such as electrical engineers, that Blumlein was responsible for inventing stereophonic sound.

Vanderlyn's choice of publication therefore served an already appreciative readership, and undoubtedly raised the awareness of an informed audience. What it did not do was to bring to attention of a wider audience the achievements of the man. For that a book was required.

# Chapter 12
# To Goodrich Castle and beyond

More than ten years had passed since the challenge of writing the 'official' Blumlein biography had been taken up. Yet still no book had been published, nor did there seem any remote chance that it might appear in the near future. It was true that through petitioning, largely by Francis Thomson, a blue plaque had been unveiled at Blumlein's last house in Ealing, in 1977. But with his health failing, his age fast approaching 70, and his attitude becoming more and more erratic, the staff at EMI, the family and friends of Alan Blumlein, all of whom had put their trust in Thomson, were becoming increasingly concerned for the long-term safety of the material, and the project as a whole.

## Another decade passes

Following the publication of the *Journal of the Audio Engineering Society* in 1978, the Blumlein publicity trail then strangely went very quiet for a long time. Then, on Thursday, 5 March 1981, Russell Burns gave a lecture entitled 'The life and work of Alan Blumlein', which several Thorn-EMI staff had attended, and which raised the question once again, where was the Blumlein biography? It had been some time since anything positive had been heard from Francis Thomson who, so it would seem, was busy engaged writing a book about tapestry. Felix Trott, Philip Vanderlyn and Maurice Harker had received an enquiry from a Mr R. Maude, principal technician at Huddersfield Polytechnic, who was conducting research into the audio developments of Alan Blumlein, and had been told that he should contact Francis Thomson, the 'official' biographer.

Maude wrote to Thomson on 14 September 1981, explaining his work, and requesting answers to two specific questions regarding Patent No. 394,325. Thomson's reply on 28 September 1981, is typical of the man's attitude at this time; he says:

> ...if you really want me to seek out any data of the sort you have requested – let me have copies of the data supplied to you by Messrs. F. Trott and P. Vanderlyn, and M. Harker, so I do not duplicate data you have already received. I have accumulated about a hundredweight-and-a-half about A.D. Blumlein and his activities, and ancestry going right back to the early 15th century and, although most of this is catalogued to facilitate biography writing there is also a great

bulk of the material which will get only passing reference or which will have to be searched for data such as you and many others have or are requesting.

In his reply of 29 September, Maude explained that Huddersfield Polytechnic had already celebrated the centenary of recorded sound in 1977, with a display, part of which was a celebration of Blumlein's work, and which he now wished to update. His specific questions related to the: '...confusion which surrounds the date at which his work on stereophonic recording and reproduction was started. The purpose of the enquiry is to "set the record straight" and satisfy my personal curiosity.'
Mr Maude had evidently not reckoned with the likes of Thomson, who sent a further letter to him on 3 October 1981:

Dear Mr. Maude, Thank you for your letter of 29.09.1981 from which I note that you are apparently not prepared to save me from seeking out and sending you data which may in some part duplicate what you already have from other sources. I do not intend to waste time and expense by sending you duplicate information or papers. Adding now to the reasonable request made in MY letter to you of recent date, I should like you to tell me (for Blumlein ancestry reasons) WHY – of all Polytechnics – that of Huddersfield should have decided to explore Sound Recording in particular and to give so much attention to A.D. Blumlein?

I must express considerable astonishment that in view of the very widespread publicity given by the general and technical media to the research and other investigative work I have undertaken with respect to Blumlein, his work, and his ancestry, since circa 1971, that I have not until the arrival of your letter of 29.09.1981, been informed of your Polytechnic's interest in Blumlein, nor even had the courtesy extended to me of being invited to your Exhibition of Machines and Recordings etc. in 1977. Since 1971, I have spent £1000s and hundreds of hours in researching and have traced the Blumlein Family back to the 14th century when, as in the 20th century, they were a powerful force in the land, and patrons of the visual art of their day, and almost certainly similarly of audio arts. Perhaps you would like to spare me the risk of duplicating what you have already?

At the end of 1981, Barry Fox, writing in *New Scientist*, published an article titled 'A hundred years of stereo: fifty of hi-fi', in which the history of dual channel sound was traced and much mention made of Blumlein's work. Then in May 1982, this time writing in *Studio Sound*, Barry Fox again wrote an article called 'Early Stereo Recording', in which he again traced the historical events leading up to modern stereo recordings, pointing out: 'A biography on Blumlein has long been promised but nothing has ever appeared.'

Once again in *New Scientist* magazine, Barry Fox re-opened the saga with his enthusiastically titled article 'The Briton who invented electronics', in the 3 June 1982 edition. Fox, who had been supportive of the long-awaited biography of Blumlein for well over a decade, pointed out the irony in the fact that Blumlein's lack of recognition was due mostly to an '...unfortunate, but fascinating set of circumstances which have conspired to deny the man the place in history which he deserves'.

Going on, explaining under a picture of Francis Thomson, taken on 1 June 1977, standing next to the blue plaque and with the heading 'Francis Thomson, whose long-awaited biography has still to appear', Barry Fox says that:

Perhaps the most relevant factor is the Official Secrets Act. Because he died while working on military hardware the act has hampered historical research. The people he was working for either kept inadequate records or have since destroyed or lost them. Many of his colleagues are now dead. So no one really knows what Blumlein was doing between 1937 and 1939. Although still on EMI's payroll when he died, Blumlein was working for the Telecommunications Research Establishment or TRE at Malvern. This is now the Royal Signals Radar Establishment. Although RSRE says that most of the papers on Blumlein would now be unclassified (under the 30-years rule), wartime TRE personnel were notorious for not keeping records and there is no original TRE material on Blumlein. The problem is exacerbated because Blumlein was employed by EMI not TRE.

From a historical point of view Blumlein suffered a disadvantage by working for EMI, now Thorn-EMI. The company is known for keeping details of its own history to itself. Although it possessed one of the best collections of old equipment, and archives, the company has done little with them. It has files, which in historical value are priceless, of original notes made by Blumlein and his colleagues in the 1930s. These are jealously guarded, but EMI has never explained why. Original equipment, such as that used to record stereo sound on film and disc was thrown away years ago. The reason? 'In the current economic climate, our first concern must be today's sales and tomorrow's products. Research into archival information cannot be given foremost priority', says Thorn-EMI's David Sowter (Public Affairs Advisor, Thorn-EMI).

The stereo-optical sound films Blumlein made in 1935, are still on nitrate stock. This is not only dangerously explosive but also becomes sticky and loses its image with the passage of time. The National Film Archive, part of the British Film Institute, believes that even under ideal storage conditions, Blumlein's material, dating back to 1935, must be on the point of irreversibly degrading. Although Thorn-EMI has been promising for five years to transfer the nitrate originals onto safe, acetate stock, the company says merely that the work is 'in hand'. The problem says Thorn-EMI, is that it is being undertaken by 'an expert' who is busy with other things.

There is another reason why Blumlein's life and work remains a secret to the world at large. Whereas most pioneers are the subject of at least one biography, there is still no book on Blumlein, something of a mystery in itself. In 1967, Basil Benzimra, an engineer who decided to immortalise Blumlein, called for reminiscences about him to be sent to his son Simon Blumlein. A year later, at a seminar on Blumlein's life organised by the British Kinematography Sound and Television Society, Simon Blumlein spoke with affection about brief memories of his father, and said how he hoped with Benzimra, to write a biography.

Ill health forced Benzimra to abandon the project and it was taken over by Francis Paul Thomson, author of books on banking, the Giro money system and tapestry. In September 1973, the authoritative magazine *Wireless World* carried a letter from Rex Baldock, organiser of the BKSTS memorial meeting, suggesting that anyone with information on Alan Blumlein should send it to Thomson at his Watford address. At the unveiling of the Blumlein plaque in 1977, Thomson told how he had been 'persuaded to write a biography'.

According to Thomson, in a letter to Huddersfield Polytechnic, he has accumulated about 80 kg of material about Blumlein and his inventions and researched the inventor's ancestry back to the early 15th century. The 1982, edition of *Who's Who* contains an entry in Thomson's name referring to a book on Blumlein called 'Engineer Extraordinary', published in 1977. But the Science Reference Division of the British Library can find no trace of any biography of Blumlein. Thomson has declined to say when he will publish his biography, or list any articles on Alan Blumlein that he has published since he started to collect biographical material nearly 10 years ago.

## Transferring the binaural films

The condition of the original nitrate films, which Blumlein had shot in 1935, when conducting his binaural experiments, had been causing concern to archive staff at EMI long before Barry Fox's letters and articles regarding the matter began in 1981. It is true that his 'prompting', certainly sped up the transferral process, but I have little doubt having researched the EMI internal memorandum archives that this action would have been taken sooner or later.

The process of transferring the binaural films to safety acetate stock took very nearly two years and involved the not inconsiderable cost of the transfer itself, the time for those given the responsibility of seeing the work through, plus of course their costs involved.

Thorn-EMI archives first refer to the transferral of the film following a letter from Barry Fox on 25 February 1982. David Sowter replied to Fox on 9 March 1982, informing him that: 'you will appreciate that in the current economic climate, our first concern must be today's sales and tomorrow's products. Consequently, research into archival information cannot be given foremost priority. Nevertheless, we value our history and the work of people such as Blumlein.'

This was not the answer Barry Fox was looking for, and in his May 1982 article which appeared in *Studio Sound*, he roundly criticized Thorn-EMI for their lack of concern on the matter of the Blumlein films. Lucas looked into the matter and reported on 4 May 1982, that:

> It is a great pity that Barry Fox does not clear his facts. The films in CRL are not dumped in a cupboard and are not deteriorating. The position is in fact reviewed regularly. The films have been offered to EMI Films and EMI Videogram Productions for internal use. They have also been offered to the British Film Institution and more recently have been discussed with ITCA. The position regarding the films is constantly reviewed. The condition of the films is in fact good. They are kept in a dry cold environment and a recent professional inspection has shown that they are well preserved.

Unfortunately for Thorn-EMI, though rather fortunately for the Blumlein films, one of Barry Fox's articles reached the attention of The Minister of State for Industry and Information Technology, then Kenneth Baker MP. He wrote to Sir William Barlow, Chairman, Thorn-EMI on 24 June 1982, to ask what the situation was: '...I saw the attached article in the *New Scientist* about Alan Blumlein, who seems to have been an archetypal example of the unsung genius. I do hope that Thorn-

EMI are taking steps to preserve Blumlein's early films, and maybe you would like to consider whether we could not do more to bring recognition of his really considerable achievements to a wider public.'

Not surprisingly, the letter from Kenneth Baker sparked a flurry of activity at Thorn-EMI with Sowter among others feeling the need to 'explain the situation' to the directors of the company who, presumably following the 'interest' and prompting of a Member of Parliament, felt obliged to act. A complete assessment of all material kept in stock relating to Blumlein seems to have taken place in late June or early July 1982, with a resultant file note appearing on 9 July 1982 which listed much of the material, its condition and whereabouts.

Humphries Film Laboratories of London were contacted with a view to getting a quote for the work of transferring the delicate nitrate stock onto a more stable acetate base. The quoted prices were for the duplication of the negative from nitrate film stock: 35mm – 5,000 feet @ £1,432.50 or 10,000 feet @ £2,865.00. To re-record the sound, 5,000 feet @ £687.50 or 10,000 feet @ £1,375.00. For a Black and White print from the nitrate material, 5,000 feet @ £761.50, 10,000 feet @ £1,523.00. The price, exclusive of VAT, was of course cheaper for 16mm stock, but the entire Blumlein binaural films were shot on 35mm.

By October 1982, the contents of the tins of nitrate film had been examined and broken down into reels, lengths in feet, even the contents of the footage had been listed. It was pointed out that: "This significant milestone has been achieved by the very helpful efforts by Norman Green (ITCA) and Reg Willard (Thorn-EMI). When, that month, Lucas and Sowter corresponded to examine the probable costs that Thorn-EMI would incur, Lucas wrote: 'The estimated cost is £1,800. We had been warned that it might be £3k or £4k. I have said go right ahead with the job.'

On 11 January 1983, Norman Green collected the films from Reg Willard at Thorn-EMI in order to examine the quality of the nitrate stock. If this was OK, he would then proceed with the transfer process. Norman Green reported on 25 February 1983, that the matter was in hand and he would be in touch shortly. By 7 April, however, with Norman Green under extreme pressure on other affairs still nothing had been done. Lucas wrote to him for an explanation. In his reply Norman Green explained: 'As you know this is a spare time activity for me and in the last three months I have been out of the country for some five weeks. I have edited approximately 75% of the films and will arrange for the remaining 25% to be done...I should therefore be able to return the films to you in the week beginning 18 April.' (See Author's note on p. 398.)

That date came, and went. Still no sign of the films. On 27 April 1983, Norman Green wrote again to Lucas explaining the situation: 'It has always been my impression regarding these films, that EMI wanted to preserve them in some way, but was not too keen on the expenditure involved ... the arrangement for the transfer to videotape has been made ... if EMI has money to spend, then I am quite happy that they should arrange to take over this work and complete it'. EMI left things in the hands of Norman Green who reported on 3 June: 'we have been delayed in the film to tape transfer. This is to be done next week (6 – 10 June), and so I should be able to let you progress on the 13th.'

Finally, by 28 June, Norman Green was able to report to John Jarrett, EMI Sound & Vision Equipment Limited:

Last night I was able to listen to the films directly off a video tape. The problem with them seem to be that they were printed for picture quality and not for sound quality. That is, instead of the sound track being black or white it is grey and white. To prove to ourselves that this is causing most of the background noise on the sound tracks could you please arrange for Humphries to print say just the sound tracks of the two playlets in the most optimised way and we then can check the improvement in dynamic range and signal to noise.

As Norman Green had no way of knowing at the time that the sound on the films he was listening to had indeed been recorded poorly in 1935, it is not surprising that he suggested the action to Jarrett. Of course, we now have written notes retrieved from the EMI archives that confirm Blumlein had also looked at the first shooting of the 'playlet', and he too had decided that it had been filmed with the sound too low. That is why the scene was filmed for a second time on 26 July 1935.

On 17 October 1983, Lucas reported to Sowter, Ingham, Jarrett and Willard that:

From the original nitrate films duplicate negatives and prints have been made. The next stage was to do transfer from film to tape. During investigation of how best to do this, it was found that although the picture quality was good, the sound quality was poor. There are two reasons for this: – 1. Non-standard position of the sound tracks on the film. 2. Lack of density in the film dupe negatives. Resolving these problems involves: – 1. Co-operation of Dolby Laboratories who have special equipment for locating non-standard sound-based positioning. 2. Co-operation of the National Film Institute who have special equipment, and a permit, to print direct from nitrate film. Both these organisations are keen to help because of the historical interest and both have offered to provide some of the funding on the undertaking that they could have copies for themselves.

By 8 December 1983, Jarrett was able to report to Lucas, Ingham and Willard that:

A satisfactory situation now exists regarding the preservation of the historic binaural films made during Blumlein's experiments of 1935. Starting from the original nitrate stock we now have: – 1. Copy duplicate negatives. 2. 1 set of prints, separated. 3. 1 set of prints, joined (1). 4. 1 U-matic VCR cassette, two runs, one with monosound on an audio track, and one with sum and difference sound on separate audio tracks. 5. 1 ¼ inch audio tape of sum and difference sound on separate tracks (2). In the course of the work, it has been possible to accurately date most of the films. One shows a few seconds of Blumlein himself: apart from the much-publicised single photograph, this is probably the only other photographic record that we have of him (and his voice). The prints on safety stock are exact replicas of the originals. However, the present day position of the sound tracks on the 35-mm films are 25 thou different. It is therefore proposed that we go one step further and make a standard 35-mm cinema print for demonstration purposes from (1) and (2) above, that has left and right (matrixed from sum and difference) stereo sound tracks in the correct positions.

## Memorial window

Over the next few years the odd article was published about Blumlein and the absence of the long-awaited biography. The book had been promised for so long that when the 50th anniversary of the opening of Alexandra Palace came and went in November 1986 without a firm publication date it was almost expected. By September 1990, the Secretary of the IEE, increasingly concerned following the adverse publicity and the lack of a publication date visited Francis Thomson worried that the biography progress had ground to a halt. They suggested that Thomson work with a collaborator and suggested Gordon Bussey, the author of a meticulously researched series of books on the history of radio and television. Although at first Thomson agreed that in the interests of the biography he would indeed work with Mr Bussey, later on he would go back on his word.

As the 50th anniversary of the crash of Halifax V9977 approached in June 1992, William Sleigh and some of the members of the Royal Signals Radar Establishment at Malvern began putting together the embryonic plans for some kind of permanent memorial near the crash-site, not only dedicated to the memory of those who had perished on 7 June 1942, but for all the men and women who had made the ultimate sacrifice during the years of radar development. While the bitter arguments between Francis Thomson, Barry Fox and now the Blumlein family continued regarding the non-publication of Blumlein's biography, at RSRE Malvern, Sleigh, who had been responsible for the production of 'Aircraft for Airborne Radar Development', which, once-and-for-all, had established the circumstances surrounding the crash of V9977, was writing to English Heritage to enquire as to the possibility of installing a memorial window at Goodrich Castle, just under a mile from the crash site itself.

Barry Fox had been one of the first interested writers to be made aware of this, and wrote an article in *New Scientist* which appeared on 27 April 1991, entitled 'How did Blumlein die?', in which the last moments of his life were supposedly traced. The article relies heavily on 'Aircraft for Airborne Radar Development', which by now had been de-classified and released to the Imperial War Museum (among others). At the end of the article, Fox pointed out that RSRE were planning to unveil some form of memorial on the 50th anniversary of the crash of Halifax V9977 in June 1992, as a tribute to all the men who had died in the pursuit of the development of radar. English Heritage, he reported, had offered a site for the memorial at Goodrich Castle, a mile north of the crash-site, and that the library at RSRE was busy collecting contributions for the memorial.

Sleigh wrote to Fox on 5 June 1991; having read his article, he pointed out certain anomalies in his work (which do not bear mention here, as the correct information is given in detail in Chapter 10 of this book). However, the key factor was the erecting of the memorial at Goodrich Castle, '...may I say how grateful my colleagues and I are for your mention of our memorial appeal for the Radar Research Squadron which is intended to commemorate those who died during the Unit's active years between 1936 and 1976, as well as remind future generations that Britain led the world during the dawn of applied radar sciences.'

Meanwhile, Francis Thomson appealed once again for more information for his 'biography', this time in the pages of *Television*, the journal of the Royal Television Society:

My biography of Britain's most prolific and diverse electronics inventor and patentee, Alan Dower Blumlein (1903–42), will be ready for publication 1992–93. May I appeal to your readers to send me for possible inclusion in appendices: 1. Copies of their learned papers or journal articles referring to Blumlein and/or, his associates' development of stereophony, telecommunications, cable, television, radar, measuring instruments, military etc. circuits, ideas, etc., in postwar computer, aerial, or electromechanical systems. It would help if print-offs or copies were suitable for direct production of printing plates. 2. Particulars of papers, articles, books etc., in any language, for inclusion in the bibliography.

The Institution of Electrical Engineers and publishers of The Gramophone and others have given me permission to reproduce papers and articles free – but subject to acknowledgements. It is hoped your readers will accord me a similar privilege. Some of your readers may be among those who expected publication in 1983–84 and have been annoyed because I gave no convincing explanation for the delay. The reason was that I was trying to shield the Blumleins from the belated discovery that a person close to them had supplied me with letters he had stolen from the late Mr A.K. van Warrington. There were also photographs belonging to Mr van Warrington, kept by this person, which ought to have been given to me for possible inclusion in the biography. From 1973, until his death in 1983, Mr van Warrington tried to persuade the thief to surrender letters written in the 1930s by A.D. Blumlein to him, and the photographs I needed, but without success. Consequently, may I appeal to your readers to loan me any letters, notes and photographs which could help to bridge those still (presumably) in the thief's haul? Please post letters to me marked on the envelope 'Blumlein Biography', to arrive here before mid-August 1991.

Similar articles were placed in the *Journal of the Audio Engineering Society* (June 1991), *Nature* (23 May 1991), and *Image Technology*, the journal of the British Kinematograph, Sound and Television Society (July 1991). If certain elements of these appeals sound familiar, it is because Thomson uses almost exactly the same form of vague persuasion as he had in his appeals of 1972, 1973, 1974, 1982, 1983 and 1984. Quite who the 'thief' that had 'stolen' the letters and papers from van Warrington was, Thomson does not say; strange then, that in all his correspondence with Lucas and Sowter at EMI, Felix Trott and Eric White, from 1981 to 1991, he never mentions this 'reason' for the delay in publication of the book.

Following this, a few weeks later, again in *New Scientist*, Barry Fox returned again to the Francis Thomson saga in a short article titled 'Alan Blumlein: a warning':

For nearly 20 years, Francis Thomson has been promising to write a biography of Alan Blumlein, the brilliant pioneer of television and radar technology who died in a plane crash in 1942. Thomson is appealing again for source material. Those who know that he started collecting original papers in 1972, but has published nothing, refuses third party access to his collection, and will not catalogue what he has obtained, have treated the latest call with suspicion. Others, apparently ignorant of the facts, are still publishing Thomson's plea. For example, the current issue of *Television*, the respected journal of the Royal Television Society, contains another appeal by Thomson. Gordon Bussey, the author of meticulously researched books on the history of radio and television, is among the many who have been upset by Thomson's behaviour. Last year, the Institution of Electrical Engineers asked Bussey if he would help Thomson write the biography. Bussey met with Thomson and the IEE's publishers in November, but Thomson later refused to co-operate.

Blumlein's son Simon is also in despair. He is now resigned to the fact that the 50th anniversary of his father's death, next June, will pass without the publication of the biography for which he and his mother permitted Thomson to collect material. His mother died without seeing any results. Her son now worries about what will happen to the collected material long term. The same fear deters those who could bring peer pressure to bear on Thomson. This leaves Thomson free to continue collecting material. And not just on Blumlein. In 1979, Thomson wrote to one of the inventors of the heart and lung machine, saying he expected to finish his biography of Blumlein soon and would like to have a go at writing their life story. Perhaps wisely, they did not accept. Thomson then appealed in the January 1980 edition of *Wireless World* for 'papers notes, photographs etc.' from anyone who knew S.G. Brown. Brown had patents on inventions such as the gyrocompass and audio headphones. There is no sign of a biography on Brown either. Now Thomson has returned to Blumlein, explaining the decades of delay with a puzzling tale of his attempts to 'shield the Blumlein's from the belated discovery that a person close to them had supplied me with the letters ... stolen from the late Mr A.K. van Warrington'. The IEE, and others, have tried to play honest broker but failed. There are no further avenues of reason, duty and obligation left to explore. It will be a tragedy for the nation if Blumlein's biography is permanently blocked.

Barry Fox repeated the message this time in August 1991 in the pages of *Hi-Fi News & Record Review*. 'In this issue ('Views' page 7) Francis Thomson, who describes himself as the biographer of Alan Blumlein (inventor of stereo and much more) publishes another appeal for material. This will come as a surprise and disappointment to those who have followed the saga of the Blumlein biography and who know Francis Thomson first put out public calls for material nearly twenty years ago and has so far declined to identify anything which he has published on Alan Blumlein. Nor has Mr Thomson yet given any firm publication date for his promised biography'; and again in September 1991, in *Everyday Electronics*: 'Next year, 1992, sees the unification of Europe. For a very few people 1992 also means the fiftieth anniversary of the death of Alan Blumlein – arguably the most significant electronics engineer to be born, live and work in Britain...despite the importance of his work on stereo sound, television, telecommunications and radar, only electronics engineers know enough of the man's achievements to respect his memory. The public at large will have heard of Edison, Bell and Baird, but not Blumlein.'

There was, however, some good news. English Heritage had approved the erection of the proposed permanent memorial to the Radar Research Squadron intended to be unveiled in June 1992. By 5 September 1991, a provisional specification for the memorial in the form of a stained glass window had been designed by William Sleigh and his wife Audrey. It had then been drawn up at RSRE; it was to be installed in the chapel at Goodrich Castle which was owned and run for the nation by English Heritage. The design and selection of the glass was to be compatible with the his-

Figure 12.1
The concept drawing of the proposed memorial window for Goodrich Castle.
(*Courtesy of William Sleigh*)

toric atmosphere of the thirteenth century chapel, and, because natural light only entered by two windows, the east window space was chosen, ensuring that 'excessive use of heavy and deep colours' be kept to a minimum. The design itself was centred on the amorial bearing of the Government's former Royal Radar Research Squadron surrounded by three Royal Air Force heraldic crests from squadrons of the period.

The remaining panels of the window were to be supplemented with a geometric cross section of the magnetron valve, the floral emblems of the four primary regions of Great Britain, signifying the collective involvement of the nation in the application of radar sciences and finally the heraldic insignia of the counties of Worcestershire and Herefordshire. In the first working drawing of the window, designed by Sleigh at his home in Malvern, the window is shown with three large glass panels at the bottom, the right hand panel having a Halifax bomber in profile above two H2S CRT displays. The middle panel simply says, 'Radar Research Squadron 1936–1976 To Those Who Gave Their Lives', and the left panel shows three towers of the Chain Home array.

On 7 September 1991, Group Captain F.C. Griffiths DFC, AFC, RAF (retired), chairman of the window committee, who had been the Air Force Commander of the RAF station at Defford, and who had known the crew of V9977 personally, visited the site at Goodrich Castle for a first-hand appraisal. By the end of September 1991, William Sleigh, who had been placed in charge of the project for RSRE, was writing to Mrs J. Duberley, resident custodian of Goodrich Castle for English Heritage, enclosing a full-colour print of the proposed stained glass window, a copy of which had been tendered to potential stained glass window contractors. There now remained the problem of raising the money to pay for the window, which was expected to cost in excess of £6000. The money was raised mostly from donations by the staff at RSRE in Malvern, which is a testimony not only to their generosity, but to the esteem in which they held the men who had paid the highest price for research and development of radar systems. The large, bronze plaque which would be placed below the window, and would explain its purpose would cost another £1000. English Heritage, upon hearing that the cost of the window had been raised by donation, decided to pay for the cost of the plaque from their funds. In all, the memorial at Goodrich Castle, had come to just over £7000.

In February 1992, an article entitled 'A.D. Blumlein – engineer extraordinary', written by Russell Burns, appeared in the first issue of the *Engineering Science & Educational Journal*, a subsidiary publication of the IEE. In the article, Burns traces the life of Blumlein with a hitherto unknown degree of accuracy, detailing events which could only have been obtained through long and exhaustive research. The fifteen page article contained photographs of the Blumlein moving coil cutter, the Blumlein-Holman moving coil binaural pick-up head from 1933, as well as large picture of a sound-locator incorporating Blumlein's VIE equipment from 1939. Several diagrams of circuits as well as the Marconi-EMI 405-line television system waveform were also included. At the conclusion of the article (in which Burns, perhaps wisely, neglects to mention either a biography or biographer; opting instead to credit Jim Lodge at EMI for much of the assistance in preparing the document), there is a comprehensive list of all of Blumlein's 128 patents though, somewhat disappointingly, this seems to be a near-verbatim copy of the Benzimra list that had appeared in June 1967.

Also in February 1992, now with the full backing of the IEE, Jim Lodge was part of a 'working party' from the Professional Group Committee (History of Technology), IEE, that were making preparation for the memorial unveiling. These working parties met regularly through March, April and May 1992, leading up to the event itself, organizing such things as who would be invited to attend the ceremony itself. The task required quite a lot of research; not since the plaque unveiling in June 1977, would so many members all linked in some way or another to Blumlein and the other ten men who had died in V9977, be called to attend one place at once.

Figure 12.2
The installed memorial window on
7 June 1992, the fiftieth anniversary
of the crash of Halifax V9977.
(*Courtesy of English Heritage*)

Figure 12.3
Goodrich Castle looking north west towards the flat ground that Pilot Berrington was hoping to use as a crash landing site. This is approximately 1½ miles from the actual impact point. (*Courtesy of English Heritage*)

Many had died in the fifteen years that had passed. It now fell to the researchers to find and trace those that remained.

The window itself was now nearing completion. The contract had gone to H. John Hobbs of Much Birch, Wormelow, Herefordshire, a stained glass maker of quite exquisite skill, who had constructed a masterpiece in glass, which was befitting its ceremonial purpose. Completed in early May 1992, the window was packed and shipped from Mr Hobbs' workshop to Goodrich Castle, and carefully installed in the east window space that had been waiting for it. The original design had been modified a little from the early working drawings; the window is essentially three, long, tall glass windows known as 'lights'. Each of these is made up of four rectangular panes of glass, criss-crossed with leading, the higher-most pane of all three being topped-off with a small, diamond-shaped pane of glass.

There is no design in the small diamond pane at the top of the left-hand window, though just below in the fourth pane, are two of the four floral emblems of Britain at the top (the leek of Wales and the rose of England); below these, across the third and second panes, the heraldic crest of A&AEE Martlesham Heath, the RAF host station for radar's first flying unit (1936–1939), designated 'D' flight which, with its successors, served the research and development programme of radar continually for forty years. Below this, a diamond shaped panel depicting three Chain Home aerial masts in perspective, with a scroll underneath on which the words 'Chain Home System 1936–1950' are written.

The middle window pane has the crest of the Royal Air Force in the very top diamond pane, immediately below which is a cross-section diagram of the cavity magnetron in the fourth pane. In the third pane, below, the coat of arms of the

Royal Radar Establishment, and below this in the second pane, the heraldic crest of the RAF Telecommunications Flying Unit (TFU). In the lowest pane of the centre window is a scroll on which apprear the words 'Radar Research Squadron 1936–1976 To Those Who Gave Their Lives'.

The final, right-hand panel of glass again has no design in the topmost diamond panel. Below this in the fourth pane are the floral emblems of Scotland (thistle) and Ireland (shamrock). Across the third and second panes, the crest of the RAF Radar Research Flying Unit (RRFU), and in the lower pane another diamond shaped panel, this time showing a Halifax bomber in profile, below the aircraft is a graphic representation of the view on an H2S CRT, based on an actual one taken during an attack on Leipzig in 1943. Below the diamond panel, a scroll bears the words 'Centimetric Airborne Radars AI – H2S – ASV 1941–1982'.

Underneath the window, the bronze plaque with the following words picked out in brushed relief was placed:

> The memorial window, unveiled on the 7th June 1992, commemorates the many Service and civilian aircrews who lost their lives in radar development flying duties between 1936 and 1976. The Radar Research Squadron's parent Establishment created between 1935 and 1939 the world's first radar managed defence system. This was fundamental to our victory in the Battle of Britain in 1940, and was one of the many British radar systems to transform air power and earn the nation's gratitude. The unveiling marks the anniversary of the worst tragedy when a Halifax bomber carrying the prototype of the first ever ground mapping radar bombing aid crashed near Goodrich Castle, killing all eleven on board. This navigational bombing aid made possible effective strategic air power, while its maritime derivative saved the British Isles from total isolation by submarines. Together both versions allowed the assembly of large military resources in Britain that enabled the Allies to liberate Europe in 1944/45. 'They applied the frontier of scientific knowledge to the salvation of their country'.

The finished window is surprisingly light and airy, in fact just as the original specification had wanted it to be. The colours, reds and blues mostly, shine through perfectly and the overall effect is most pleasing on the eye. William Sleigh is still very pleased with the overall effect though he explained, 'I was disappointed with just one thing – they omitted the word "50th" before "anniversary" on the plaque. After all, that's why we were to be there, on the 50th anniversary, on a Sunday, at exactly the same time as the Halifax had left Defford all those years ago.'

While work on the memorial window at Goodrich Castle was being finished, Barry Fox's persistence, continuing to use the pages of the many periodicals he regularly writes for to chip away at Francis Thomson, finally bore some fruit. In May 1992, the IEE and the Royal Society, after years of doing nothing, directed Thomson, quite specifically, to hand over all and any papers that he might hold on Alan Blumlein in order that future generations might benefit from this source of knowledge and that, at long last, a comprehensive biographical tribute to Alan Blumlein might be written. Thomson, at first acknowledged the IEE and Royal Society directives by saying that he would agree to hand over all the papers in his possession just as soon as the fiftieth anniversary of Blumlein's death was over.

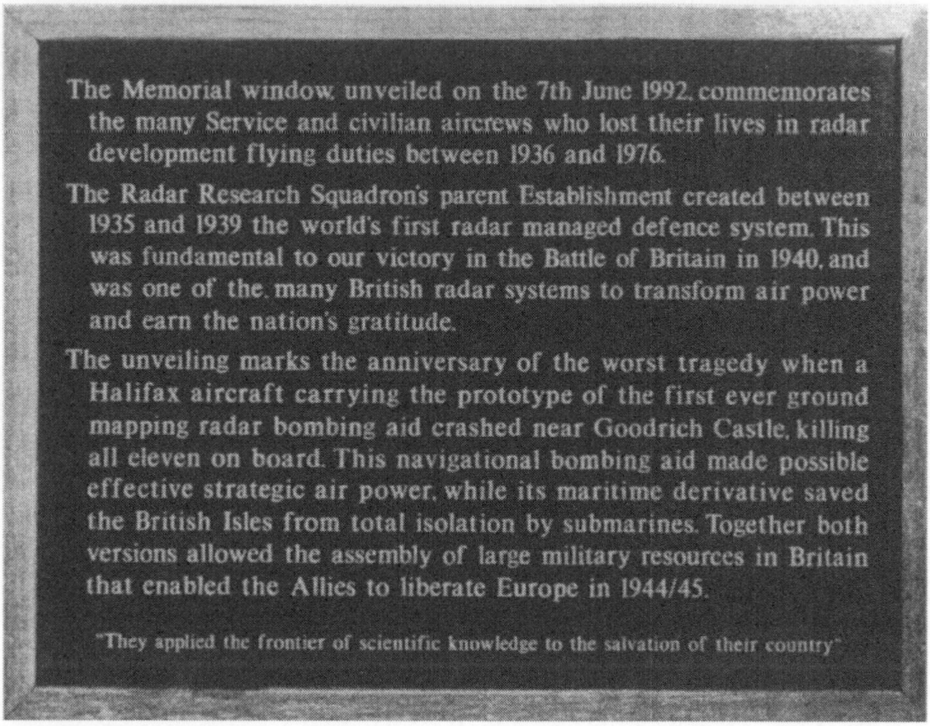

The Memorial window unveiled on the 7th June 1992, commemorates the many Service and civilian aircrews who lost their lives in radar development flying duties between 1936 and 1976.

The Radar Research Squadron's parent Establishment created between 1935 and 1939 the world's first radar managed defence system. This was fundamental to our victory in the Battle of Britain in 1940, and was one of the many British radar systems to transform air power and earn the nation's gratitude.

The unveiling marks the anniversary of the worst tragedy when a Halifax aircraft carrying the prototype of the first ever ground mapping radar bombing aid crashed near Goodrich Castle, killing all eleven on board. This navigational bombing aid made possible effective strategic air power, while its maritime derivative saved the British Isles from total isolation by submarines. Together both versions allowed the assembly of large military resources in Britain that enabled the Allies to liberate Europe in 1944/45.

"They applied the frontier of scientific knowledge to the salvation of their country"

Figure 12.4 The memorial plaque at Goodrich Castle. While the anniversary of the crash of Halifax V9977 is recorded, sadly the plaque does not point out that it was erected and unveiled on the fiftieth anniversary of that date. (*Courtesy of English Heritage*)

## The fiftieth anniversary ceremony, Goodrich Castle

The fiftieth anniversary of the crash of V9977, the 7 June 1992, was a Sunday, just as it had been all those years before, when the eleven aboard the Halifax lost their lives. Just as 50 years earlier, it was a bright sunny, though somewhat breezy day at Goodrich Castle. The invited dignitaries and guests began to arrive around 1.30 p.m., and assembled at Goodrich Hall, which the villagers of this tiny Herefordshire community had kindly made available for the day. At around 2.10 p.m. the congregation, now numbering in excess of 200, moved slowly across the carpark and along the grass path to the entrance of the castle itself which sits upon a hill overlooking the English countryside. The chapel, where the memorial window had been erected, was full to capacity as the officiating dignitaries arrived and also entered the castle across a wooden bridge, spanning the long-since-drained moat surrounding Goodrich.

The official ceremony itself began with a dedication service from the Royal Air Force Chaplain, after which an RAF trumpeter sounded the 'Last Post', followed by a short silence and then 'Reveille'. It then fell to Sir Bernard Lovell, to officially donate the window as a permanent reminder to future generations of the influence of British scientific ingenuity on the nation's history. A memorial register was brought forward in which was recorded the names of all 71 scientists and engineers who had died during 40 years of radar research and development. The register is kept at the Royal Air Force Museum, Hendon. At the conclusion of the ceremony, a single RAF Nimrod aircraft, the modern-day equivalent of the radar research aircraft such as Halifax V9977, flew past Goodrich Castle at 3.15 p.m. The entire event passed without a hitch, and all who were in attendance, concluded that the ceremony, and memorial window formed a fitting tribute to those who had died. While the castle had been closed to the public during the ceremony itself, it was opened up to all soon after and remained open, as it does to this day, until 6.00 p.m. each evening. Francis Thomson was not invited to the ceremony, and did not attend.

Following the proceedings at Goodrich Castle, *The Times*, reporting on the events the next day, summed up exactly what the lack of a biography of Alan Blumlein meant; they completely omitted to mention him by name in the article. Thankfully, in the days and weeks that followed, most of the press did cover the event accurately, reporting that the memorial window, while dedicated to the lives of all who had died, had been brought about because of the persistence of a group of individuals, determined to see a permanent memorial to those who had died in Halifax V9977. The *Hereford Times* reported: 'People came from Australia and the United States to the ceremony. Among those attending was Sir Bernard Lovell, a leading scientist of the day who was on the project team and later became director of Jodrell Bank radio astronomy station.'

The *DRA News* of June 1992 (RSRE had become DRA, Defence Research Agency, some months before the unveiling ceremony), carried an intriguing personal recollection of Geoffrey Hensby, the scientific officer attached to TRE, who had of course also died in V9977. In the article, written by Bob Rose, who had known Hensby during the war as he had been his neighbour, a brief biographical history of Hensby was given, and the activities he carried out at the Royal Aircraft Establishment and later the Telecommunications Radar Establishment at Malvern. The article was read and a response arrived from Geoffrey Hensby's cousin, Patrick Benger, who supplied a photograph of Hensby which was subsequently printed in *DRA News*, August 1992.

Despite the pressure that had been applied to him, Francis Thomson did not produce the papers as he had promised he would, and still the IEE and the Royal Society refused to take any action against him. They gave their reasons as 'confidential', and refused Barry Fox's attempts to lift the restriction on taking any further action. They did however state to Barry that, 'The investigating panel received assurances from Mr Thomson that the archival material he has collected will be safeguarded and (the panel) has no reason to doubt this assurance.' At this time, Barry Fox wrote to the Royal Society's librarian, Sheila Edwards, to ask if they had received any further news from Thomson as to when he might give up the papers. Mrs Edwards replied: 'I am happy to confirm that I have been in correspondence with Mr Thomson, and that the Officers of the Society have agreed to accept Mr Thomson's generous offer to donate his collection of Blumlein papers to the Society.'

Still articles continued to appear about Blumlein. Having evidently been pleased with the response to the initial publication of Russell Burns work in *Engineering Science & Educational Journal*, a follow-up article, titled 'The life and work of

A.D. Blumlein', this time written by Burns, Jim Lodge, A.C. Lynch, Eric White, K.R. Thrower and R.M. Trim, appeared in the June 1993 issue. Covering specific aspects of his work, the article has sections on Blumlein's transformer-bridge network (Lynch), audio engineering (Lodge), television (White), Blumlein's contribution to electronics (Thrower), and radar (Trim), with an introduction by Burns. It is, once again, a comprehensive and well-researched piece of work, though it is debatable how wide an audience would have read it.

The publication returned once more to the subject of Blumlein, and more specifically the memorial window at Goodrich Castle a year later in an article by Dr Ernest Putley, in the June 1994 issue. It had been decided that the chapel at Goodrich Castle would be used for another commemorative ceremony, this time on Sunday, 5 June 1994, to honour the 50th anniversary of D-Day and again in memory of the radar researchers killed in V9977. The event was organized by Gillian Duberley and English Heritage with representatives from the Radar Research Squadron again present for the dedication. William Sleigh, writing to thank Mrs Duberley on 16 June 1994, explained:

> You will be well aware that most of our former colleagues live considerable distances from Goodrich, compounded with us not growing any younger over the passing years. However, I wrote to Sir Bernard Lovell the following day to advise him of the Service and flowers, including those from the area where his Halifax aircraft was so tragically destroyed, and to my surprise, Sir Bernard in his busy life, replied by return to say how touched he was with the thoughts of those involved. It makes us feel very content to know that the location of this British Service Memorial of Goodrich Castle, could not have been in a more appropriate location in the United Kingdom, and we hope its presence in the Castle long after we have departed the scene will have continued meaning.

Throughout 1995, as Britain, Europe and America commemorated the ending of the Second World War fifty years earlier, the nagging question of what had happened to the Blumlein archives held by Francis Thomson returned to the fore once again. Barry Fox pointed out in his article in *Everyday with Practical Electronics*, September 1995:

> The Royal Society's rules decree that papers are kept secret until forty years after the death of the subject. As Alan Blumlein died over fifty years ago, this means that the Blumlein papers will be available as soon as The Royal Society has them. Unfortunately, the promise of seeing the Blumlein papers safe in the library of The Royal Society proved to be a false dawn. At first, The Royal Society would say only that it was 'still negotiating with Mr Thomson'. At a seminar held at the IEE in October 1992, I answered a question on archiving progress with the suggestion that those who are concerned should phone The Royal Society and ask for themselves. This suggestion did not however appeal to the IEE. The Institution's Assistant Secretary, Philip Secker, rose from the audience at the Savoy Hill meeting and warned that to make such approaches would be 'counter productive' because 'negotiations are at a delicate stage'.
>
> Honouring this request I left it several years before contacting The Royal Society again. The time seemed right when I heard that the new Centre for the History of Defence Electronics in Bournemouth University refers to Blumlein's work. New generations will be wondering why there is no biography they can

read. In early July (1995), I phoned The Royal Society Library to ask if it had ever received the Blumlein archives. 'It doesn't ring any bells', a librarian told me, coming back later to add that 'there is no entry in the archive catalogue'. Head Librarian, Sheila Edwards, then confirmed the bad news, 'nothing has come. I must admit, I wasn't surprised', she added.

Barry Fox concluded his 1995 article with the following:

The ball is now firmly back in the IEE's court. The Institution said previously that it saw no reason to doubt the assurance given by Francis Thomson that the collected papers would be safely archived. But after three years this has still not happened. I suggest it is now the clear responsibility of the IEE to approach Francis Thomson and ask for the promise to be honoured. This historical material is too important to risk losing. I have written to the Secretary of the IEE to ask for the Institution's attributable view on Francis Thomson's apparent failure to do what he promised three years ago.

In July 1996, again in *Everyday Practical Electronics*, Barry wrote that he had still not heard anything from the IEE, and that Simon Blumlein was now becoming 'very concerned' as he had received notification of legal proceedings against him by Francis Thomson, who, by now, seemed to be bordering on the hysterical in his determination not to give up the Blumlein papers he said he had in his possession. Simon Blumlein was as anxious as everybody else to ensure that the mass of original material which Thomson had been collecting since 1972 found its way to a safe archive. Fox pointed out that: 'Although Mr Thomson is still active in writing curiously worded letters, and threatening legal actions, there is clearly no hope of his ever finishing the biography he has for so long now been promising to write. The wider moral of this unsavoury story is that under current laws if someone wants to bury the reputation of a dead genius, they can do so by jackdawing all available material and continuing to promise a biography until they themselves die. This, of course, raises the all-important question, what happens to that material when the jackdaw dies?'

By the autumn of 1996, having begun the work which would eventually result in this book (in January 1995), I felt it was necessary to contact Francis Thomson myself for the first time, though Barry Fox (and later on Simon Blumlein), warned me: 'He won't help you, believe me, he won't help you'. Undeterred, I wrote to Thomson on 5 July 1996, deciding it was better to introduce myself formally, rather than call on him (at the time we both lived in Watford, by a curious coincidence in fact, just a few streets away from each other). My letter was the precursor to a series of articles that I had in mind, and which were eventually published in the spring of 1997, in *Audio Media* magazine. I explained this to Thomson, but as Barry Fox had foretold, he rebuffed my requests for help in two letters which contained such vicious outbursts of anger towards me that I was completely taken aback. Needless to say, that was the end of my direct correspondence with Francis Thomson.

In March of 1997, the first of my four-part series titled 'A.D. Blumlein – The Audio Patents', appeared in *Audio Media* magazine, and ran concurrently until June. Despite this publication being one of the foremost of its kind for the professional audio industry, it evidently did not reach Francis Thomson, or, if it did, he chose to remain silent. My correspondence with Barry Fox and Simon Blumlein continued, and the remainder of the research necessary for this book was carried out. As I pro-. gressed with the work on this book, I discovered through correspondence with Felix Trott that Francis Paul Thomson died in February 1998. At the request of the Thomson family, his death was not widely announced at the time, and so the large majority of people who had pursued him were unaware that he had passed away. The archive had, according to Felix Trott, been distributed widely amongst the Thomson family as far away as Canada, and little more other than that is known of it. There is one last, small and rather sad point to add to this tale.

Having failed to comply with the May 1992 directive from the IEE (and The Royal Society), to hand over all and any papers he might hold on Alan Blumlein, and evidently unaware that Francis Thomson had died in February, Thomson's membership of the IEE was revoked in April 1998 as *New Scientist* reported:

> Readers regularly ask when the historical papers on Alan Blumlein, pioneer in stereo sound, television, radar and electrical engineering, will be safely archived for future generations. A small paragraph in the latest edition of *IEE News*, the monthly bulletin from the Institution of Electrical Engineers, brings some hope. British inventor Alan Blumlein died in an air crash in 1942, while testing airborne radar. His work was so important to the war effort that the accident was for a while kept secret. In the early 1970s, Francis Paul Thomson, an author of books on banking and tapestry, started to collect biographical material on Blumlein. Initially, Thomson had the support of Simon Blumlein, the inventor's son. However, 25 years on there was still no sign of a biography and not even Simon can access his father's papers.
>
> The IEE took an interest because Alan Blumlein had been a member of the institution and Thomson had used his own membership to help in collecting papers. In 1992, the IEE halted a first investigation when Thomson agreed to donate the papers to The Royal Society. But the papers never arrived and last October the IEE's disciplinary board met to review the case. In April, *IEE News* revealed its findings. In an unprecedented move, the IEE has decided that Thomson's behaviour over the Blumlein archive material is 'improper' and ruled that he should be suspended from membership of the institution. Simon Blumlein hopes that the IEE's action will do the trick. 'I do hope that Mr Thomson will now place the papers in an academic archive with free public access' he says. Having followed the case for the past 25 years, Feedback will believe this only when it happens.

I am certain that originally Francis Thomson had the most genuine of intentions to write a biography of Alan Blumlein – of that, I feel there is little doubt. As the years went by however, his work branched off into far too many facets of Blumlein's ancestry, his distant relatives and their lives. This meant that he became bogged down in what was undoubtedly too massive an undertaking for a man of advancing years, ill health and who was trying to be too much of a perfectionist. It is also the case that the persistent and public harassment of Francis Thomson by various people only served to make him even more defensive of his position. That Thomson felt his investigations warranted a near-thirty year wait is now beside the point. What is so sad is that if he had only concentrated on the task at hand, rather than the enormous undertaking which he followed, his own legacy might well have been that of 'biographer of Blumlein'. His passing in February 1998, only leads to one inevitable question – where is the archive he so passionately guarded?

# Epilogue

When I set out to write this book, I had no intention of investigating the activities of an old man, or the circumstances surrounding his inactivity, and the reasons for his lack of co-operation with the established academic society. However, as I investigated the astonishing set of circumstances that have led to the lack of a satisfactory tribute to Alan Blumlein, a man of which I think the world will undoubtedly say, 'Yes. He was a genius', it became apparent to me that any such volume of work, regardless of my personal dislike of the publishing of such sorry matters, would be incomplete, inaccurate and confusing without pointing the attention of the reader to these drawn-out events. I therefore credit the last two chapters not so much to myself, but to the set of circumstances that produced them.

At least at the end of this decade we will, finally, at last, have a tribute to Alan Blumlein that he so richly deserves. This book is a work which, I hope, will interest and help academics and researchers for many years to come. These are people who, rightly so, are interested in the achievements of the man himself, the colleagues he worked with, and the fascinating times in which he lived. They will not concern themselves with the vagaries of his thirteenth, fourteenth or fifteenth century ancestry, nor the co-directors on the boards of the various companies his father began. Nor will they be interested in who was who in Scotland, who died where at Culloden, or for that matter whether tapestries were the television sets of the past. It is to the work and achievements completed in this, the twentieth century, done for the generations of the future, that I should like to dedicate part of this book.

The past however, should not, and must not be forgotten. The further away in time that we get from those dark days of World War Two, the more people pass away. Sadly, most of the key people in this book have died, some of them in the time that I alone have been researching. Most, however, were still alive in 1970, when Francis Thomson 'began' his work. It is, I believe, saddest of all, that they did not live to see their wishes fulfilled.

To those left alive, and in the memory of those who have died, all of whom have given their time, their recollections, their personal documents and memories, I would like to dedicate another part of this book.

The largest tribute however, must of course go to Alan Dower Blumlein himself.

Will the world ever see his like again? Probably not. The world is a very different place to that in which Blumlein lived and worked. Today, he would probably be what we would call a systems engineer, an individual who can see beyond the boundaries of the obvious, to the extreme; to the whole.

Blumlein will be remembered by his family, friends and colleagues for many characteristics: his modesty, his brilliance, patience and inventiveness. Sometimes his compulsive pipe smoking, his bright green eyes and his almost child-like love of fun, sometimes his rather loud voice and ungainly walk and loping stride.

In all cases however, everyone felt the title 'genius' applied.

Philip Vanderlyn once said of Blumlein 'The world is the poorer for the fact that he did not live to try his hand with transistors or to contribute, as he surely would have done, to the development of computers.'

In several of the periodicals over the last thirty years, tributes to Alan Blumlein have often ended with the words of Isaac Shoenberg. I am sure you will forgive me if I also end with these words:

*"There was not a single subject to which he turned his mind, that he did not enrich extensively"*.

## Author's note

It has been pointed out that at no time was Mr Norman Green 'engaged' by EMI for the work of transferring the Binaural films to safety stock. Mr Green gave his time entirely voluntarily having been approached by EMI for his advice. The time taken to transfer the volatile nitrate stock and its soundtracks to a modern safety stock was determined by the volume of his work as Head of Engineering at the ITCA. The Blumlein films were primarily completed in his own time, outside that of his normal work. It should also be pointed out that EMI did not pay Norman Green for his time spent on conservation of the films.

# References

'A Brief Chronological History of EMI', EMI100 publication, Web edition, 1997

'A Guide To Events In Telecommunications History', British Telecom Archives and Historical Information Centre Publication, August 1991, Issue 1, pp. 3–51

'A Palace for the People – The Story of Alexandra Palace', 1998, KLA Film and Video Communication

Abramson, A. 'The History Of Television, 1880 to 1941', 1987, pp. 1–256

Alexander, R.C. 'Alan Blumlein – The Audio Patents', *Audio Media*, Vol. 7, March–June 1997, Nos. 76–9, Pt.1, pp. 168–76; Pt.2, pp. 142–7; Pt.3, pp. 128–32; Pt.4, pp. 102–9

Annett, W. 'John Logie Baird', *Jones Telecommunications and Multimedia Encyclopaedia*, Web edition, 1998

Ardenne, M. von. 'Cathode-Ray Television', *Wireless Magazine*, September 1931, pp. 158–61

Ariadne, Letters page, *New Scientist*, 26 May 1977, p. 517

Ariadne, Letters page, *New Scientist*, 22 September 1977, p. 776

Ashbridge, N. 'Broadcasting and Television', *Journal of the Institute Electrical Engineers*, 1939, Vol. 84, pp. 380–87

Atherton, W.A. 'Pioneers – 14. Alan Dower Blumlein (1903–1942): the Edison of electronics', *Electronic & Wireless World*, February 1988, pp. 184–6

Baird, J.L. 'An Account of some Experiments in Television', *The Wireless World and Radio Review*, Vol. 14, 7 May 1924, pp. 153–5

Baird, M.H.I. 'Eye of the World: John Logie Baird and Television – Part Two', *Kinema*, Web edition, Fall 1996

Baldock, R.N. Letters to the Editor, *Journal of the Royal Television Society*, July/August 1971, Vol. 13, No. 10, p. 27

Baldock, R.N. 'Quantities and qualities', Letters page, *Wireless World*, September 1973, p. 451

Baldwin, N. 'Edison: Inventing', *The Century*, 1995, pp. 200–46

Benzimra, B.J. 'A.D. Blumlein – an electronics genius', *Electronics & Power*, June 1967, pp. 218–24

Bidwell, S. 'Telegraphic Photography and Electric Vision', *Nature*, Vol. 78, 4 June 1908, pp. 105–6

Birkinshaw, D.C., McGee, J.D., White, E.L.C., Nind, E.A., Coulson, J., Clifford, P.M. and Bedford, L.H. 'The work of Alan Blumlein', *Journal of British Kinematography Sound and Television*, July 1968, Vol. 50, No. 7, pp. 206–18

Birkinshaw, D.C. 'The Birth of Modern Television', *Journal of the Royal Television Society*, Nov/Dec 1971, Vol. 13, No. 12, pp. 32–7

Blake, T. 'Visioneer – John Logie Baird and Mechanical Television', Web edition, 1998

Blumlein, A.D., Browne, C.O., Davis, N.E. and Green, E. 'The Marconi-EMI television system', *Journal of the Institute Electrical Engineers*, 1938, Vol. 83, pp. 758–801

'Blumlein Commemorated', *IEE News*, July 1977, p. 16

Boltz, C.L. 'An Unsung British Genius', *Spectrum*, Science Unit – Central Office of Information, October 1977, No. 151, pp. 15–16

'Bombing "eye" device that helped win air war', *Royal Air Force News*, 1 March 1975, p. 2

'Bomber crash recalled', *Monmouthshire Beacon*, 2 June 1994, p. 3

Borwick, J. 'Basic Blumlein', *Microphones – Technology and Technique*, 1990, Butterworth-Heinemann, Oxford, Ch. 5.2, pp. 118–23

Borwick, J. 'Microphones', *Newnes Audio & Hi-Fi Handbook*, Second Edition, 1993, Butterworth–Heinemann, Oxford, Ch. 2, pp. 29–56

Bowen, E.G. *Radar Days*, 1987, Institute of Physics Publishing Limited, pp. 1–46, 83–170, 209–214

Brettingham, L. 'Royal Air Force Beam Benders, No.80 (Signals) Wing, 1940–1945', 1997, pp. 9–26, 71–82

Briggs, G.A. *Audio Biographies*, 1961, Wharfedale Wireless Works Limited, pp. 72–6, 157–63, 193–200

'British Patent 291,511', *The Illustrated Official Journal (Patents)*, 25 July 1928, pp. 3399–400

'British Patent 323,037', *The Illustrated Official Journal (Patents)*, 12 February 1930, pp. 7727–8

'British Patent 334,652', *The Illustrated Official Journal (Patents)*, 29 October 1930, p. 4528

'British Patent 335,935', *The Illustrated Official Journal (Patents)*, 26 November 1930, pp. 5017–18

'British Patent 337,134', *The Illustrated Official Journal (Patents)*, 17 December 1930, p. 5471

'British Patent 338,588', *The Illustrated Official Journal (Patents)*, 14 January 1931, p. 6030

Brown, V. 'Service Accident Report No. W-1251', Accidents Investigation Branch, Air Ministry, Gloucester, 1 July 1942 (Limited circulation document, formerly Secret)

Browne, C.O. 'Multi-Channel Television', *Journal of the Institute Electrical Engineers*, 1932, Vol. 70, pp. 340–53

Buderi, R. *The Invention that Changed the World – The story of radar from war to peace*, 1996, Abacus, pp. 27–97, 451–78

Burns, R.W. 'J.L. Baird: success and failure', *Proceedings of the Institute Electrical Engineers*, Vol. 126, No. 9, September 1979, pp. 921–8

Burns, R.W. 'Wireless pictures and the Fultograph', *Proceedings of the Institute Electrical Engineers*, Vol. 128, Pt. A, No. 1, January 1981, pp. 78–88

Burns, R.W. 'A.D. Blumlein – engineer extraordinary', *Engineering Science and Educational Journal*, February 1992, pp. 19–33

Burns, R.W., Lynch, A.C., Lodge, K.R., White, E.L.C., Thrower, K.R. and Trim, R.M. 'The life and work of A.D. Blumlein', *Engineering Science and Educational Journal*, June 1993, pp. 115–36

Buskin, R. '60 Years of Abbey Road Studios', *Audio Media*, Vol. 2, October 1991, pp. 56–61

Campbell-Swinton, A.A. 'The Possibilities of Television – with wire and wireless', *The*

*Wireless World and Radio Review*, Vol. 14, 9 April 1924, Pt.1, pp. 51–6, 16 April 1924, Pt.2, pp. 82–4, 23 April 1924, Pt.3, pp. 114–18

Cave, W. *The London Television Studio – A Tour in Words and Pictures*, Alexandra Palace Television Society, Web edition, 1998

Churchill, W.S. *The Second World War, The Gathering Storm*, Vol. One, 1948, Cassell, pp. 3–17, 72, 99–142, 341–75

Churchill, W.S. *The Second World War, Their Finest Hour*, Vol. Two, 1948, Cassell, pp. 12–16, 247–65, 281–300, 337–52, 640–44

Churchill, W.S. *The Second World War, The Grand Alliance*, Vol. Three, 1948, Cassell, pp. 34–43, 716–17

Churchill, W.S. *The Second World War, The Hinge of Fate*, Vol. Four, 1948, Cassell, pp. 248–59, 336–50

Churchill, W.S. *The Second World War, Closing the Ring*, Vol. Five, 1948, Cassell, pp. 456–69

Churchill, W.S. *The Second World War, Triumph and Tragedy*, Vol. Six, 1948, Cassell, pp. 9–10, 34–49, 666–73

Clark, H.A.M. and Vanderlyn, P.B. 'Double-ratio AC bridge with inductively-coupled ratio arms', Proc. Institute Electrical Engineers, 1949, Vol. 96 Pt.III, pp. 189–202, 210–13

Clark, H.A.M., Dutton, G.F. and Vanderlyn, P.B. 'The "Stereosonic" Recording and Reproduction System', *Proceedings of the Institute of Electrical Engineers*, Sept 1957, Vol. 104, Part B, No.17, pp. 417–32

Cork, E.C. and Pawsey, J.L. 'Long Feeds for Transmitting Wide Side-Bands, with Reference to the Alexandra Palace Aerial-Feeder System', *Journal of the Institute Electrical Engineers*, 1939, Vol. 84, pp. 448–67

Curran, S. 'Personal memories of Alan Dower Blumlein and the development of H2S radar', *Electrotechnology*, December 1996/January 1997, pp. 20–21

Davies, I.F. 'Alan Dower Blumlein 1903–1942', *The Cholmeleian*, September 1977, pp. 8–9

'Dedication day at Goodrich Castle', *The Ross Gazette*, 2 June 1994, p. 4

'Discussion on "The London Television Service" and "The Marconi-EMI Television System"', Mersey and North Wales (Liverpool) Centre, at Liverpool, 28th November 1938, South Midlands Centre, at Birmingham, 5th December 1938, North Midland Centre, at Leeds, 17th January 1939, Western Centre, at Newport, 13th March 1939, *Journal of the Institute Electrical Engineers*, 1939, Vol. 85, pp. 271–9

'Early Development of Metric A.S.V.', 'A.S.V. Mark I and A.S.V. Mark II', 'Development of Long Range A.S.V.', 'Installations of A.S.V. Mark II in Anti-U-Boat Aircraft', 'Early Development of Centimetric A.S.V. in the United Kingdom', 'Development of Centimetric A.S.V. in the United Kingdom', 'Preparation of H2S for A.S.V. Role', The Second World War 1939–1945 Royal Air Force, Signals, Volume VI, Radio in Maritime Warfare, Issued by the Air Ministry (A.H.B.), 1954, (Limited circulation document, formerly Secret), Chapters 1–4, 7, 9–10, 3–37, 77–89, 100–35

'Entry into crashed aircraft – The Hampden & The Halifax', *The Journal of the Royal Observer Corps Club*, November 1942, p. 460

Fox, B. 'Early Stereo Recordings', *Studio Sound*, May 1982, Vol. 24, No. 5, pp. 36–42

Fox, B. 'The Briton who invented electronics', *New Scientist*, 3 June 1982, pp. 641–643

Fox, B. 'Alan Blumlein – Where Stereo Began', *Hi Fi for Pleasure*, January 1984, pp. 29–37

Fox, B. 'The Missing Life of Alan Blumlein', *Electronics World + Wireless World*, November 1990, pp. 973–6

Fox, B. 'How did Blumlein die?' *New Scientist*, 27 April 1991, pp. 62–3

Fox, B. 'Technology', *Hi-Fi News & Record Review*, August 1991, p19

Fox, B. 'Blumlein – The Forgotten Genius', *Everyday Electronics*, September 1991, pp. 576–80

Fox, B. 'Blumlein biography; Churchill tapes authenticity', *Studio Sound*, October 1991, Vol. 33, No. 10, p. 78

Fox, B. 'Blumlein Archives Scandal', *Everyday with Practical Electronics*, September 1995, p. 700

Fox, B. 'The Blumlein Scandal', *Everyday Practical Electronics*, July 1996, p. 559

Garrett, G.R.M. and Mumford, A.H. 'The History Of Television', *Proceedings of the Institute of Electrical Engineers*, 1952, Vol. 99, Part 3a, pp. 25–42

Gay, K. 'Palace On The Hill – A History of Alexandra Palace and Park', 1992, pp. 26–35

Gayford, M. 'Microphones', *Microphone Engineering Handbook*, 1994, Focal Press, pp. 110–14, 205–9

Gerzon, M.A. 'Applications of Blumlein Shuffling to Stereo Microphone Techniques', *Journal of the Audio Engineering Society*, Vol. 42, No. 6, June 1994, pp. 435–53

Gold, A. 'Microphones', *Newnes Audio & Hi-Fi Handbook*, Second edition, 1993, Butterworth–Heinemann, Oxford, Ch. 9, pp. 332–33

'H2S Mark I', H2S Marks II and III', The Second World War 1939–1945 Royal Air Force, Signals, Volume IV, Aircraft Radio, issued by the Air Ministry (A.H.B.), 1954, (Limited circulation document, formerly Secret), Chapters 2–3, pp. 21–69

'Historical Facts on EMI', EMI100 publication, Web edition, 1997

Hills, A.R. 'Eye of the World: John Logie Baird and Television', *Kinema*, Web edition, Spring 1996

Hope, A. 'Blumlein – A.D. Blumlein (1903–1942)', *Hi-Fi News & Record Review*, August 1977, Pt. 1, pp. 57-61

Huber, D.M. *Microphone Manual – Design and Application*, 1988, Focal Press, pp. 110–11, 115–17

Ivall, T. 'The First Radio Patent', *Electronics World*, June 1996, pp. 461–4

James, E.G., Polgreen, G.R. and Warren, G.W. 'Instruments Incorporating Thermionic Valves, and their Characteristics', *Journal of the Institute Electrical Engineers*, 1939, Vol. 85, pp. 242–62, 265 (Blumlein, A.D. 'Discussion Before The Meter And Instrument Section, 6th January 1939')

Jones, P. 'Royal Television Society holds first annual Memorial Lecture to commemorate famous EMI scientist', *EMI Electronics News*, February 1971, p. 1

Jones, R.V. 'Scientists at War – Lindemann v Tizard', *The Times*, Thursday, 6 April 1961, Pt.1 p. 13; Friday, 7 April 1961, Pt.2 p. 15; Saturday, 8 April 1961, Pt.3 p. 9

Jones, R.V. *Most Secret War*, 1978, Wordsworth Editions Limited, pp. 13–20, 34–52, 78–110, 120–85, 233–49, 318–31

Kilburn, T. and Piggott, L.S. *Frederic Calland Williams*, Biographical Memoirs of the Fellows of the Royal Society, 1978, Vol. 24, pp. 583–90

Kipping, N.V. and Blumlein, A.D. 'Introduction To Wireless Theory', *Wireless World*, Vol. 27, October-December 1925, Pt.1 pp. 536–538, Pt.2 pp. 596–7, Pt.3 pp. 636–8, Pt.4 pp. 675–78, Pt.5 pp. 711–13, Pt.6 pp. 781–2, Pt.7 pp. 812–14

Latham, C. and Stobbs, A. *Radar – A Wartime Miracle*, 1996, Sutton Publishing Limited, pp. 10–59

Lewis, P. *The British bomber since 1914*, 1974, Purnell Book Services Limited, pp.

189–91, 216–19, 309–13

Lodge, J.A. 'History Of Research Laboratories: 10, Thorn EMI Central Research Laboratories – an anecdotal history', *Journal of Physics Technology*, Vol. 18, 1987, pp. 258–64

Lovell, B. 'Blumlein Crash', *New Scientist*, 8 December 1977, p. 659

Lovell, B. 'There's a war on!', *New Scientist*, Monitor, 21 October 1982, p. 189

Lovell, B. 'An astonishing strong echo', *New Scientist*, Forum, 28 October 1982, pp. 246–7

Lovell, B. 'Turning point of the war', *New Scientist*, Forum, 4 November 1982, pp. 315–16

Lovell, B. 'Some bombs in the right place', *New Scientist*, Forum, 11 November 1982, pp. 377–8

Lovell, B. 'A tragic crash', *New Scientist*, Forum, 18 November 1982, p. 433

Lovell, B. 'Any news, any problems?', *New Scientist*, Forum, 25 November 1982, pp. 525–6

Lovell, B. 'Coded "Scent Spray"', *New Scientist*, Forum, 2 December 1982, pp. 595–6

Lovell, B. 'Fishpond, U-boats and peace', *New Scientist*, Forum, 9 December 1982, pp. 677–8

Lovell, B. *Astronomer By Chance*, 1991, Oxford University Press, pp. 44–104

Lovell, B. *Echoes of War*, 1991, Oxford University Press, pp. 1–136

Macnamara, T.C. and Birkinshaw, M.A. 'The London Television Service', *Journal of the Institute Electrical Engineers*, December 1938, Vol. 83, No. 504, pp. 729–57

Mallett, E. and Blumlein, A.D. 'A new method of high-frequency resistance measurement', *Journal of the Institute Electrical Engineers*, 1925, Vol. 63, pp. 397–414

Martin, D. 'Thorn EMI 50 Years Of Radar – EMI at War, 1939–1945 The First Electronic War', Thorn EMI publications, 1985, Ch. 1.1–1.9, pp. 1–17

McGee, J.D. 'Campbell Swinton and Television', *Nature*, 17 October 1936, Vol. 138, pp. 674–6

McGee, J.D. and Lubszynski, H.G. 'E.M.I. Cathode-Ray Television Transmission Tubes', *Journal of the Institute Electrical Engineers*, 1939, Vol. 84, pp. 468–82

McGee, J.D. 'The life and work of Sir Isaac Shoenberg, 1880–1963', *Journal of the Royal Television Society*, May/June 1971, Vol. 13, No. 9, pp. 207–16

McLean, D.M. 'The Earliest Known Recording of Broadcast Television', Web edition, 1998

McLean, D.M. 'Phonovision – The First TV Recording Studio?', Web edition, 1998

McLean, D.M. 'The First Commercial TV Recording – "Major Radiovision" Recording: 1934', Web edition, 1998

Mondey, D. *British Aircraft of World War II*, 1982, Chancellor Press, pp. 47–56, pp. 61–7, pp. 127–31

'One Foot In The Past – Sound Mirrors', 1996, BBC TV

'Pre War Development Trends', The Second World War 1939–1945 Royal Air Force, Signals, Volume III, Aircraft Radio, issued by the Air Ministry (A.H.B.), 1954, (Limited circulation document, formerly Secret), Chapter 1, pp. 1–19

Price, A. 'Doubts and Decisions', *Instruments of Darkness*, 1966, Macdonald and Jane's Publishers Limited, pp. 121–24

Preston, S.J. 'The Birth of a High Definition Television System', *Journal of the Television Society*, 1953, Vol. 7, No. 3, pp. 115–25

Randall, J.T. 'The Cavity Magnetron', *Journal of the Physical Society*, Vol. 58, Pt. 3, 1946, pp. 246–52

Rapier, B.J. *Halifax at war*, 1987, The Promotional Reprint Company Limited, pp. 16–35

Ratcliff, J. and Papworth, N. *Single Camera Stereo Sound*, 1992, Focal Press, Oxford, pp. 6–19, 21–8, 95–117

'Report of the Television Commission', HMSO, Parliamentary Papers 1933–34, 14 January 1935, Vol. XI, Cmd. 4793, pp. 921–47

Roberts, N. 'Handley Page Hampden & Hereford, Crash Log', 1980, p. 27

Roberts, R.N. 'The Halifax File', Air Britain (Historical) Limited & The British Aviation Archaeological Council, 1982, pp. 29–46

Rogers, K., Scroggie. M.G., Kendall, G.P. and Graham, G.A. *The Encyclopaedia Of Radio And Television*, 1950, Odhams Press Limited, pp. 13–16, 17–24, 31–7, 78–84, 97–105, 204–6, 355–7, 365–7, 441–2, 486–90, 505–6, 589–92, 612–14

Rose, R. 'A Tribute to Geoffrey Hensby – Great young scientist lost during vital wartime work', *DRA News*, June 1992, p5

Rose, R. 'Letter to the Editor – Tribute's appearance was excellently timed', *DRA News*, August 1992, p10

Rose, R. 'Geoffrey Spencer Hensby', *Electrotechnology*, December 1996/January 1997, pp. 17–19

Rumsey, F. 'Microphones in stereophonic applications', *Microphone Engineering Handbook*, 1994, Focal Press, Oxford, Ch. 9, pp. 360–67, 369–72, 377–97

*School For Secrets*, 1946, (film), Rank Film Organization

Scroggie, M.G. 'The Genius of A.D. Blumlein', *Wireless World*, September 1960, pp. 451–6

'Sir Isaac Shoenberg', Obituary, *Wireless World*, March 1963, Vol. 69, p119

Sleigh, W.H. 'Aircraft for Airborne Radar Development' (unpublished), RSRE Internal Document, Historical Collection TRE H 785, June 1985

Strachan, J. 'The Early History of Television', *The Wireless World and Radio Review*, Vol. 14, 11 June 1924, pp. 305–7

'Television', *Nature*, Vol. 115, 4 April 1925, pp. 505–6

'Television Transmissions – Details of the Baird and Marconi-EMI systems', *Wireless World*, 4 October 1935, pp. 371–3

'The Life and work of the late Sir Isaac Shoenberg', *EMI News*, Jan/Feb 1971, No. 76, p. 1

'The Secret War – To See for a Hundred Miles/Battle of the Beams', 1977, BBC TV

Thomson, F.P. 'Blumlein', Letters page, *IEE News*, June 1972

Thomson, F.P. 'A.D. Blumlein', Letters page, *Wireless World*, December 1973, p. 580

Thurbin, K. 'Sir Isaac Shoenberg 1880–1963', *EMI News*, February 1971, p. 1

Tucker, J.D. 'Shoenberg', *International Broadcast Engineer*, May 1972, pp. 35–8

Turnbull, I.L. and Clark, H.A.M. 'The Marconi-E.M.I. Audio-Frequency Equipment At The London Television Service', *Journal of the Institute Electrical Engineers*, 1939, Vol. 85, pp. 439–67

Vanderlyn, P.B. 'In Search Of Blumlein: The Inventor Incognito', *Journal of the Audio Engineering Society*, September 1978, Vol. 26, No. 9, pp. 660–70

Virgo. R. 'Unsung – the boffin whose flying radar helped to save Britain during the war', *Headline*, October 1982, pp. 6–7

Waddell, P. 'Alan Blumlein', Letters page, *New Scientist*, 1 July 1982

Webster, C., Noble, F. and Butler, J. (eds) 'Radar Aids', *The History Of The Second World War, The Strategic Air Offensive Against Germany*, 1961, Vol. IV, Annexes & Appendices, Section One, pp. 11–15

Willard, R.A. 'Television Sound Equipment', Thorn-EMI Central Research Laboratories, 1976

Yonge, M. 'Blumlein looking at stereo', *Line Up, Journal of the Institute of Broadcast Sound*, September 1988, pp. 51–53.

# Patent chronology

| No. | Patent. No. | Original Application | Co-writer (1) | Co-writer (2) | Co-writer (3) |
|---|---|---|---|---|---|
| 1. | 291,511 | 1 March 1927 | J.P. Johns | ST&C | |
| 2. | 323,037 | 13 September 1928 | ST&C | | |
| 3. | 334,652 | 29 June 1929 | | | |
| 4. | 337,134 | 29 June 1929 | ST&C | | |
| 5. | 335,935 | 3 July 1929 | ST&C | | |
| 6. | 338,588 | 21 August 1929 | ST&C | | |
| 7. | 350,954 | 10 March 1930 | II.E. Holman | | |
| 8. | 359,998 | 10 March 1930 | H.E. Holman | | |
| 9. | 357,229 | 19 June 1930 | ST&C | | |
| 10. | 362,472 | 30 July 1930 | | | |
| 11. | 361,468 | 12 September 1930 | P.W. Willans | | |
| 12. | 363,627 | 12 September 1930 | | | |
| 13. | 368,336 | 1 December 1930 | | | |
| 14. | 369,063 | 13 May 1931 | | | |
| 15. | 394,325 | 14 December 1931 | EMI | | |
| 16. | 400,976 | 4 April 1932 | EMI | | |
| 17. | 402,483 | 10 June 1932 | | | |
| 18. | 417,718 | 7 March 1933 | EMI | | |
| 19. | 419,284 | 7 March 1933 | EMI | | |
| 20. | 421,546 | 16 June 1933 | EMI | | |
| 21. | 422,914 | 11 July 1933 | C.O. Browne | J. Hardwick | EMI |
| 22. | 425,553 | 18 September 1933 | H.A.M. Clark | EMI | |
| 23. | 429,022 | 23 October 1933 | EMI | | |
| 24. | 432,485 | 29 December 1933 | | | |
| 25. | 434,876 | 5 February 1934 | | | |
| 26. | 432,978 | 7 February 1934 | J.L. Pawsey | | |
| 27. | 429,054 | 10 February 1934 | EMI | | |
| 28. | 436,734 | 17 April 1934 | C.O. Browne | | |
| 29. | 446,661 | 3 August 1934 | J.D. McGee | | |
| 30. | 445,968 | 17 August 1934 | E.C. Cork | E. L. C. White | |
| 31. | 446,663 | 4 September 1934 | | | |
| 32. | 448,421 | 4 September 1934 | | | |
| 33. | 449,242 | 18 September 1934 | C.O. Browne | F. Blythen | |
| 34. | 449,533 | 24 October 1934 | M. Bowman-Manifold | | |
| 35. | 447,754 | 26 October 1934 | H.E. Holman | | |
| 36. | 447,824 | 26 October 1934 | H.E. Holman | | |
| 37. | 452,713 | 6 December 1934 | E.C. Cork | | |
| 38. | 456,444 | 7 February 1935 | | | |
| 39. | 452,772 | 25 February 1935 | E.C. Cork | | |
| 40. | 452,791 | 28 February 1935 | | | |
| 41. | 455,492 | 7 March 1935 | J. Hardwick | | |
| 42. | 458,585 | 20 March 1935 | | | |

| No. | Patent. No. | Original Application | Co-writer (1) | Co-writer (2) | Co-writer (3) |
|---|---|---|---|---|---|
| 43. | 456,135 | 3 April 1935 | E.A. Nind | | |
| 44. | 455,858 | 24 April 1935 | | | |
| 45. | 461,004 | 4 July 1935 | | | |
| 46. | 462,530 | 8 July 1935 | | | |
| 47. | 462,583 | 8 July 1935 | | | |
| 48. | 462,584 | 8 July 1935 | | | |
| 49. | 461,324 | 12 August 1935 | | | |
| 50. | 462,823 | 12 August 1935 | EMI | | |
| 51. | 463,111 | 17 September 1935 | E.C. Cork | J.L. Pawsey | |
| 52. | 464,443 | 19 October 1935 | E.C. Cork | J.L. Pawsey | |
| 53. | 470,495 | 14 November 1935 | | | |
| 54. | 466,418 | 22 November 1935 | | | |
| 55. | 470,408 | 13 February 1936 | | | |
| 56. | 473,276 | 10 March 1936 | | | |
| 57. | 475,729 | 19 March 1936 | | | |
| 58. | 474,607 | 29 April 1936 | E.L.C. White | | |
| 59. | 479,113 | 29 April 1936 | | | |
| 60. | 476,935 | 15 May 1936 | | | |
| 61. | 477,392 | 22 May 1936 | E.L.C. White | | |
| 62. | 482,740 | 4 July 1936 | | | |
| 63. | 479,599 | 8 July 1936 | W.H. Connell | | |
| 64. | 480,355 | 18 July 1936 | | | |
| 65. | 482,370 | 27 August 1936 | E.L.C. White | | |
| 66. | 482,195 | 21 September 1936 | | | |
| 67. | 483,650 | 20 October 1936 | | | |
| 68. | 490,205 | 10 November 1936 | C.O. Browne | | |
| 69. | 491,490 | 2 December 1936 | | | |
| 70. | 489,950 | 5 February 1937 | | | |
| 71. | 490,150 | 4 March 1937 | EMI | | |
| 72. | 496,139 | 24 May 1937 | | | |
| 73. | 496,883 | 5 June 1937 | | | |
| 74. | 495,724 | 7 June 1937 | | | |
| 75. | 501,966 | 7 June 1937 | R.E. Spencer | | |
| 76. | 497,060 | 9 June 1937 | | | |
| 77. | 505,079 | 4 October 1937 | | | |
| 78. | 503,765 | 8 October 1937 | | | |
| 79. | 503,555 | 14 October 1937 | | | |
| 80. | 505,480 | 6 November 1937 | E.L.C. White | | |
| 81. | 507,239 | 6 November 1937 | E.L.C. White | | |
| 82. | 507,665 | 18 November 1937 | C.S. Bull | | |
| 83. | 507,417 | 8 December 1937 | | | |
| 84. | 508,377 | 24 December 1937 | | | |
| 85. | 512,109 | 24 December 1937 | E.L.C. White | | |
| 86. | 515,684 | 7 March 1938 | | | |
| 87. | 515,044 | 23 March 1938 | J. Hardwick | C.O. Browne | |
| 88. | 515,348 | 30 March 1938 | | | |
| 89. | 514,825 | 13 April 1938 | E.L.C. White | | |

| No. | Patent. No. | Original Application | Co-writer (1) | Co-writer (2) | Co-writer (3) |
|---|---|---|---|---|---|
| 90. | 514,065 | 25 April 1938 | | | |
| 91. | 515,360 | 30 May 1938 | | | |
| 92. | 515,361 | 30 May 1938 | EMI | | |
| 93. | 515,362 | 30 May 1938 | F. Blythen | J. Hardwick | |
| 94. | 515,364 | 30 May 1938 | | | |
| 95. | 517,516 | 28 June 1938 | H.E. Kallmann | W.S. Percival | |
| 96. | 520,646 | 27 October 1938 | | | |
| 97. | 528,310 | 25 February 1939 | | | |
| 98. | 529,590 | 19 May 1939 | A.H. Cooper | | |
| 99. | 543,602 | 21 June 1939 | | | |
| 100. | 581,920 | 20 July 1939 | E.L.C. White | | |
| 101. | 534,465 | 28 July 1939 | F. Blythen | | |
| 102. | 536,089 | 28 July 1939 | | | |
| 103. | 577,817 | 13 October 1939 | E.A. Nind | | |
| 104. | 585,906 | 3 November 1939 | | | |
| 105. | 554,715 | 16 November 1939 | | | |
| 106. | 585,907 | 1 December 1939 | E.L.C. White | | |
| 107. | 589,228 | 13 December 1939 | E.L.C. White | | |
| 108. | 585,908 | 4 December 1939 | | | |
| 109. | 581,161 | 10 January 1940 | | | |
| 110. | 581,164 | 10 January 1940 | EMI | | |
| 111. | 589,229 | 10 January 1940 | EMI | | |
| 112. | 567,227 | 26 January 1940 | | | |
| 113. | 579,725 | 27 January 1940 | E.L.C. White | EMI | |
| 114. | 579,154 | 29 March 1940 | | | |
| 115. | 541,947 | 17 June 1940 | | | |
| 116. | 563,462 | 17 June 1940 | | | |
| 117. | 581,561 | 17 June 1940 | E.L.C. White | | |
| 118. | 587,878 | 17 June 1940 | EMI | | |
| 119. | 574,708 | 23 September 1940 | | | |
| 120. | 587,562 | 31 March 1941 | | | |
| 121. | 581,167 | 5 May 1941 | E.C. Cork | | |
| 122. | 589,127 | 10 October 1941 | | | |
| 123. | 577,942 | 28 October 1941 | | | |
| 124. | 585,789 | 21 November 1941 | W.H. Connell | D.G. Holloway | |
| 125. | 580,527 | 5 June 1942 | | | |
| 126. | 574,133 | 17 June 1942 | C.O. Browne | | |
| 127. | 582,503 | 15 October 1943 | F.C. Williams | E.L.C. White | |
| 128. | 595,509 | 12 May 1945 | E.L.C. White, EMI | | |

## Associated web site

The following web sites contain all the above 128 patents in full, with associated diagrams, all binaural audio recordings, together with all binaural films:
http://www.focalpress.com (choose the 'audio' section from the main menu)
http://www.doramusic.com
Web site constructed by http://www.jayjaybee.com

# Index